POLYMER BRUSHES

Substrates, Technologies, and Properties

POLYMER BRUSHES

Substrates, Technologies, and Properties

Edited by
Vikas Mittal

CRC Press
Taylor & Francis Group
Boca Raton London New York

CRC Press is an imprint of the
Taylor & Francis Group, an **informa** business

CRC Press
Taylor & Francis Group
6000 Broken Sound Parkway NW, Suite 300
Boca Raton, FL 33487-2742

First issued in paperback 2017

© 2012 by Taylor & Francis Group, LLC
CRC Press is an imprint of Taylor & Francis Group, an Informa business

No claim to original U.S. Government works

ISBN-13: 978-1-4398-5794-6 (hbk)
ISBN-13: 978-1-138-07497-2 (pbk)

Library of Congress Cataloging-in-Publication Data

Polymer brushes : substrates, technologies, and properties / editor, Vikas Mittal.
    p. cm.
"A CRC title."
Includes bibliographical references and index.
ISBN 978-1-4398-5794-6 (hardcover : alk. paper)
    1. Paintbrushes. 2. Polymers--Mechanical properties. 3. Polymers--Surfaces. 4. Finishes and finishing--Equipment and supplies. I. Mittal, Vikas.

TT305.3.P65 2012
667'.60284--dc23   2011049567

Visit the Taylor & Francis Web site at
http://www.taylorandfrancis.com

and the CRC Press Web site at
http://www.crcpress.com

Contents

Preface

Organic-inorganic hybrids find applications in a large number of areas. Compatibilization of the organic and inorganic phases is often required to ensure their nanoscale interactions at the interface. One novel way to achieve such compatibilization is the attachment of polymer brushes on the surface, which can then compatibilize with the other surface. Apart from that, a number of applications of the various materials require an accurate control on the surface properties. Such control is also achieved by the generation of polymer brushes from these surfaces. The polymer brushes either can be physically bound on these surfaces or can be chemically grafted to them. Both systems have their own advantages and limitations; thus the choice is dependent on the requirement. A number of substrates like organic-inorganic, flat or spherical, etc., have been reported for the generation of brushes, and a large number of synthesis methods and materials for the brushes have been developed in recent years. This book intends to summarize the various aspects of the polymer brushes technology, starting from synthesis and properties to performance and applications. It also provides hands-on information to the reader in order to decide the synthesis methods, materials, etc., for the system at hand. Apart from the experimental details, theoretical insights into the system to provide better understanding are also provided.

Chapter 1 provides an overview of the polymer brush systems. A number of systems involving different polymers, substrates and polymerization or grafting methods have been reviewed. Chapter 2 details the grafting of organic brushes from the surface of clay platelets. Different methodologies like grafting to, grafting from, as well as physical adsorption have been presented. Other chemical reactions like esterification to enhance the brush length have also been reported. Chapter 3 presents a systematic investigation of the effect of grafting density on the thermomechanical properties of poly(N-isopropylacrylamide) (PNIPAm) brushes. By using plasma activation of mica surfaces, which allows for covalent grafting of PNIPAm brushes by surface-initiated atom transfer radical polymerization (SI-ATRP), the authors have been able to take advantage of the surface forces apparatus technique to probe in detail the collapse and adhesion of such polymer layers. Chapter 4 provides a comprehensive summary of the syntheses and applications of the ferrocene polymer brushes. Ferrocene-functionalized polymer brushes covalently bonded or physisorbed on solid substrate surfaces could be readily prepared via well-known surface modification techniques, such as surface-initiated polymerizations, reaction of the end functional groups of polymer chains with substrate surfaces, and host-guest interaction. Chapter 5 focuses on the preparation and characterization of nonfouling brushes on the poly(ethylene terephthalate) (PET) film surfaces. Chapter 6 focuses on the case of concave geometry and more specifically on brushes formed on the inner surface of cylindrical pores that have a characteristic diameter in the nanometer micrometer range, i.e., close to the characteristic length of polymer brushes. Chapter 7 reports the "zipper brush" approach, which presents a way to produce polymer brushes by adsorption that circumvents all the normal problems associated with diblock-copolymer adsorption. The driving force for the adsorption is very large, while there is no crowding of anchor blocks or a large steric barrier that prevents the brush from becoming dense. Chapter 8 reports the use of scanning electrochemical microscopy (SECM) for the analysis of brushes. SECM is a new powerful tool valuable for a wide range of local interrogation, characterization, and chemical or electro-chemical transformations at very diverse interfaces. It is then a very prom-ising tool for the inspection, imaging, or patterning of polymer brushes. Chapter 9 compares the surface-controlled atom transfer radical polymer-ization (SC-ATRP) and surface-controlled single-electron transfer (SC-SET) living radical polymerizations by reviewing prior methods used to prepare (meth)acrylate polymer brushes—including a series of poly(amino (meth) acrylate)s and PNIPAm brushes—from various substrates. A wide range of catalytic systems have been discussed, including a variety of copper catalyst, ligand, and solvent combinations. Chapter 10 further reports the grafting of PNIPAm brushes on the surface of latex particles along with their appli-cation in chromatographic separations of viruses and proteins. Chapter 11 demonstrates well-defined concentrated polymer brushes of hydrophilic polymers with applications related to suppression of proteins and cell adhe-sions. Chapter 12 further describes synthesis of polymer brushes by surface-initiated iniferter-mediated polymerization.

I am grateful to CRC Press for kind acceptance to publish the book. I dedicate this book to my mother for being a constant source of inspiration. I express heartfelt thanks to my wife, Preeti, for her continuous help in coediting the book as well as for her ideas to improve the manuscript.

Vikas Mittal

Editor

Dr. Vikas Mittal studied chemical engineering from Punjab Technical University in Punjab, India. He later obtained his master's of technology in polymer science and engineering from Indian Institute of Technology, Delhi, India. Subsequently, he joined the polymer chemistry group of Professor U. W. Suter at the Department of Materials at the Swiss Federal Institute of Technology, Zurich, Switzerland. There he worked for his doctoral degree with a focus on the subjects of surface chemistry and polymer nanocomposites. He also jointly worked with Professor M. Morbidelli at the Department of Chemistry and Applied Biosciences for the synthesis of functional polymer latex particles with thermally reversible behaviors.

After completion of his doctoral research, Mittal joined the Active and Intelligent Coatings section of the Sun Chemical Group Europe in London. He worked for the development of water- and solvent-based coatings for food packaging applications. He later joined as polymer engineer at BASF Polymer Research in Ludwigshafen, Germany, where he worked as laboratory manager responsible for the physical analysis of organic and inorganic colloids. Currently, he is an assistant professor in the Chemical Engineering Department at the Petroleum Institute, Abu Dhabi.

Mittal's research interests include organic-inorganic nanocomposites, novel filler surface modifications, thermal stability enhancements, polymer latexes with functionalized surfaces, etc. He has authored more than 40 scientific publications, book chapters, and patents on these subjects.

Contributors

Lionel Bureau
University of Grenoble 1/CNRS
Grenoble, France

Martien A. Cohen Stuart
Laboratory of Physical Chemistry
and Colloid Science
Wageningen University
Wageningen, The Netherlands

Wiebe M. de Vos
Polymers at Interfaces Group
University of Bristol
Cantock's Close
Bristol, United Kingdom

Shijie Ding
College of Life Science and
Chemical Engineering
Jiangsu Provincial Key Laboratory
of Palygorskite Science and
Applied Technology
Huaiyin Institute of Technology
Huaian, People's Republic
of China

Fu Guo Dong
School of Chemistry and Chemical
Engineering
Southeast University
Nanjing, People's Republic of China

Kang En-Tang
Department of Chemical and
Biomolecular Engineering
National University of Singapore
Kent Ridge, Singapore

Qiang Fu
College of Polymer Science and
Engineering
State Key Lab Polymer and
Materials Engineering
Sichuan University
Chengdu, China

Frédéric Kanoufi
Physicochimie des Electrolytes
Colloïdes et Sciences Analytiques
CNRS UMR 7195
ESPCI ParisTech
Paris, France

S. Michael Kilbey II
Departments of Chemistry and
Chemical and Biomolecular
Engineering
University of Tennessee
Knoxville, Tennessee
and
Center for Nanophase Materials
Sciences
Oak Ridge National Laboratory
Oak Ridge, Tennessee

J. Mieke Kleijn
Laboratory of Physical Chemistry
and Colloid Science
Wageningen University
Wageningen, The Netherlands

Hisatoshi Kobayashi
Biomaterials Center
National Institute for Materials
Science
Ibraki, Japan

Alexandros G. Koutsioubas
Laboratoire Leon Brillouin
CEA/CNRS UMR 12
Gif-sur-Yvette, France

Jiehua Li
College of Polymer Science and
    Engineering
State Key Lab Polymer and
    Materials Engineering
Sichuan University
Chengdu, China

V. Mittal
Chemical Engineering Department
Petroleum Institute
Abu Dhabi, United Arab Emirates

Xu Li Qun
Department of Chemical and
    Biomolecular Engineering
National University of Singapore
Kent Ridge, Singapore

Santosh B. Rahane
School of Polymers and High
    Performance Materials
University of Southern Mississippi
Hattiesburg, Mississippi

Hong Tan
College of Polymer Science and
    Engineering
State Key Lab Polymer and
    Materials Engineering
Sichuan University
Chengdu, China

Muriel Vayssade
CNRS UMR 6600
Université de Technologie de
    Compiègne
Compiègne, France

Keisha B. Walters
Dave C. Swalm School of Chemical
    Engineering
Mississippi State University
Mississippi State, Mississippi

Chiaki Yoshikawa
WPI Research Center
International Center for Materials
    Nanoarchitectonics
National Institute for Materials
    Science
Ibraki, Japan

1

Polymer Brushes: An Overview

V. Mittal

Chemical Engineering Department, Petroleum Institute,
Abu Dhabi, United Arab Emirates

CONTENTS

1.1 Introduction

Surface properties of substrates are among the most important contributors to their applications, especially when the application involves the interaction of the surfaces with foreign media or entities. As an example, the surface properties are of tremendous importance when the chromatographic separations of proteins, polymers, or any mixture must be handled. There is always a requirement or desire to achieve extra functionalities on the surface by controlling the surface properties of the substrates, which can be used for one application or another or a combination of a few. The substrates can also be of any nature, e.g., flat, spherical, colloidal, etc. These functionalized surfaces represent a novel class of materials, which at the forefront of technology can lead to revolutionary changes in the conventional processes.

Generation of the polymer brushes on the surface is the most common means of controlling the substrate surface properties. Many techniques to functionalize the surfaces with polymer brushes have been reported in the literature, such as physical adsorption of oligomeric or polymeric chains on the surface [1–3], grafting of polymer chains to the surface [4,5], and grafting of polymer chains from the surface [6,7]. Although physical adsorption has the advantage of adsorption of log polymer chains of well-defined length, molecular weight distribution, and composition to the surface, this approach suffers from the limitations of steric hindrance posed by the long polymer chains, thus seriously limiting the final surface density of the grafts. Apart from that, as the chains are only physically bound to the surfaces, the efficiency of these linkages for load transfer or resistance toward washing or cleaving off may not be sufficient. Grafting of polymer chains to the surface approach focuses on chemical immobilization of end-functionalized polymer chains to the surface by reacting the ends with the reactive groups already immobilized on the surface. It can also be achieved by attaching a monomer on the surface and subsequently polymerizing the monomer in the presence of an externally added monomer and initiator. In both cases, the preformed long chains approach the surfaces; therefore, the steric hindrance concerns may still hinder the generation of high-density grafts on the surface. Another important and extensively studied approach is grafting of polymer brushes from the surface. In this approach, an initiator is covalently bound to the surface, which is subsequently polymerized with the external monomer without the addition of any further initiator. This approach leads to the generation of densely packed polymer brushes along with the advantage of covalently bound polymer chains on the surface.

Such structures are referred to as polymer brushes, when the distance between the grafted chains is less than twice the radius of gyration of the unperturbed polymer chain [8]. A number of different controlled polymerization methods have also been used to graft polymer brushes on the surfaces apart from conventional free radical polymerization. Such controlled polymerization techniques include ionic [9,10], nitroxide-mediated polymerization (NMP) [11], atom transfer radical polymerization (ATRP) [12–14], and reversible addition fragmentation chain transfer (RAFT) [15–17]. Conventional polymer systems sometimes suffer from unwanted bimolecular termination reactions, thereby leading the grafting process to an abrupt end. However, the growth of living polymer chains from the surface generated by controlled living polymerization techniques ensures better control over the molecular weight distribution and the amount of grafted polymer. A number of systems involving different polymers, substrates, and polymerization or grafting methods have been reported in the literature to obtain polymer brushes, the examples of a few of which are reviewed in the following.

1.2 Functional Polymer Brushes: Examples

Brzozowska et al. [18] reported on the stability of the polymer brushes formed by adsorption of ionomer complexes on hydrophilic and hydrophobic surfaces. The ionomer complexes or micelles consisted of oppositely charged polyelectrolyte blocks (poly(acrylic acid) and poly(N-methyl-2-vinyl pyridinium iodide)), and a neutral block (poly(vinyl alcohol)) or neutral grafts (poly(ethylene oxide)). The results showed that adsorbed micellar layers were relatively weakly attached to hydrophobic surfaces and much stronger to hydrophilic surfaces, which led to a significant impact on their stability. Examples of the measured force curves for silica, polystyrene, and polysulfone substrates with various polymer brushes using silica probe are demonstrated in Figure 1.1.

It was observed that the adsorbed layers lowered the attraction between the surface and the hydrophilic probe. Friction between the coated surfaces and the colloidal probe also remained low, indicating that adsorbed micellar layers were mobile and had lubricating properties.

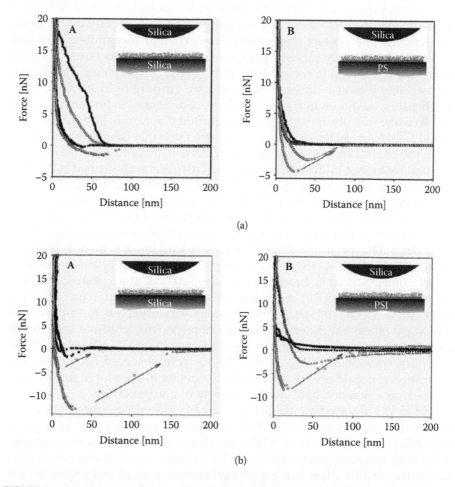

(a)

(b)

FIGURE 1.1
Force curves measured between (A) silica (a) and polystyrene (b) substrates coated with C3M-PEO$_{204}$/PAA$_{139}$ and a silica probe at 10 mM NaCl, pH 7; (B) silica (a) and polysulfone (b) substrates coated with C3M-PVA$_{445}$/P2MVPI$_{228}$, and a silica probe at 10 mM NaCl, pH 7. Closed symbols correspond to approach and open to retraction. C3M corresponds to complex coacervate core micelles. (Reproduced from Brzozowska, A. M., et al., *Journal of Colloid and Interface Science*, 353, 380–91, 2011. With permission from Elsevier.)

Pan et al. [19] described a new approach to obtaining molecularly imprinted polymers (MIPs) with both pure water-compatible and stimuli-responsive binding properties. The proof of principle was demonstrated by the facile modification of the preformed MIP microspheres via surface-initiated reversible addition fragmentation chain transfer (RAFT) polymerization of N-isopropylacrylamide (NIPAAm). The authors observed that on introduction of PNIPAAm brushes onto the MIP, microspheres significantly improved their surface hydrophilicity and imparted stimuli-responsive properties to them. Figure 1.2b and c shows the profiles of water drops on the ungrafted and PNIPAAm-grafted MIP films. Static water contact angles at 20°C were determined to be 66.7° and 124.8° for the grafted and ungrafted MIP films, respectively. The contact angle results showed that the grafted MIP film exhibited significantly higher hydrophilicity at room temperature than the ungrafted one owing to the swelling of the PNIPAAm brushes. The dispersion stability was also affected by the surface grafting of hydrophilic polymer brushes, as shown in Figure 1.2a. More sedimentation for the ungrafted MIP in water was observed than for the grafted dispersion. The authors also mixed PNIPAAm polymer with the ungrafted polymer dispersion to understand the generation of dispersion stability. The stability in this case was also comparable to that in the case of ungrafted dispersion confirmed by significant sedimentation, indicating that the mere presence of PNIPAAm did not lead to the dispersion stability. PNIPAAm was present as covalently grafted brushes that imparted the observed characteristics to the MIP particles.

Mizutani et al. [20] reported the synthesis of poly(N-isopropylacrylamide-co-N-tert-butylacrylamide) [P(IPAAm-co-tBAAm)] brushes on poly(hydroxy methacrylate) (PHMA) [hydrolyzed poly(glycidyl methacrylate-co-ethylene glycol dimethacrylate)] beads by surface-initiated atom transfer radical polymerization (SI-ATRP) as shown in Figure 1.3. The copolymer brush of P(IPAAm-co-tBAAm)-grafted PHMA beads was observed to improve the stationary phase characteristics of thermoresponsive chromatography for the all-aqueous separation of peptides and proteins. PHMA beads with the optimum amount of P(IPAAm-co-tBAAm) were observed to perform the separation of peptides with high resolution owing to their large surface area for the interaction and hydrophilic property of graft interface.

Li et al. [21] reported the surface grafting of nonfouling poly(ethylene glycol) methyl ether acrylate (PEGMA) brushes on poly(ethylene terephthalate) (PET) carried out via SI-ATRP. To achieve such brushes, the coupling agent with hydroxyl groups for the ATRP initiator was first immobilized on the surface of PET films using a photochemical method. Subsequently, the hydroxyl groups were esterified by bromoisobutyryl bromide, from which PET grafted with various main chain lengths of PEGMA was prepared. It was observed that for all PET-grafted poly(PEGMA), the topographic and phase images of the surfaces were completely changed compared to a pristine PET surface due to the coverage by poly(PEGMA) layers, as shown in Figure 1.4. The authors studied the protein adsorption resistance on the surfaces of PET

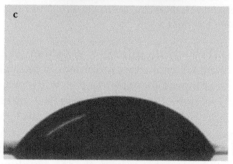

FIGURE 1.2
(a) The images of the dispersion of the ungrafted (a2) and grafted (a3) microspheres and the mixture of ungrafted microspheres with PNIPAAm (a1) in pure water at 20°C. (b) Water drop on the ungrafted MIP films and (c) water drop on the grafted MIP films. (Reproduced from Pan, G., et al., *Biosensors and Bioelectronics*, 26, 976–82, 2010. With permission from Elsevier.)

by an enzyme-linked immunosorbent assay (ELISA). It was observed that the protein adsorption could be well suppressed by poly(PEGMA) brush structure on the surface of PET.

Chen et al. [22] reported a combination of controlled ring-opening polymerization (CROP) and click reaction to synthesize linear poly(e-caprolactone) brushes at attapulgite surface. To achieve this, self-assembly of 3-chloropropyltrimethoxysilane from the surfaces of attapulgite was achieved to generate chlorine-terminated attapulgite, which was subsequently substituted with azido groups. Linear propargyl-terminated

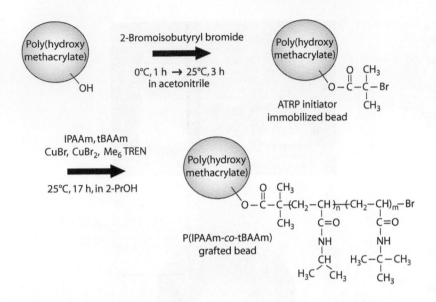

FIGURE 1.3
Schematic representation of the preparation of poly(N-isopropylacrylamide-co-tert-butyl-acrylamide) (P(IPAAm-co-tBAAm)) brush-grafted poly(hydroxy methacrylate) (PHMA) beads by SI-ATRP. (Reproduced from Mizutani, A., et al., *Journal of Chromatography A*, 1217, 5978–85, 2010. With permission from Elsevier.)

poly(ε-caprolactones) (PCLs) with different molecular weights were synthesized by the CROP of a ε-CL monomer in toluene separately. In the last step, the azido-terminated attapulgite was reacted with propargyl-terminated PCLs via the click reaction. Figure 1.5 shows the transmission electron microscopy (TEM) micrographs of the ungrafted and grafted attapulgite surfaces. The rod-like structure of the mineral was retained after grafting, but the diameter of the rods increased owing to grafting of polymer chains.

Mi et al. [23] reported a tunable mixed-charge copolymer brush containing positively charged quaternary amine monomers ([2-(acryloyloxy)ethyl] trimethyl ammonium chloride (TMA)) and negatively charged carboxylic acid monomers (2-carboxy ethyl acrylate (CAA)). The nonfouling properties of this copolymer brush were observed to depend on environmental pH. Bacteria adhered to the surface under acidic conditions, but could be easily released as bulk pH increased, as shown in Figure 1.6. The authors attributed this response of the mixed polymer brush to the increased spatial freedom and reduced interference of the oppositely charged groups in the TMA:CAA copolymer. The number of bacteria adhered to the TMA:CAA polymer brush coated surface showed a sixfold difference between acidic and neutral pH test conditions, while no significant differences were observed in both the positive (bare gold) and negative (pSBMA) controls. Such surfaces were proposed to be effective for surface enrichment, detection, and removal of microorganisms.

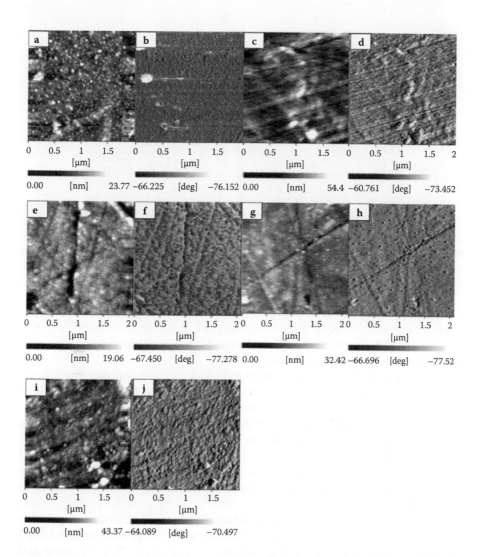

FIGURE 1.4
Topography and phase images of PET (a and b), PET10 (c and d), PET50 (e and f), PET100 (g and h), and PET200 (i and j) at 2 μm scan sizes. The PET 10, 50, 100, and 200 signify an increase in chain length of the grafted polymer. (Reproduced from Li, J., et al., *Colloids and Surfaces B: Biointerfaces*, 78, 343–50, 2010. With permission from Elsevier.)

Hou at al. [24] reported the synthesis of poly(methyl methacrylate) (PMMA) nanobrushes on silicon based on localized surface-initiated polymerization. To achieve this, self-assembled monolayers (SAMs) of octadecyltrichlorosilane (OTS) were first generated on a silicon surface. Introduction of nanostructures was achieved on these SAM surfaces using a conductive atomic force microscopy (AFM) tip, which led to the oxidation of OTS SAMs. These

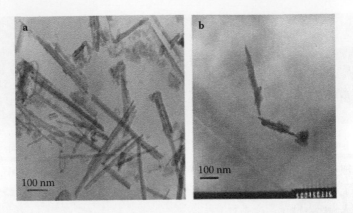

FIGURE 1.5
TEM images of (a) the bare attapulgite and (b) poly(caprolactone)-grafted attapulgite. (Reproduced Chen, J. C., et al., *Chinese Chemical Letters*, 21, 496–500, 2010. With permission from Elsevier.)

were subsequently covered with hydroxyl-terminated silane SAMs and were reacted sequentially with 2-bromoisobutyryl bromide to connect initiator groups. PMMA brushes were then formed on the initiator functionalized silicon surface, as shown in Figure 1.7.

Farhan et al. [25] reported surface-initiated polymerizations from polymeric surfaces of commercially important polyester films, poly(ethylene terephthalate) (PET), and poly(ethylene naphthalate) (PEN). Patterned self-assembled monolayers (SAMs) of the trichlorosilane initiator were first immobilized on the surface through a soft lithographic method of micro-contact printing (lCP). Grafting from the surface was initiated via controlled ATRP, under aqueous conditions, to create patterned brushes of the thermoresponsive polymer poly(N-isopropylacrylamide) (PNIPAm), as shown in Figure 1.8.

Chang et al. [26] generated roughness-enhanced thermal-responsive brushes by surface-initiated polymerization of a polymer on ordered ZnO pore array films. The effects of thickness of the grafted PNIPAm layer and surface morphology on the thermally responsive switching behavior of the PNIPAm-modified films were studied considering the three-dimensional (3D) capillary effect (Wenzel's model) and air trapping effect (Cassie's model). Figure 1.9 shows AFM images of E9D and E12D ZnO films and the films after PNIPAm grafting. As compared to the virgin ZnO film, the roughness of the PNIPAm-modified substrate was observed to decrease after grafting. The 3D capillary effect was observed to be dominant when the thin PNIPAm layer was grafted on the pore array film. On the other hand, when the PNIPAm layer became thicker, the air trapping effect dominated and the thermal responsive wettability switching of the brushes was enhanced. The authors observed that the thermally responsive switching of the brushes between hydrophilicity and hydrophobicity was enhanced with increasing molecular weight of

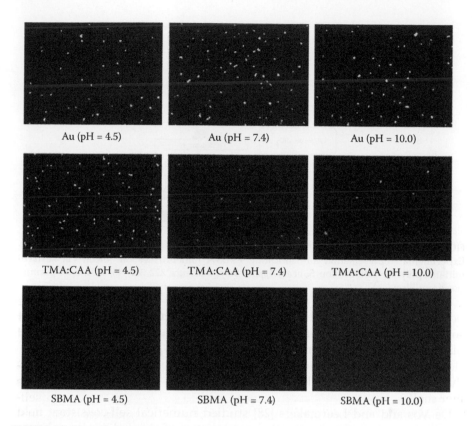

FIGURE 1.6 (See color insert.)

Fluorescence microscopy images showing *S. epidermidis* attachment to uncoated Au, TMA:CAA copolymer, and pSBMA (N-(3-sulfopropyl)-N-(methacryloxyethyl)-N,Ndimethylammonium betaine (SBMA)) surfaces at pH values of 4.5, 7.4, and 10.0 following a 3 h flow chamber adhesion assay. (Reproduced Mi, L., et al., *Biomaterials*, 31, 2919–25, 2010. With permission from Elsevier.)

PNIPAm. The authors concluded that when the pore structure of substrate was completely filled, the switching efficiency of the brushes decreased.

Mu et al. [27] reported synthesis of cross-linked polymeric nanocapsules with an inner diameter of about 20–50 nm by posttreatment of the poly(methyl acrylate) (PMA) brush-grafted silica nanoparticles (SN-PMA) generated using the SI-ATRP technique. The PMA brushes were modified with amino groups by treating with ethanediamine (EDA) followed by cross-linking of PMA chains with hexamethylene diisocyanate (HDI) and etching of the silica nanoparticle (SN) templates. Figure 1.10 shows the TEM images of the silica nanoparticles, silica nanoparticles with grafted PMA brushes, silica nanoparticles with cross-linked brushes, and nanocapsules after etching of silica, respectively.

The inner diameter of nanocapsules was observed to be larger than the sizes of the primary particles (10–20 nm). The authors attributed this to the

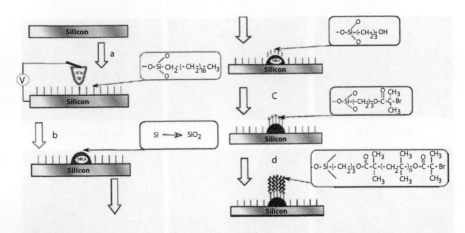

FIGURE 1.7
Experimental representation for fabricating PMMA nanobrushes on silicon via surface-initiated polymerization. (Hou, S., et al., *Applied Surface Science*, 222, 338–45, 2004. With permission from Elsevier.)

possible aggregation of the primary particles. It was also suggested that some cross-linking reactions of the amino side groups with HDI might take place between the different poly(methyl acrylate)-grafted silica nanoparticles, leading to the formation of the relatively big core-shell structures (more than one silica nanoparticle template encapsulated in the cross-linked polymer shells).

De Vos and and Leermakers [28] studied numerical self-consistent field theory to analyze the structural characteristics of a polydisperse polymer brush. The authors considered the case of a Schulz–Zimm distribution and found that even a small degree of polydispersity completely destroyed the parabolic density profile. The first moment (average height) of the brush was observed to increase with polydispersity, whereas the average stretching in the brush decreased. The authors also observed that the short chains were found to be compressed close to the grafting interface, whereas longer chains had a characteristic flower-like distribution. Figure 1.11 shows the effect of polydispersity on the volume fraction and endpoint distribution profiles. Even for polydispersities considered low from a synthetic point of view, the effect on the profiles was significant. The profile changed from convex to linear upon going from a polydispersity of unity (homodisperse brush) to a polydispersity of Mw/Mn 1.1. At higher polydispersities the profile became completely concave. A maximum in endpoint distribution existed at the edge of a homodisperse brush. On increasing polydispersity this maximum moved closer to the grafting interface, even though the overall height of the brush increased.

Zhang et al. [29] reported well-defined poly(2-(dimethylamino) ethyl methacrylate) (PDMAEMA) brushes with a high density on the surface of polystyrene latex particles by ATRP using acetone/water as the solvent and

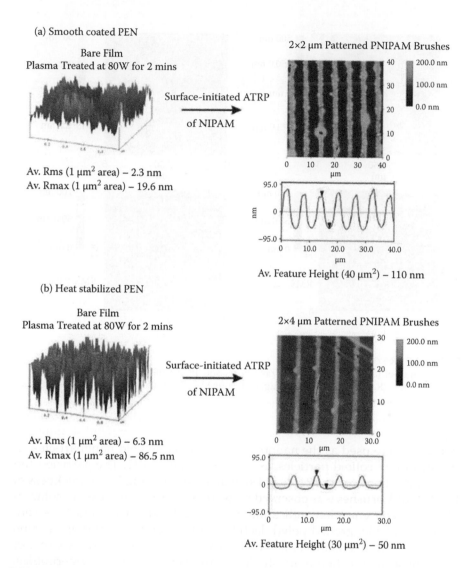

FIGURE 1.8
The left-hand column represents topographical 3D AFM images of the bare films (postplasma) of (a) smooth-coated heat-stabilized PEN and (b) heat-stabilized PEN. Right-hand column shows the patterned PNIPAm brushes. (Reproduced from Farhan, T., and Huck, W. T. S., *European Polymer Journal*, 40, 1599–604, 2004. With permission from Elsevier.)

CuCl/CuCl$_2$/bpy as the catalyst. The ATRP initiator layer on the surface of polystyrene particles was prepared by seed emulsion polymerization of 2-(2-bromoisobutyryloxy) ethyl methacrylate (BIEM) using polystyrene (PS) latex as seeds.

It was observed that the polydispersity of PDMAEMA brushes decreased with the increasing external CuCl$_2$ concentration. Subsequently, PDMAEMA

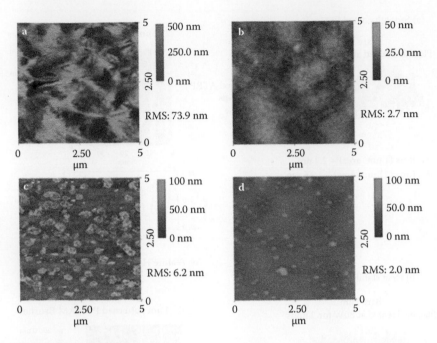

FIGURE 1.9 (See color insert.)
AFM images of (a) E9D and (c) E12D ZnO films and PNIPAm-modified (b) E9DPN and (d) E12DPN films. (Reproduced from Chang, C.-J., and Kuo, E.-H., *Thin Solid Films*, 519, 1755–60, 2010. With permission from Elsevier.)

domains were used as the nanoreactors to generate gold nanoparticles on the surface of colloid particles, as shown in Figure 1.12. The particles had an average diameter of 4.2 nm and size distribution of 1.12. The thickness of PDMAEMA brushes was observed to decrease with the increase of solution pH or salt concentration, indicating pH- and salt-sensitive characteristics.

Yang et al. [30] grafted poly(L-lactide) (PLLA) comb polymer brushes on poly(hydroxyethyl methacrylate) (PHEMA) backbone on the surface of clay layers by a combination of in situ ATRP and ring-opening polymerization. An ATRP initiator with a quaternary ammonium salt end group was ionically exchanged on the surface of clay. In situ ATRP was used to generate PHEMA polymer brushes on the surface of clay, whereas PLLA comb polymers on PHEMA backbone were prepared by ring-opening polymerization. Figure 1.13 shows the TEM images of the grafted clay with increasing length of the comb. It is evident that with the growth of comb chain length a significant degree of silicate exfoliation was achieved.

Figure 1.13a shows, at the lower length of the comb, large clay particles and their aggregation. Figure 1.13b and c indicates that on increasing the comb length, most of the big clay particles disappeared. Figure 1.13d shows that further increment in the comb length achieved highest exfoliation, thus

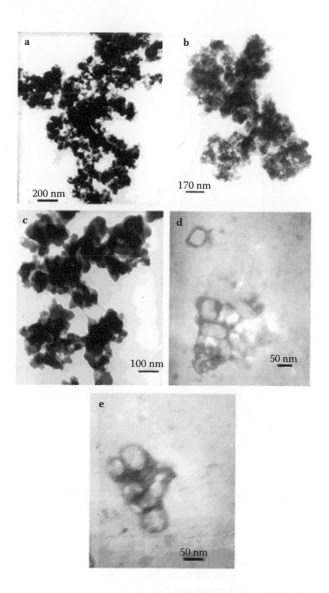

FIGURE 1.10
TEM images of SN (a), SN-PMA (b), SN-CP (cross-linked particles) (c), and cross-linked polymeric nanocapsules (d and e). (Reproduced from Mu, B., et al., *Colloids and Surfaces B: Biointerfaces*, 74, 511–15, 2009. With permission from Elsevier.)

indicating that with increase of comb chain length, more and more exfoliated structure was created.

Wang et al. [31] used a surface-initiated reverse atom transfer radical polymerization (reverse ATRP) technique to synthesize the well-controlled nanostructure of polymer brushes from silicon wafer. PMMA brushes were prepared by modifying the surface of the silicon substrate with peroxide

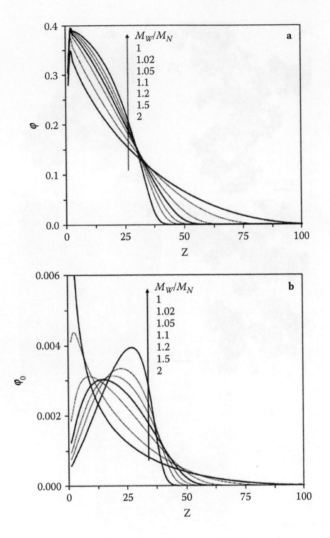

FIGURE 1.11

(a) Overall volume fraction profile for brushes with increasing polydispersity M_w/M_n, but still low in extent as indicated in the plots. (b) The corresponding overall distribution of endpoints. (Reproduced de Vos, W. M., and Leermakers, F. A. M., *Polymer*, 50, 305–16, 2009. With permission from Elsevier.)

groups. Figure 1.14a shows the plot of the thickness of the PMMA brushes grown on the silicon surface using reverse ATRP versus time. The figure also shows the first-order kinetic plot vs. time. A linear increase in the thickness of the PMMA graft layer on the silicon surface with polymerization time was observed. Apart from that, a linear relationship between ln ([M0]/ [M]) and time, where [M0] is the initial monomer concentration and [M] is monomer concentration at any time, was also observed, thus confirming that the concentration of the growing species remained constant. The authors

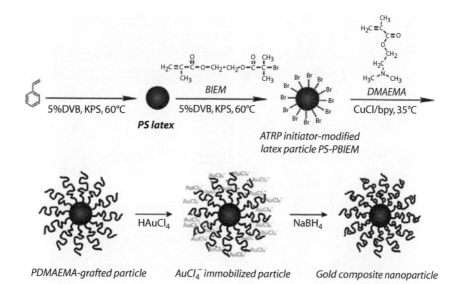

FIGURE 1.12
Schematic representation of the synthesis of PDMAEMA brushes on the surface of colloid particles by ATRP. (Reproduced Zhang, M., et al., *Polymer*, 48, 1989–97, 2007. With permission from Elsevier.)

suggested that the molecular weight of the graft polymer was expected to be proportional to that of the polymer formed in the solution. Figure 1.14b also shows the plot of the graft layer thickness vs. Mn of the homopolymer, again indicating a proportional relationship confirming that the process of surface-initiated reverse ATRP of methyl methacrylate (MMA) was controlled.

Parvole et al. [32] described formation of well-defined polymer brushes based on the nitroxide-mediated polymerization (NMP) of n-butyl acrylate (BA), which were initiated from self-assembled monolayers (SAMs) of an azoic initiator in the presence of a stable nitroxide radical. The grafting of the brushes was carried out from a silica surface. It was observed that the polymer monolayers exhibited all the features of the polymer brushes, i.e., strong crowding of neighboring grafted chains and extension of the chains in the direction perpendicular to the grafting surface with respect to the Gaussian dimensions. Figure 1.15 shows the AFM images of grafted poly(butyl acrylate) (PBA) monolayers. In Figure 1.15a, the holes were observed on the extreme surface, which could be associated to the presence of pinned micelles or other types of nano-segregated structures. In this case, the degree of polymerization of PBA was around 26, leading to an average molecular weight of 3,700 g/mol. In such a case, the degree of overlapping of the chains in the layer led to the appearance of individual pinned globules or pinned micelles in the dry (collapsed) state.

Wang et al. [33] prepared polymer brushes on a palygorskite surface by the RAFT polymerization technique using palygorskite as a RAFT chain

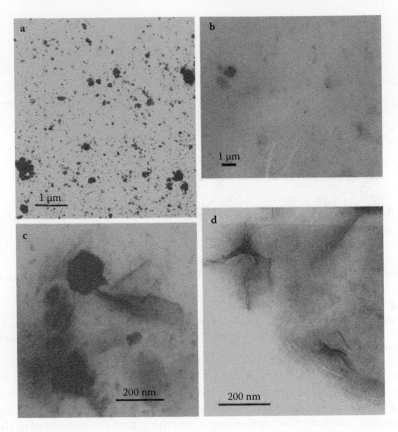

FIGURE 1.13
TEM images of (a) PLLA10-PHEMA/clay at a low magnification, (b) PLLA92-PHEMA/clay at a low magnification, (c) PLLA92-PHEMA/clay at a high magnification, and (d) PLLA120-PHEMA/clay at a high magnification. (Reproduced Yang, Y., et al., *Polymer*, 47, 7374–81, 2006. With permission from Elsevier.)

transfer agent (CTA) support. Figure 1.16 shows the synthetic pathway for palygorskite-supported benzyl dithiopropyltrimethoxysilane (P–Si-BDPM) and the RAFT polymerization of MMA using P–Si-BDPM as the RAFT agent.

The organic content of the P-g-PMMA was determined by thermogravimetric analysis (TGA). The value of organic content obtained was relatively high in all cases (17–21 wt%), which agreed well with the relatively high grafting rates. A linear relationship in the first-order rate plot for the polymerization of MMA using P–Si-BDPM as the RAFT agent was observed, which indicated that the kinetics was of first order in monomer, and the concentration of propagating radicals remained constant during the polymerization. A linear relationship was also observed between the monomer conversion and the molecular weight of the free polymer formed in the solution, indicating the living nature of the polymerization. Narrow molecular weight of the polymer further confirmed the living nature of the polymerization.

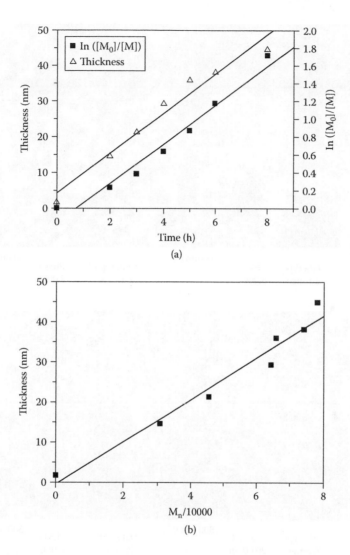

FIGURE 1.14
(a) First-order kinetic plot (right) and plot of ellipsometric polymer brush thickness vs. time (left) for the reverse ATRP of MMA from initiator-immobilized silicon surface. (b) The linear relationship between the thickness of the PMMA layer and the molecular weight of the "free" polymer formed in the solution. (Reproduced from Wang, Y.-P., et al., *European Polymer Journal*, 41, 737–41, 2005. With permission from Elsevier.)

Hoven et al. [34] used a chemically grafted tris(trimethylsiloxy)silyl (tris(TMS)) monolayer on a silicon oxide substrate as a template for creating nanoclusters of polymer brushes. Surface-initiated polymerization of 2-methacryloyloxyethyl phosphorylcholine (MPC) and tert-butyl methacrylate (t-BMA) was used to generate polymer brushes via ATRP from α-bromoester groups tethered to the residual silanol groups on the silicon surface, as

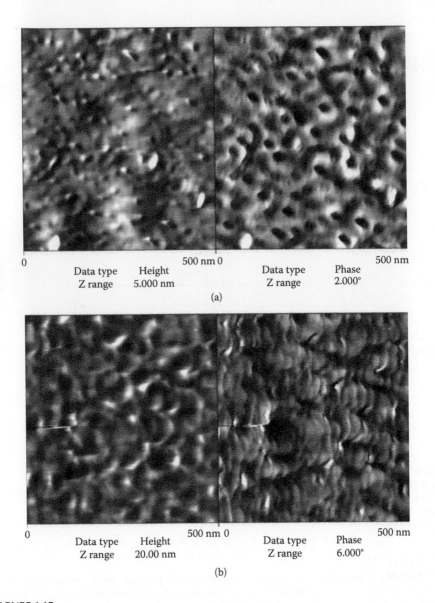

FIGURE 1.15
AFM images of PBA polymer brushes grafted on silicon wafer: (a) M_n = 3,700 g/mol and
(b) M_n = 86,000 g/mol. (Reproduced Parvole, J., et al., *Polymer*, 47, 972–81, 2006. With permission
from Elsevier.)

depicted in Figure 1.17. The percentage of tris(TMS) coverage on the surface
was observed to significantly influence the thickness and morphology of
the resulting polymer brushes. The thickness of PMPC and Pt-BMA brushes
increased with time, suggesting that polymerization time could be used as
a tool for controlling the growth of polymer brushes. Both advancing and

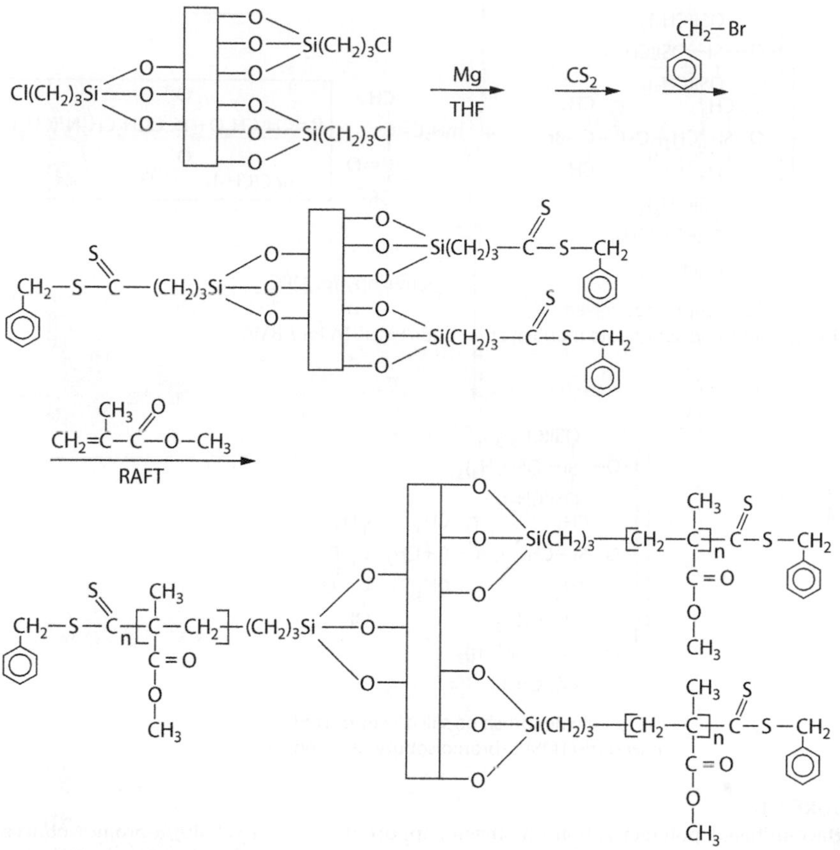

FIGURE 1.16
Synthetic pathway for palygorskite-supported benzyl dithiopropyltrimethoxysilane (P–Si-BDPM) and the RAFT polymerization of MMA using P–Si-BDPM as the RAFT agent. (Reproduced Wang, L.-P., et al., *Reactive and Functional Polymers*, 68, 643–48, 2008. With permission from Elsevier.)

receding water contact angles of the silicon-supported α-bromoisobutyrate monolayer were observed to drastically change from 72°/68° to 20°/1° for the hydrophilic silicon-supported PMPC brushes and to 100°/80° for the hydrophobic silicon-supported Pt-BMA brushes.

Liu et al. [35] investigated the grafting of thiol-terminated poly[(2-dimethylamino)ethyl methacrylate] (HS-PDMEM) chains to a gold surface from a solution. Quartz crystal microbalance with dissipation (QCM-D) in real time together with atomic force microscopy was used to study the grafting process. The studies revealed a three-regime kinetics of the grafting of the polymer chains to the surface. In the first regime, the chains quickly grafted on the surface forming a random mushroom. In regime II, the grafted chains underwent a rearrangement and formed an ordered mushroom structure. In regime III,

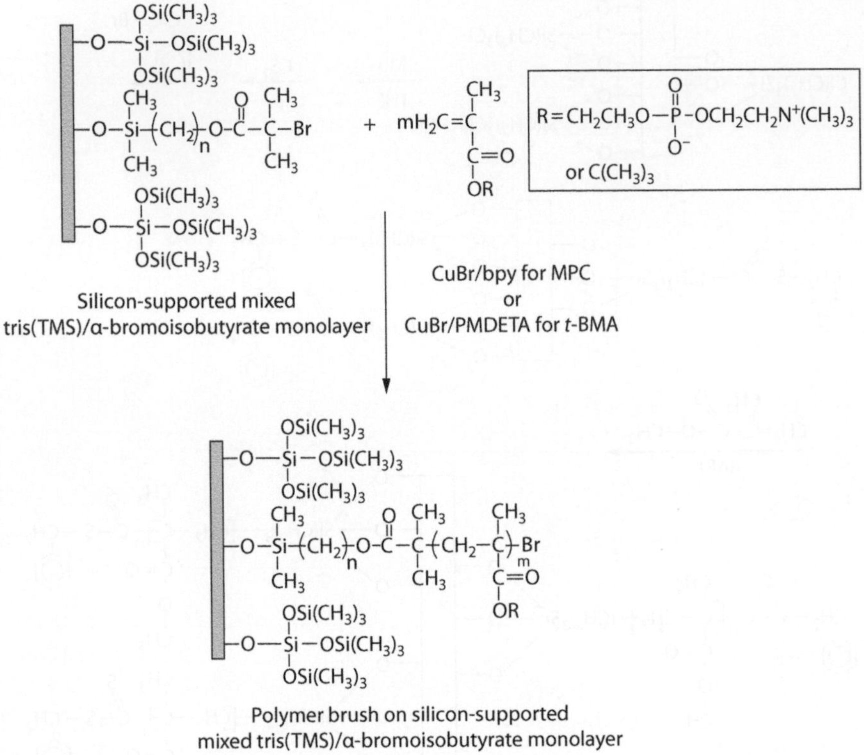

FIGURE 1.17
Surface-initiated polymerization on silicon-supported mixed tris(TMS)/α-bromoisobutyrate monolayer. (Reproduced Hoven, V. P., et al., *Journal of Colloid and Interface Science*, 314, 446–59, 2007. With permission from Elsevier.)

the grafting was accelerated and as the grafting density increased, the chains formed brushes. The authors observed a mushroom-to-brush transition from regime II to III. Figure 1.18 shows the AFM images of PDMEM layers at different pH values. For the bare gold crystal surface, the surface was very smooth with roughness approximately 0.75 nm. Figure 1.18a and b shows the AFM images of the PDMEM layer in regimes II and III at pH 2. The roughness values were 1.47 and 2.80 nm, respectively. At pH 6, the roughness value increased and increased further at pH 10. As a higher degree of charging led to a more stretched brush and as the roughness increased with the stretching of the chains, the fact that the roughness increased from regime II to III at each pH confirmed that the mushroom-to-brush transition occurred.

Suzuki et al. [36] modified silica particles (SiP) with a 2-bromoisobutyryl group-carrying silane coupling reagent, and a polymer brush of carboxy methylbetaine, poly[1-carboxy-N,N-dimethyl-N-(2′- methacryloyloxyethyl) methanaminium inner salt] (PolyCMB), was generated onto the surface of the particles using SI-ATRP.

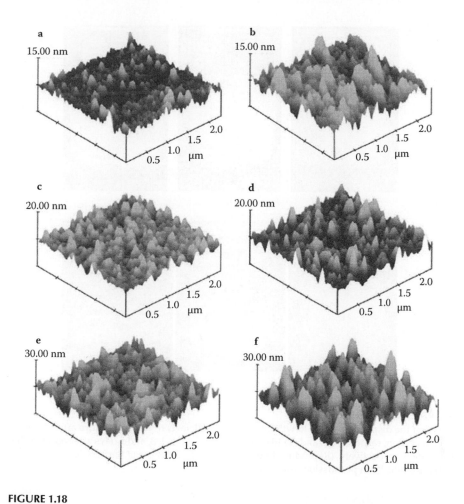

FIGURE 1.18

AFM images of PDMEM layers at different pH: (a) in regime II at pH 2, (b) in regime III at pH 2, (c) in regime II at pH 6, (d) in regime III at pH 6, (e) in regime II at pH 10, (f) in regime III at pH 10. (Reproduced from Liu, G., et al., *Polymer*, 47, 3157–63, 2006. With permission from Elsevier.)

SiP-PolyCMBs particles dispersed in water showed neither salt- nor freeze-thaw cycle-induced coagulation. These particles were observed to be resistant against the nonspecific adsorption of proteins such as bovine serum albumin and egg white lysozyme in contrast to the significant adsorption of the proteins onto the bare silica particles.

As shown in Figure 1.19A, the polymer-modified silica particles showed iridescence indicating that the colloidal crystal was formed. This behavior was in contrast with the rapid sedimentation of bare SiP. It was attributed to the lower density of SiP-PolyCMB as compared to bare SiP.

FIGURE 1.19
Images of the dispersion of (A) 10 wt% of SiP-PolyCMB: (a–c) 6, 18, and 36 days after onset of the observation, respectively; and (B) 10 wt% of bare SiP: (d–f) 1, 3, and 5 days after onset of the observation, respectively. (Reproduced Suzuki, H., et al., *Colloids and Surfaces B: Biointerfaces,* 84, 111–16, 2010. With permission from Elsevier.)

References

1. Chaudhury, M. K., and Whitesides, G. M. 1992. How to make water run uphill. *Science* 256:1539–41.
2. Nagasaki, Y., and Kataoka, K. 1996. An intelligent polymer brush. *Trends in Polymer Science* 4:59–64.
3. Mansky, P., Liu, Y., Huang, E., Russell, T. P., and Hawker, C. J. 1997. Controlling polymer-surface interactions with random copolymer brushes. *Science* 275:1458–60.
4. Fytas, G., Anastasizdis, S. H., Seghrouchni, R., Vlassopoulos, D., Li, J., Factor, B. J., et al. 1996. Probing collective motions of terminally anchored polymers. *Science* 274:2041–44.
5. Mir, Y., Auroy, P., and Auvray, L. 1995. Density profile of polyelectrolyte brushes. *Physical Review Letters* 75:2863–66.

6. Prucker, O., and Ruhe, J. 1998. Synthesis of poly(styrene) monolayers attached to high surface area silica gels through self-assembled monolayers of azo initiators. *Macromolecules* 31:592–601.
7. Prucker, O., and Ruhe, J. Mechanism of radical chain polymerizations initiated by azo compounds covalently bound to the surface of spherical particles. *Macromolecules* 31:602–13.
8. Milner, S. T. 1991. Polymer brushes. *Science* 251:905–14.
9. Jordan, R., Ulman, A., Kang, J. F., Rafailovich, M. H., and Sokolov, J. 1999. Surface-initiated anionic polymerization of styrene by means of self-assembled monolayers. *Journal of the American Chemical Society* 121:1016–22.
10. Ingall, M. D. K., Honeyman, C. H., Mercure, J. V., Bianconi, P. A., and Kunz, R. R. 1999. Surface functionalization and imaging using monolayers and surface-grafted polymer layers. *Journal of the American Chemical Society* 121:3607–13.
11. Mittal, V. 2007. Polymer chains grafted "to" and "from" the layered silicate clay platelets. *Journal of Colloid and Interface Science* 314:141–51.
12. Zhao, B., and Brittain, W. J. 2000. Synthesis, characterization, and properties of tethered polystyrene-*b*-polyacrylate brushes on flat silicate substrates. *Macromolecules* 33:8813–20.
13. Kim, J., Bruening, M. L., and Baker, G. L. 2000. Surface-initiated atom transfer radical polymerization on gold at ambient temperature. *Journal of the American Chemical Society* 122:7616–17.
14. Sun, T., Wang, G., Feng, L., Liu, B., Ma, Y., Jiang, L., et al. 2004. Reversible switching between superhydrophilicity and superhydrophobicity. *Angewandte Chemie International Edition* 43:357–60.
15. You, Y. Z., Hong, C. Y., Pan, C. Y., and Wang, P. H. 2004. Synthesis of a dendritic core-shell nanostructure with a temperature-sensitive shell. *Advanced Materials* 16:1953–57.
16. D'Agosto, F., Charreyre, M.-T., Pichot, C., and Gilbert, R. G. 2003. Latex particles bearing hydrophilic grafted hairs with controlled chain length and functionality synthesized by reversible addition-fragmentation chain transfer. *Journal of Polymer Science, Part A: Polymer Chemistry* 41:1188–95.
17. Lamb, D., Anstey, J. F., Fellows, C. M., Monteiro, J. M., and Gilbert, R. G. 2001. Modification of natural and artificial polymer colloids by "topology-controlled" emulsion polymerization. *Biomacromolecules* 2:518–25.
18. Brzozowska, A. M., Spruijt, E., de Keizer, A., Cohen Stuart, M. A., and Norde, W. 2011. On the stability of the polymer brushes formed by adsorption of ionomer complexes on hydrophilic and hydrophobic surfaces. *Journal of Colloid and Interface Science* 353:380–91.
19. Pan, G., Zhang, Y., Guo, X., Li, C., and Zhang, H. 2010. An efficient approach to obtaining water-compatible and stimuli-responsive molecularly imprinted polymers by the facile surface-grafting of functional polymer brushes via RAFT polymerization. *Biosensors and Bioelectronics* 26:976–82.
20. Mizutani, A., Nagase, K., Kikuchi, A., Kanazawa, H., Akiyama, Y., Kobayashi, J., Annaka, M., and Okano, T. 2010. Preparation of thermo-responsive polymer brushes on hydrophilic polymeric beads by surface-initiated atom transfer radical polymerization for a highly resolutive separation of peptides. *Journal of Chromatography A* 1217:5978–85.

21. Li, J., Tan, D., Zhang, X., Tan, H., Ding, M., Wan, C., and Fu, Q. 2010. Preparation and characterization of nonfouling polymer brushes on poly(ethylene terephthalate) film surfaces. *Colloids and Surfaces B: Biointerfaces* 78:343–50.

22. Chen, J. C., Wang, Y. P., Wang, H. D., Wei, S. Y., Li, H. J., and Yan, X. D. 2010. Preparation of polymer brushes on attapulgite surfaces via a combination of CROP and click reaction. *Chinese Chemical Letters* 21:496–500.

23. Mi, L., Bernards, M. T., Cheng, G., Yu, Q., and Jiang, S. 2010. pH responsive properties of non-fouling mixed-charge polymer brushes based on quaternary amine and carboxylic acid monomers. *Biomaterials* 31:2919–25.

24. Hou, S., Li, Z., Li, Q., and Liu, Z. F. 2004. Poly(methyl methacrylate) nanobrushes on silicon based on localized surface-initiated polymerization. *Applied Surface Science* 222:338–45.

25. Farhan, T., and Huck, W. T. S. 2004. Synthesis of patterned polymer brushes from flexible polymeric films. *European Polymer Journal* 40:1599–604.

26. Chang, C.-J., and Kuo, E.-H. 2010. Roughness-enhanced thermal-responsive surfaces by surface-initiated polymerization of polymer on ordered ZnO porearray films. *Thin Solid Films* 519:1755–60.

27. Mu, B., Shen, R., and Liu, P. 2009. Crosslinked polymeric nanocapsules from polymer brushes grafted silica nanoparticles via surface-initiated atom transfer radical polymerization. *Colloids and Surfaces B: Biointerfaces* 74:511–15.

28. de Vos, W. M., and Leermakers, F. A. M. 2009. Modeling the structure of a polydisperse polymer brush. *Polymer* 50:305–16.

29. Zhang, M., Liu, L., Wu, C., Fu, G., Zhao, H., and He, B. 2007. Synthesis, characterization and application of well-defined environmentally responsive polymer brushes on the surface of colloid particles. *Polymer* 48:1989–97.

30. Yang, Y., Wu, D., Li, C., Liu, L., Cheng, X., and Zhao, H. 2006. Poly(L-lactide) comb polymer brushes on the surface of clay layers. *Polymer* 47:7374–81.

31. Wang, Y.-P., Pei, X.-W., He, X.-Y., and Lei, Z.-Q. 2005. Synthesis and characterization of surface-initiated polymer brush prepared by reverse atom transfer radical polymerization. *European Polymer Journal* 41:737–41.

32. Parvole, J., Montfort, J.-P., Reiter, G., Borisov, O., and Billon, L. 2006. Elastomer polymer brushes on flat surface by bimolecular surface-initiated nitroxide mediated polymerization. *Polymer* 47:972–81.

33. Wang, L.-P., Wang, Y.-P., Wang, R.-M., and Zhang, S.-C. 2008. Preparation of polymer brushes on palygorskite surfaces via RAFT polymerization. *Reactive and Functional Polymers* 68:643–48.

34. Hoven, V. P., Srinanthakul, M., Iwasaki, Y., Iwata, R., and Kiatkamjornwong, S. 2007. Polymer brushes in nanopores surrounded by silicon-supported tris(trimethylsiloxy)silyl monolayers. *Journal of Colloid and Interface Science* 314:446–59.

35. Liu, G., Yan, L., Chen, X., and Zhang, G. 2006. Study of the kinetics of mushroom-to-brush transition of charged polymer chains. *Polymer* 47:3157–63.

36. Suzuki, H., Murou, M., Kitano, H., Ohnob, K., and Saruwatari, Y. 2010. Silica particles coated with zwitterionic polymer brush: Formation of colloidal crystals and anti-biofouling properties in aqueous medium. *Colloids and Surfaces B: Biointerfaces* 84:111–16.

2

Grafting of Organic Brushes on the Surface of Clay Platelets

V. Mittal

*Chemical Engineering Department, Petroleum Institute,
Abu Dhabi, United Arab Emirates*

CONTENTS

2.1 Introduction

Clay or layered aluminosilicate (especially montmorillonite) has been a filler of choice for most of the studies on polymer nanocomposites where the thin aluminosilicate platelets are dispersed in the polymer matrix at a nanometer level, thus leading to significant enhancements in the polymer properties at low filler fractions. The general formula of montmorillonites is $M_x(Al_{4-x}Mg_x)Si_8O_{20}(OH)_4$. Its particles consist of stacks of 1 nm thick aluminosilicate layers (or platelets) with a regular gap in between (interlayer). Each layer consists of a central Al-octahedral sheet fused to two tetrahedral silicon sheets. In the tetrahedral sheets, silicon is surrounded by four oxygen atoms, whereas in the octahedral sheets, an aluminum atom is surrounded by eight oxygen atoms. Isomorphic substitutions of aluminum by magnesium in the octahedral sheet generate negative charges, which are compensated for by alkaline-earth or hydrated alkali-metal cations, as shown in Figure 2.1 [1,2]. The majority of these cations are present in the interlayers between the sheets, but some percentage of them is present on the edges of the sheets. Based on the extent of the substitutions in the silicate crystals, the term *layer charge density* is defined. Montmorillonites have a mean layer charge density of 0.25–0.5

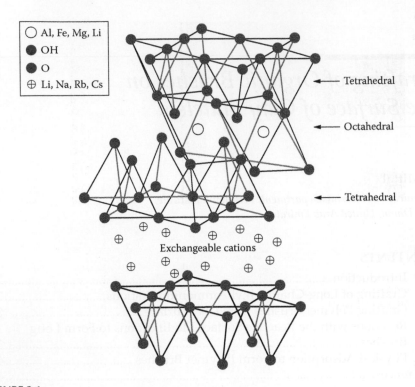

FIGURE 2.1
Structure of the layered aluminosilicates. (Reproduced from Pavlidoua, S., and Papaspyrides, C. D., *Progress in Polymer Science*, 33, 1119–98, 2008, and Beyer, G., *Plastics Additives and Compounding*, 4, 22–27, 2002. With permission from Elsevier.)

equiv.mol⁻¹. The layer charge is also not constant and can vary from layer to layer; therefore, it should be considered more of an average value. The electrostatic and van der Waals forces holding the layers together are relatively weak in these materials, and the interlayer distance varies depending on the radius of the cation present and its degree of hydration. As a result, the stacks swell in water and the 1 nm thick layers can be easily exfoliated by shearing, giving platelets with a high aspect ratio. Unfortunately, their high energetic hydrophilic surfaces are incompatible with many polymers, whose low energetic surfaces are hydrophobic. Compatibility is required between these organic and inorganic phases for the nanoscale dispersion of the platelets in the polymer. In the absence of compatibility, uniform dispersion of the filler would not be achieved. To achieve this, the inorganic cations are exchanged with organic ions (e.g., alkyl ammonium) to give organically modified montmorillonite (OMMT) that does not suffer from this problem of incompatibility with the polymer matrix [3,4]. An exchange of inorganic cations with organic cations renders the silicate organophilic and hydrophobic, lowers the surface energy of the platelets, and increases the basal plane or interlayer spacing (d-spacing). This improves the wetting, swelling, and exfoliation of

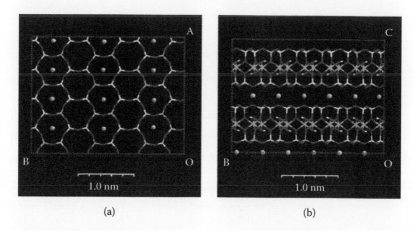

(a)   (b)

FIGURE 2.2 (See color insert.)
A simulated representation of mica unit cells. Al atoms are depicted in blue, Si in yellow, O in red, H in white, and K (or Li) ions in violet. (a) Top view showing only the upper tetrahedral layer and (b) side view showing that the cavities in the lamellae are fully occupied with alkali ions and stacked on each other. (Reprinted with permission from Heinz, H., et al., *Journal of American Chemical Society*, 125, 9500–10, 2003. Copyright (2003) American Chemical Society.)

the aluminosilicate in the polymer matrix. Figure 2.2 also shows the molecular dynamics images of the top and side views of the aluminosilicates [5].

Modification of the clay surface by ion exchange of the inorganic cations with the long-chain ammonium cations leads to the generation of oligomeric brushes on the surface. However, it is not the only way to achieve such brushes. Several methods used to achieve organic brushes on the surface of clay are:

1. Ion exchange with the long-chain ammonium cations
2. Exchange of monomer or initiator moiety on the surface of clay platelets followed by polymerization of externally added monomer
3. Chemical reactions with the surface modification, which has reactive functional groups
4. Physical adsorption of polymer chains on the clay surface

2.2 Grafting of Long-Chain Alkyl Ammonium Cations

Figure 2.3 shows the representation of the exchange of long-chain alkyl ammonium cations on the filler surface [6]. Depending on the area available per cation as well as the cross section of the ammonium cation, the chains either lie on the surface or radiate away from the surface, forming an angular spatial arrangement with the surface. A number of ammonium cations

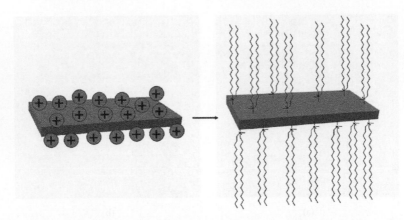

FIGURE 2.3
Exchange of long-chain alkyl ammonium ions on the surface. (Reproduced from Mittal, V., in *Optimization of Polymer Nanocomposites Properties*, ed. V. Mittal, Wiley VCH, Weinheim, Germany, 2010, pp. 1–20. With permission from Wiley.)

like octadecyltrimethylammonium, dioctadcyldimethylammonium, triocta-decylmethylammonium, benzylhexadecyltrimethylammonium, etc., have been used to form grafts on the clay surface. Figure 2.4 also demonstrates the impact of cation exchange capacity as well as chain density in surface modification on the grafting [6].

Three different surface modifications like octadecyltrimethylammonium, dioctadecyldimethylammonium, and trioctadecylmethylammonium were used to form grafts on two clay platelet surfaces with cation exchange capacities of 680 and 880 μ.eq g^{-1}. As shown in Figure 2.4a, the amount of grafted material increased on increasing the chain density in surface modification, as indicated by the increase in the amount of organic matter obtained from thermogravimetric analysis (TGA). The increase in basal spacing with increasing the chain density in the surface modification was further confirmation of the increased grafting as the chains were immobilized in a more straight fashion as the chain density increased and more crowding of the modification chains occurred on the surface. For the same modification, the filler with higher cation exchange capacity had a higher extent of grafting owing to the lesser are per cation available on the surface with higher cation exchange capacity. This again led to more straight positioning of the modification chains on the surface. Figure 2.4b also shows the TGA thermograms of the modified fillers for the filler with a cation exchange capacity of 880 μ.eq g^{-1}. The increase in the amount of grafted material is visible on increasing the chain density in the surface modification. The peak degradation temperature of the grafts was also observed to shift to higher temperature on increasing the chain density.

Figure 2.5 also shows an interesting finding in the case of trioctadecyl-methylammonium-grafted clay platelets. The order-disorder transition of these brushes was observed to be near 50°C. The same order-disorder

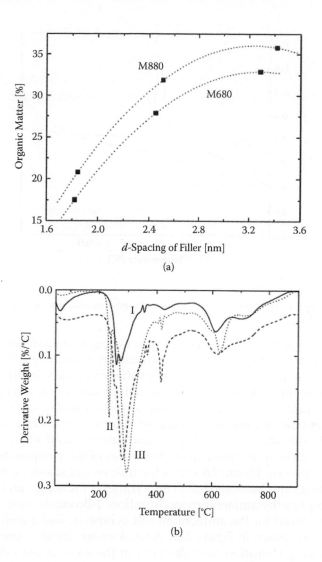

FIGURE 2.4
(a) Basal plane spacing as a function of increasing chain density in the modification as well as filler CEC. (b) TGA of the filler modified with (I) octadecyltrimethylammonium, (II) dioctadecyldimethylammonium, and (III) trioctadecylmethylammonium (with a CEC of 880 μ.eq g^{-1}). (Reproduced from Mittal, V., in *Optimization of Polymer Nanocomposites Properties*, ed. V. Mittal, Wiley VCH, Weinheim, Germany, 2010, pp. 1–20. With permission from Wiley.)

transition of the monolayer was also observed to take place at the same temperature in polypropylene composites containing 3 vol% of the modified filler, indicating that the grafts did not mix well with the polymer, indicating their resistance to penetration by the polypropylene chains.

Apart from chain density in the surface modification, the length of the chains also affects the grafting process. The preformed long chains are

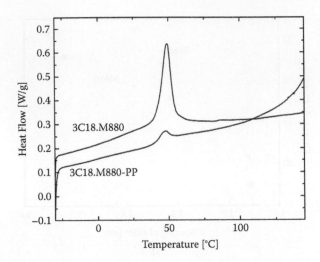

FIGURE 2.5
DSC thermograms of trioctadecylmethylammonium-grafted clay and its 3 vol% composite with polypropylene.

difficult to be grafted on the filler surface owing to insolubility. However, with suitable reagents, the surface modification molecules with optimum chain length can be generated and do not suffer from the solubility concerns. In one such example, docosyltriethylammonium with 22 carbon atoms in the chain was synthesized and used to form grafts on the clay surface [7].

The basal plane spacing of the modified filler increased to a small extent compared to the octadecyltrimethylammonium-modified filler, and significant improvement in the gas barrier properties of the composites with this filler were observed. Figure 2.6 shows the differential scanning calorimetry (DSC) thermograms associated with the synthesis, exchange, and composite of the docosyltriethylammonium modification. 1-Bromodocosane, which is a starting material for the ammonium salt synthesis, had a melting transition at 45°C, as shown in Figure 2.6a. A weak order-disorder transition and a strong melting transition were observed in the ammonium salt at 44 and 96°C, respectively, as shown in Figure 2.6b. The modified filler also indicated similar transitions but at higher temperatures of 55 and 129°C, as shown in Figure 2.6c. This is due to improved thermal behavior after tethering one end of the ammonium salt with the clay surface. Finally, the polypropylene composites, as shown in Figure 2.6d, also showed the presence of similar peaks with the lower temperature transition shifting by 2°C and higher temperature transition shifting by 17°C. The behavior was different from that in the above-mentioned case of grafts formed from trioctadecylmethylammonium modification, where no change in the transition temperature was observed on making the composites. An increase of 2 and 17°C in the transitions associated with docosyltriethylammonium grafts is only possible if the modifier chains on the surface mix well with the polypropylene chains, thus forming

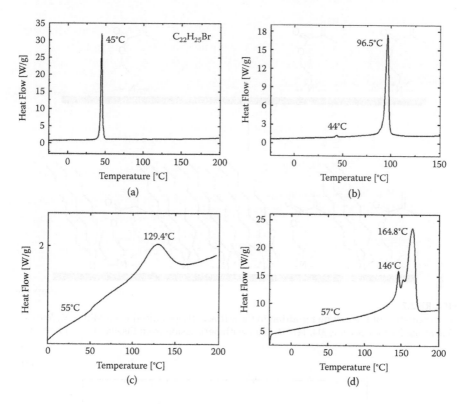

FIGURE 2.6
DSC thermograms of (a) 1-bromodocosane, (b) docosyltriethylammonium bromide, (c) clay exchanged with docosyltriethylammonium, and (d) polypropylene nanocomposite incorporated with docosyltriethylammonium-modified clay. (Reproduced from Mittal, V., *Philosophical Magazine*, 90, 2489–506, 2010. With permission from Taylor and Francis.)

a high melting layer at the interface, which helps to generate better interfacial morphology in the composite.

2.3 Grafting Polymer Brushes from Clay Surface

Grafting of polymer chains either to or from the surface is a common method to generate polymer brushes from the surfaces. In the polymerization to the surface method, the filler can be modified with a monomer, as shown in Figure 2.7 [8]. The surface can also be partially covered with the monomer, while the rest of the surface can have nonreactive brushes, thus generating mixed brushes. In the polymerization from the surface, the surface can be partially modified with an initiator, thus again generating a mixed brush.

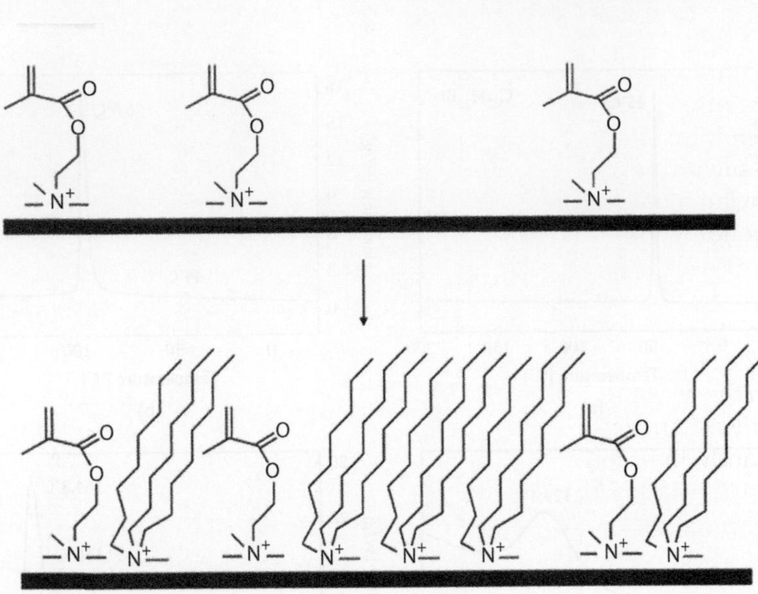

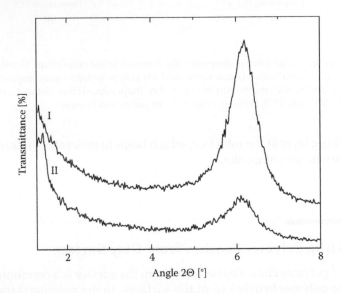

FIGURE 2.7
Representation of the polymer grafting to the surface. (Reproduced from Mittal, V., *Journal of Colloid and Interface Science*, 314, 141–51, 2007. With permission from Elsevier.)

FIGURE 2.8
X-ray diffractogram of (I) clay modified by using methacryloxyethyltrimethylammonium chloride (MOETMAC) in an amount corresponding to 70% of the CEC and methyltrioctylammonium in an amount corresponding to 30% of the CEC and (II) clay of I after reaction at 60°C (solution polymerization). (Reproduced from Mittal, V., *Journal of Colloid and Interface Science*, 314, 141–51, 2007. With permission from Elsevier.)

The polymerization in both cases can be initiated by adding external monomer. Polymerization from the surface generates more controlled brushes, and even the controlled living polymerization can be applied to this system. In one such example of polymerization to the surface, methacryloxyethyltrimethylammonium chloride was exchanged on the surface in an amount corresponding to 70% of cation exchange capacity (CEC), with the rest being nonreactive methyltrioctylammonium. Polymerization of lauryl methacrylate in the presence of initiator azo bis(iso butyro nitrile) (AIBN) led to the generation of weak peaks at 1.46 and 6.40 nm in the x-ray diffractogram shown in Figure 2.8. The clay before the reaction had a basal plane spacing of 1.43 nm. Though the peak at 6.40 nm was on the border of sensitivity of the wide-angle x-ray diffraction device, the intensity of the reacted clay was also diminished significantly, indicating grafting of polymer brushes on the surface.

Similarly, trials of polymerization of lauryl methacrylate from the surface were also reported [7,8]. For this purpose, two initiators were synthesized, one being monocationic and the other being dicationic. Figures 2.9 and 2.10

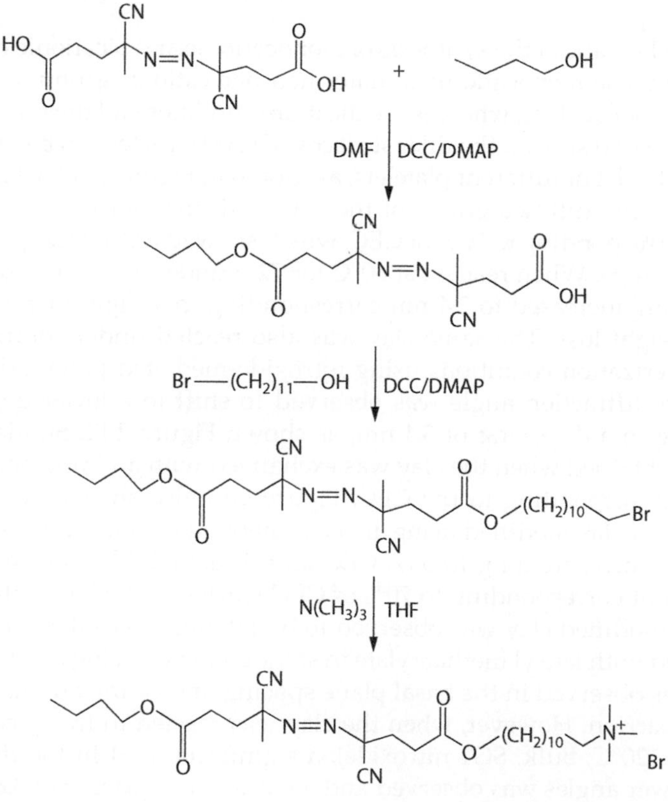

FIGURE 2.9
Schematic of the synthesis of monocationic initiator. (Reproduced from Mittal, V., *Philosophical Magazine*, 90, 2489–506, 2010. With permission from Taylor and Francis.)

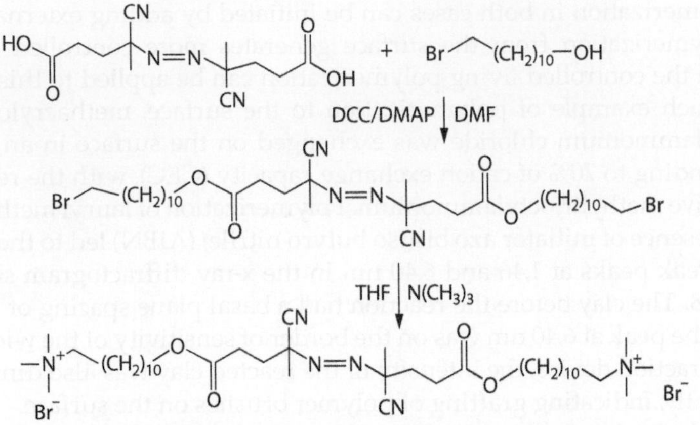

FIGURE 2.10
Schematic of the synthesis of dicationic initiator. (Reproduced from Mittal, V., *Journal of Colloid and Interface Science*, 314, 141–51, 2007. With permission from Elsevier.)

show the schematic of the synthesis of monocationic and bicationic initiators, respectively. The monocationic initiator had one cationic group to immobilize on the filler surface, whereas the dicationic initiator had the possibility to exchange on two sites on the filler surfaces. These two sites can be present on the same platelet or different platelets, as shown in Figure 2.11 [9]. Figure 2.12 shows the x-ray diffractograms of the clay with initiator exchanged in an amount corresponding to 30% of CEC, which showed a 001 basal plane spacing of 2.2 nm [8]. When reacted at 60°C for 72 h under nonliving conditions, the *d*-spacing increased to 2.4 nm corresponding to a slight increase in the organic weight loss. The same clay was also reacted under controlled living polymerization conditions using nitroxide-mediated polymerization at 120°C. The diffraction angle was observed to shift to a lower degree corresponding to a *d*-spacing of 3.4 nm, as shown Figure 2.12. Similar results were also obtained when the clay was exchanged with an initiator in higher amounts corresponding to the CEC. Figure 2.13 also shows the x-ray diffractogram of clay modified using an asymmetrical monocationic initiator at an amount corresponding to 30% CEC and dimethyldidodecylammonium at an amount corresponding to 70% of CEC (curve I). The basal plane spacing of the modified clay was observed to be 1.9 nm. Curve II represents the clay reacted with lauryl methacrylate in solution in nonliving conditions. No change was observed in the basal plane spacing, indicating an unsuccessful grafting reaction. However, when the clay was reacted in living conditions (curve III, 120°C, bulk, SG1 nitroxide), a significant shift in the diffraction peak to lower angles was observed and a basal plane spacing of 3.8 nm was achieved. This indicates that in the presence of a living polymerization agent, termination reactions could be significantly avoided and polymer chains could be successfully grafted.

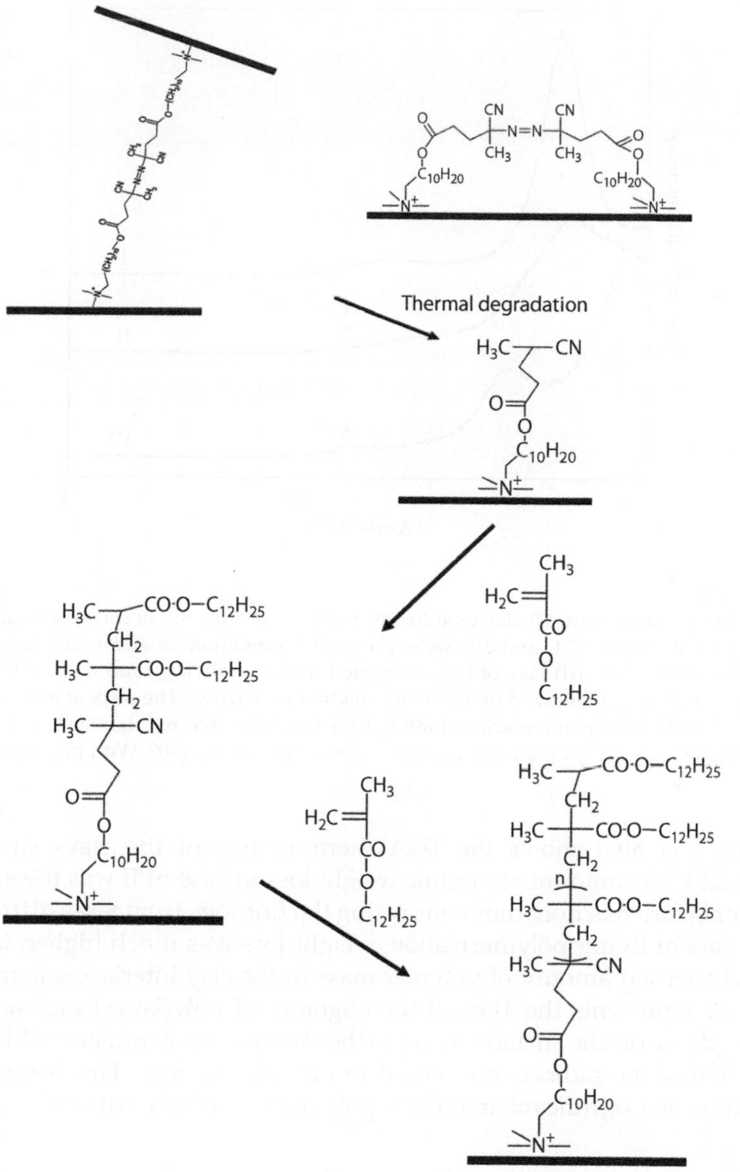

FIGURE 2.11
Representation of the polymerization from the surface approach with bicationic initiator bound to the clay surface. It should be noted that the bicationic initiator may be attached either to one platelet or to two platelets. (Reproduced from Mittal, V., in *Encyclopedia of Polymer Composites*, ed. M. Lechkov and S. Prandzheva, Nova Science Publishers, New York, 2009, pp. 76–99. With permission.)

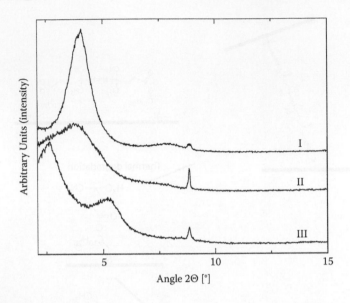

FIGURE 2.12

X-ray diffractograms of the (I) clay modified by using bicationic initiator in an amount corresponding to 30% of the CEC and dihexadecyldimethylammonium in an amount corresponding to 70% of the CEC, (II) clay of I when reacted under nonliving conditions at 60°C, and (III) clay of I when polymerized under living conditions at 120°C. The peak at 8.84° 2θ corresponds to the 001 basal plane spacing of mica platelets used as internal standard. (Reproduced from Mittal, V., *Journal of Colloid and Interface Science*, 314, 141–51, 2007. With permission from Elsevier.)

Figure 2.14 also shows the TGA thermograms of the clays shown in Figure 2.13. The amount of organic weight loss in case of II was the same as the filler before reaction, thus confirming the findings from x-ray diffraction. In the case of living polymerization, weight loss was much higher, indicating an increased amount of polymer mass in the clay interlayers (curve III). Curve IV represents the TGA of the oligomer of poly(lauryl methacrylate) produced free off the surface owing to the monocationic initiator, which produces half of the radicals not bound to the filler surface. This leads to the generation of a significant amount of polymer free off the surface.

2.4 Reactions with the Reactive Surface Modifications to Form Long Brushes

Surface reactions with the reactive surface modifications are another important route to form long brushes grafted from the surface. In one such study, esterification reactions were carried out on the surface of clay platelets, which

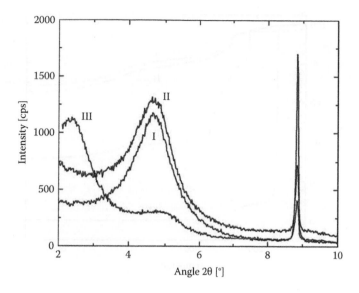

FIGURE 2.13

X-ray diffractograms of clay (I) modified using asymmetrical monocationic initiator in an amount corresponding to 30% of CEC and dimethyldidodecylammonium in an amount corresponding to 70% of CEC, (II) the clay of I when reacted with lauryl methacrylate in nonliving conditions, and (III) the clay of I when reacted with lauryl methacrylate in living conditions. The peak at 8.84° 2θ corresponds to mica platelets used as internal standard. (Reproduced from Mittal, V., *Philosophical Magazine*, 90, 2489–506, 2010. With permission from Taylor and Francis.)

were modified with carrying terminal hydroxyl groups [10]. Figure 2.15a shows the schematic of a general esterification reaction on the clay surface and Figure 2.15b is the example of the reaction of the montmorillonite modified with benzyldibutyl(2-hydroxyethyl)ammonium chloride (BzlOH) with dotriacontanoic acid to form long organic brushes on the surface. Figure 2.16 also shows the effect of increasing the amount of octadecanoic acid compared to the OH groups available on the surface of filler platelets on the surface modification molecules. An increase in organic weight loss in the TGA thermograms of the montmorillonite when acid-to-alcohol molar ratios of 1:1, 4:1, and 8:1 were used was observed. The increase in the organic weight loss on increasing the molar ratio indicated that the extent of surface reaction was enhanced. The surface modification in this case did not have any long alkyl chain in the chemical structure. The basal plane spacing in the filler before the esterification reaction was observed to be 1.53 nm, which was enhanced to 2.26 nm after surface esterification, indicating grafting of octadecanoic chains on the already present surface modification molecules. As the montmorillonites modified with ammonium ions carrying long alkyl chains result in higher initial basal plane spacing and can help in better swelling of the clay during esterification reactions, it was desired to analyze the effect of the presence of these alkyl chains along with hydroxyl groups

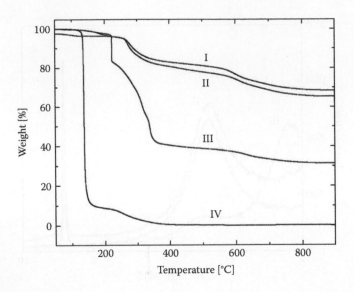

FIGURE 2.14
TGA thermograms of clay (I) modified using asymmetrical monocationic initiator in an amount corresponding to 30% of CEC and dimethyldidodecylammonium in an amount corresponding to 70% of CEC, (II) the clay of I when reacted with lauryl methacrylate in nonliving conditions, and (III) the clay of I when reacted with lauryl methacrylate in living conditions. The TGA thermogram of pure lauryl methacrylate is shown in curve IV. (Reproduced from Mittal, V., *Philosophical Magazine*, 90, 2489–506, 2010. With permission from Taylor and Francis.)

on the extent of surface esterification reactions. Therefore, montmorillonite modified with benzyl(2-hydroxyethyl)methyloctadecylammonium chloride (BzC18OH) was reacted with dotriacontanoic acid on the surface. A significant increase of 0.9 nm in the basal plane spacing was also observed, as shown in x-ray diffractograms in Figure 2.17. The extent of conversion of OH groups to the corresponding ester groups was also higher in this case compared to the Bz1OH-modified clay, thus confirming that the presence of a long alkyl chain in the surface modification led to effective enlargements of the brushes grafted on the surface.

It is further important to stress the uniformity of the brushes of surface modification molecules before the surface reactions can be carried out. Commercially modified montmorillonites are known to contain an excess of surface modification molecules forming pseudo bilayers on the surface that are not ionically bound to the surface. These excess surface molecules are thermally less stable and can interfere negatively with the surface reaction. To confirm this, commercially treated Cloisite 30B, surface modified with bis(2-hydroxyethyl)methyl-hydrogenated tallow ammonium (C18 2OH), was used for esterification reactions before and after washing in a methanol-water mixture [7]. Figure 2.18a shows TGA thermograms of the unwashed clay before washing (curve I) and the clay after esterification with dotriacontanoic acid (curve II). No or marginal increase in the weight loss was

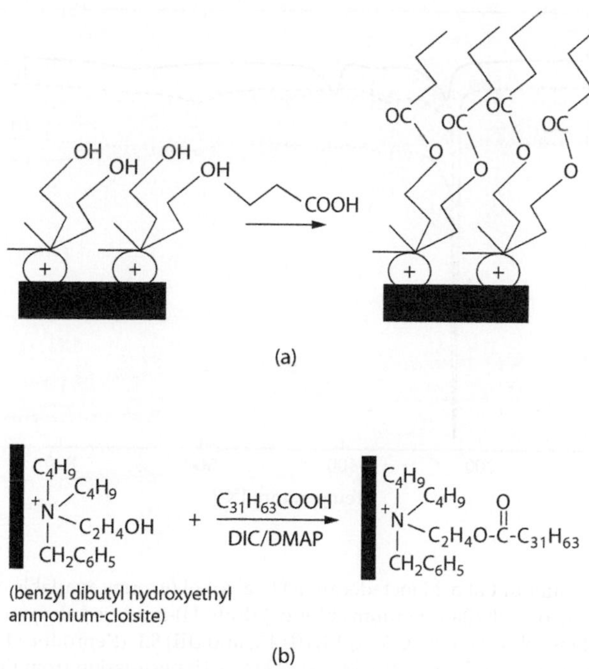

(a)

(benzyl dibutyl hydroxyethyl
ammonium-cloisite)

(b)

FIGURE 2.15

(a) Schematic of esterification reaction on the filler surface and (b) schematic of surface esterification reaction of montmorillonite modified with benzyldibutyl(2-hydroxyethyl)ammonium chloride (Bz1OH) with dotriacontanoic acid. (Reproduced from Mittal, V., *Journal of Colloid and Interface Science*, 315, 135–41, 2007. With permission from Elsevier.)

observed, which confirmed that the grafting did not take place. On the other hand, Figure 2.18b shows the derivative TGA thermograms of washed clay (curve I) and clay esterified with dotriacontanoic acid (curve II). The cumulative weight loss as a function of temperature is also demonstrated. Much higher increase in weight loss was observed in this case. The free surface modification molecules unbound to the clay surface are probably more accessible to the acid chains, thus reacting with them and subsequently getting washed in the cleaning process after the esterification reaction.

In another example [7] vermiculite with brushes of benzyl(2-hydroxyethyl) methyloctadecylammonium on the surface was reacted with epoxy prepolymer with the hope of tethering epoxy chains to the filler surface through the hydroxyl groups present on the filler surface. The whole process of composite generation was studied through x-ray diffraction studies as shown in Figure 2.19.

The diffraction pattern of the filler powder was depicted as curve I and a basal plane spacing of 3.4 nm was observed. The filler was suspended in dimethyl formamide (DMF) in order to generate a composite by solution polymerization. The basal plane spacing of the filler in DMF was observed to increase to 3.8 nm,

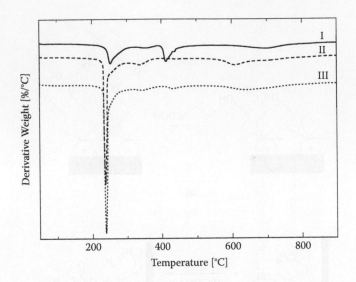

FIGURE 2.16
Effect of increasing amount of acid (octadecanoic) to alcohol (groups available on surface with benzyldibutyl(2-hydroxyethyl)ammonium chloride) (Bz1OH-modified clay) molar ratio on the increase in weight loss observed in TGA: (I) 1:1, (II) 4:1, and (III) 8:1. (Reproduced from Mittal, V., *Journal of Colloid and Interface Science*, 315, 135–41, 2007. With permission from Elsevier.)

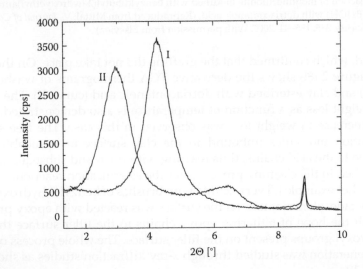

FIGURE 2.17
X-ray diffractograms of (I) BzC18OH-modified clay and (II) clay of I after surface esterification reaction with dotriacontanoic acid. The peak at 8.84 2θ corresponds to the 001 basal plane spacing of mica platelets used as the internal standard. (Reproduced from Mittal, V., *Journal of Colloid and Interface Science*, 315, 135–41, 2007. With permission from Elsevier.)

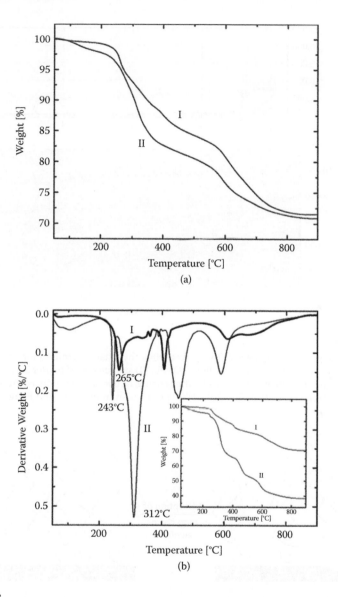

FIGURE 2.18

(a) TGA thermograms of (I) commercially treated Cloisite 30B surface modified with bis(2-hydroxyethyl)methyl-hydrogenated tallow ammonium (C182OH) and (II) clay of I esterified with dotriacontanoic acid. (b) Derivative TGA thermograms of (I) washed Cloisite 30B and (II) clay of I esterified with dotriacontanoic acid. The inset also shows the cumulative weight loss as a function of temperature. (Reproduced from Mittal, V., *Philosophical Magazine*, 90, 2489–506, 2010. With permission from Taylor and Francis.)

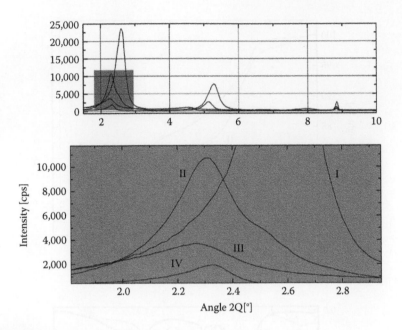

FIGURE 2.19
X-ray diffractograms of (I) benzyl(2-ydroxyethyl)methyloctadecylammonium-modified vermiculite, (II) suspension of modified filler in DMF, (III) suspension of II after addition of epoxy prepolymer, and (IV) the nanocomposite after curing of epoxy. (Reproduced from Mittal, V., *Philosophical Magazine*, 90, 2489–506, 2010. With permission from Taylor and Francis.)

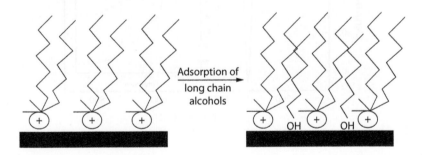

FIGURE 2.20
Schematic of physical adsorption process on the clay surface. (Reproduced from Mittal, V., and Herle, V., *Journal of Colloid and Interface Science*, 327, 295–301, 2008. With permission from Elsevier.)

as shown in curve II. The basal plane spacing remained unchanged after addition of the epoxy prepolymer to the filler suspension in DMF (curve III). The basal plane spacing of the filler decreased marginally after polymerization compared to curves II and III, but increased compared to curve I for the modified filler, indicating intercalation of the polymer in the interlayers (curve IV).

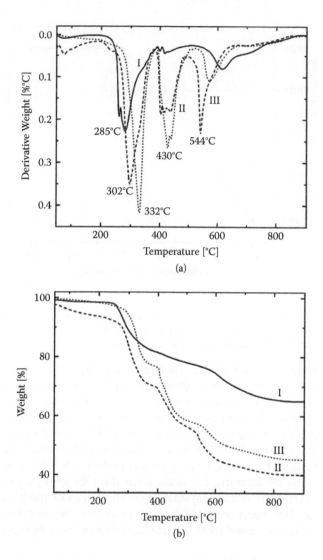

FIGURE 2.21

(a) Derivative weight and (b) weight loss thermograms of organically modified montmorillonite before adsorption (I), after adsorption with 1.5 g of initially added PVP (for 2 g of organically modified montmorillonite) (II), and after adsorption with 1 g of initially added PVP (for 2 g of organically modified montmorillonite) (III). (Reproduced from Mittal, V., and Herle, V., *Journal of Colloid and Interface Science*, 327, 295–301, 2008. With permission from Elsevier.)

2.5 Physical Adsorption to Form Polymer Brushes

The area available per cation to the surface modification molecules on the montmorillonite surface is generally larger than the cross section of the

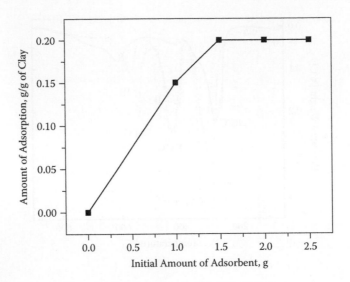

FIGURE 2.22
Extent of PVP adsorption on the clay surface as a function of initial amount of PVP used for the adsorption process. (Reproduced from Mittal, V., and Herle, V., *Journal of Colloid and Interface Science*, 327, 295–301, 2008. With permission from Elsevier.)

molecules. Thus, this leaves the surfaces partially uncovered and residual forces between the surfaces therefore exist. To overcome these attractive forces, it is possible to add another variety of brush into the existing structure by physical adsorption. Long-chain polar polymers or long-chain alcohol molecules are some of the examples of the adsorbents that can be used for such purpose. These molecules (e.g., long alkyl chain alcohols) can absorb physically in between the gaps generated after modification with ammonium ions by forming H bonds with the OH groups present either in the inside structure of clay crystals or on the edges of the platelets. Also, the adsorption has been reported to take place on the preadsorbed water molecules in the clay interlayers. Figure 2.20 shows the schematic of such a process [11].

Figure 2.21 shows the TGA thermograms of the system where a mixed brush with poly(vinylpyrrolidone) (PVP) was formed with the dimethyl ditallow ammonium-modified clay. The amount of physical adsorption corresponded to the amount of PVP initially used for the process. Figure 2.22 also confirmed an increase in the amount of physically adsorbed material as a function of the initial amount of PVP. However, a maximum adsorption was observed at 1.5 g of initial PVP, at which 0.2 g PVP was observed to adsorb on the surface. On further increasing the amount of PVP, no additional increase in the amount of adsorption was observed. The thermal performance of the brushes was also observed to be enhanced in increasing the amount of physically adsorbed PVP, as shown in Figure 2.21, where the peak degradation temperatures were enhanced significantly after PVP absorption.

Thus, the physical adsorption led to the generation of mixed brushes with reduced attractive forces between the platelets and more thermal stability of the system.

References

1. Pavlidoua, S., and Papaspyrides, C. D. 2008. A review on polymer-layered silicate nanocomposites. *Progress in Polymer Science* 33:1119–98.
2. Beyer, G. 2002. Nanocomposites: A new class of flame retardants for polymers. *Plastics Additives and Compounding* 4:22–27.
3. Theng, B. K. G. 1974. *The chemistry of clay-organic reactions*. New York: Wiley.
4. Jasmund, K., and Lagaly, G. 1993. *Tonminerale und Tone Struktur*. Darmstadt, Germany: Steinkopff.
5. Heinz, H., Castelijns, H. J., and Suter, U. W. 2003. Structure and phase transitions of alkyl chains on mica. *Journal of American Chemical Society* 125:9500–10.
6. Mittal, V. 2010. Polymer nanocomposites: Synthesis, microstructure and properties. In *Optimization of polymer nanocomposites properties*, ed. V. Mittal, 1–20. Weinheim, Germany: Wiley VCH.
7. Mittal, V. 2010. Clay exfoliation in polymer nanocomposites: Specific chemical reactions and exchange of specialty modifications on clay surface. *Philosophical Magazine* 90:2489–506.
8. Mittal, V. 2007. Polymer chains grafted "to" and "from" layered silicate clay platelets. *Journal of Colloid and Interface Science* 314:141–51.
9. Mittal, V. 2009. Advances in grafting of polymer chains "to" and "from" the layered-silicate clay surface. In *Encyclopedia of polymer composites*, ed. M. Lechkov and S. Prandzheva, 76–99. New York: Nova Science Publishers.
10. Mittal, V. 2007. Esterification reactions on the surface of layered silicate clay platelets. *Journal of Colloid and Interface Science* 315:135–41.
11. Mittal, V., and Herle, V. 2008. Physical adsorption of organic molecules on the surface of layered silicate clay platelets: A thermogravimetric study. *Journal of Colloid and Interface Science* 327:295–301.

Thus, the physical adsorption led to the generation of mixed brushes with reduced attractive forces between the platelets and more thermal stability of the system.

3

Collapse, Compression, and Adhesion of Poly(N-Isopropylacrylamide) Brushes

Lionel Bureau

Laboratory for Interdisciplinary Physics,

UMR 5588 University of Grenoble 1–CNRS, Grenoble, France

Muriel Vayssade

Biomechanics and Bioengineering,

UMR 6600 Université Technologie de Compiègne–CNRS, Compiègne, France

CONTENTS

3.1 Introduction

Stimuli-responsive polymers attract an ever-growing interest for the design of "smart" surfaces or interfaces, where properties (surface energy (Yakushiji et al. 1998), adhesion (Jones et al. 2002), friction (Ikeuchi et al. 1996, Chang et al. 2007)) can be strongly altered upon application of an external stimulus, such as a change in temperature, light, pH, etc. (Luzinov et al. 2004, Gil and Hudson 2004, Cole et al. 2009). Among such smart systems, thermosensitive coatings generally exploit the properties of polymers that exhibit a lower critical solution temperature (LCST) when mixed with a solvent (Freeman and Rowlinson 1960). Below the LCST, polymer chains are in good solvent conditions and adopt a swollen coil conformation. As temperature is increased across the LCST, the solvent quality shifts from good to poor, and macromolecules collapse into a dense globular state. Such a phase separation is commonly encountered in binary mixtures in which specific interactions (e.g., hydrogen bonding) exist between constituents (Hirschfelder et al. 1937, Goldstein 1984), as it is the case in aqueous solutions of poly(N-isopropylacrylamide) (PNIPAM), one of the most extensively studied thermosensitive systems (Schild 1992, Wu and Zhou 1995).

PNIPAM in water exhibits a LCST at about 32°C, above which polymer chains become hydrophobic. The coil-to-globule transition in bulk solutions occurs over a narrow temperature range and involves large variations in the overall volume occupied by the chains (Wu and Zhou 1995, Wu and Wang 1998). These features, which have been exploited in microfluidics for the design of temperature-controlled actuators or valves (Lokuge et al. 2007, Harmon et al. 2003), have found their main applications in bioengineering (Gil and Hudson 2004, Cole et al. 2009, Ernst et al. 2007). Indeed, numerous issues in this field are related to the control of adsorption/adhesion of proteins on surfaces. In many instances, protein binding to surfaces is undesirable, as it leads, e.g., to biofouling, reduced sensitivity of biosensors, or thrombosis of implants. On the other hand, adsorption of proteins is vital for biocompatibility of the synthetic surfaces of implants, which requires good cell proliferation. The sharp hydrophilic-hydrophobic transition of PNIPAM, along with the close proximity of its LCST to physiological temperature, has made it a choice material for surface coatings in view of designing surfaces that are protein resistant at low temperatures and protein adsorbent above the LCST. Surface-grafted PNIPAM thin films have thus been successfully used for the control of adhesion in applications ranging from biomolecule purification and separation in temperature-selective chromatography (Nagase et al. 2009) to harvesting of cultured cells (Mizutani et al. 2008, Nagase et al. 2009). The latter application is of paramount importance. It relies on thermoresponsive coatings for the modification of surface properties of cell culture dishes. Since the pioneer work performed in the group of Okano (Nagase et al. 2009), various studies have shown that mammalian cells adhere and proliferate on hydrophobic PNIPAM surfaces at 37°C, and can subsequently be detached by cooling below the LCST (Cole et al. 2009). Such a temperature-triggered detachment therefore allows for cell harvesting without the need for harsh methods like mechanical scrapping or enzymatic digestion of the extracellular matrix (ECM) proteins secreted during proliferation. Thus, when cells are grown until confluence on a PNIPAM surface, in order to form a "cell sheet," lowering the temperature allows for the lift-off of this sheet along with the major constituents of the extracellular matrix to which it adheres. The ability to preserve the contact between cell sheets and ECM during harvesting further permits their transfer to other substrates onto which they can readhere. Stacking of such sheets has allowed for the fabrication of three-dimensional functional tissue successfully transplanted in vivo (Ohashi et al. 2007). This emerging cell sheet engineering technology, based on thermal harvesting from PNIPAM-modified substrates, thus opens extraordinary perspectives in the field of tissue reconstruction for regenerative medicine.

Cell-substrate adhesion is mediated by proteins, secreted by cells during their growth, which constitute the ECM. Fibronectin is a protein present in the ECM that plays a pivotal role: it binds to the cell membrane through specific transmembrane receptors (integrins), and adheres to the facing substrate via nonspecific or hydrogen bond interactions, thus ensuring cell attachment. A qualitative argument commonly invoked to explain cell proliferation/ detachment on a PNIPAM substrate is the variation of surface hydrophobicity with temperature, which is believed to affect the binding of fibronectin and thus cell adhesion (Cole et al. 2009, Nagase et al. 2009). Now, expertise in elaborating functional thermoresponsive culture substrates largely stems from empirical research. Various techniques have been employed to produce PNIPAM coatings for cell culture. The most widely used are electron beam polymerization and vapor phase plasma deposition, both methods yielding surface-bound films of 1–100 nm in thickness. All of these films exhibit the expected change in surface energy when temperature crosses the LCST, as probed by, e.g., water contact angle measurements. However, successful cell proliferation and detachment is observed only on a limited number of these substrates (Cole et al. 2009).

This underlines that the sole criterion of surface hydrophobicity is far from being sufficient to rationalize observations and optimize coating design. For instance, it has been shown that cell adhesion and proliferation at 37°C depends significantly upon film thickness. This point is still a matter of debate: Okano and coworkers observed that cell proliferation is strongly favored on very thin thermoresponsive films (Akiyama et al. 2004), whereas other groups report either a weak effect (Li et al. 2008) of thickness or even better proliferation on thicker films (Xu et al. 2004). These studies explored cell adhesion on PNIPAM coatings elaborated by different techniques (electron beam irradiation, plasma deposition, surface-initiated polymerization), which most likely produce polymer layers exhibiting very different molecular structures. There is now a general agreement on the fact that the PNIPAM film architecture obviously plays a crucial role in cell adhesion, but the mechanisms by which it affects binding and detachment remain to be identified (Cole et al. 2009). Further improvement and rational design of thermoswitchable substrates, and further success of cell sheet engineering, must now come from a deeper understanding of the mechanisms controlling interactions at smart biointerfaces, in particular of the role of coatings molecular structure of the coatings.

PNIPAM brushes are considered promising surface modifiers to accurately tune cell-substrate interactions. The structure of such brushes in water has been the subject of several experimental studies. On the one hand, force measuring techniques (atomic force microscopy and surface forces apparatus) have been used to probe the repulsive forces resulting from brush compression (Ishida and Biggs 2007, Kidoaki et al. 2001, Kaholek et al. 2004, Plunkett et al. 2006, Zhu et al. 2007), as well as adhesion forces between PNIPAM and various countersurfaces, including monolayers of proteins relevant to bioadhesion (Kidoaki et al. 2001). Most of these works have been performed at two temperatures, on both sides of the bulk LCST, but did not address the detailed temperature dependence of brush mechanical properties. On the other hand, more systematic investigations have been made, using neutron reflectivity (Yim et al. 2006), quartz crystal microbalance (Ishida and Biggs 2007, Liu and Zhang 2005, Annaka et al. 2007), or surface plasmon resonance (Balamurugan et al. 2003), in order to track the evolution with temperature of the thickness of brushes of different densities and molecular weights. From these studies, two important results can be highlighted:

1. Densely grafted PNIPAM chains collapse over a broader tempera-ture range than that of dilute chains in bulk solution or brushes with low grafting density.
2. The magnitude of the collapse transition, i.e., the ratio of the swol-len to the collapsed thickness, is affected by both chain length and grafting density.

The former result has been obtained in different studies and is in qualita-tive agreement with numerical and theoretical predictions about the collapse of brushes when the solvent quality goes from good to poor (Dimitrov et al. 2007, Zhulina et al. 1991, Wijmans et al. 1992). However, open questions remain regarding the latter point, due to the rather limited number of systematic stud-ies that investigated the effect of molecular parameters on the magnitude of brush collapse.

This prompted us to investigate in further detail the effect of grafting den-sity on the properties of PNIPAM brushes. We have used the surface forces apparatus to study the thermal response of PNIPAM brushes grown on mica substrates by the so-called grafting-from method (Malham and Bureau 2010). Our main results show that, in the range of temperature explored (20°C–40°C):

1. The compression of well-solvated brushes of high density, as those investigated here, cannot be fully described within the classical the-oretical framework developed by Alexander and de Gennes, most likely because of the breakdown of the semidilute assumption on which it relies.
2. Irrespective of the density, the brush thickness decreases slightly as the temperature is raised up to 30°C, then exhibits a steeper decrease between 30°C and 35°C, the temperature above which the collapsed thickness is reached. This confirms the broadening of the collapse transition previously reported for dense brushes.
3. The swelling ratio decreases noticeably at larger surface coverage. This trend, which we observe with an unprecedented resolution

over a large range of grafting densities, is in good qualitative agreement with predictions of Mendez et al. (2005).

4. Adhesive forces between two identical brushes, which build up for temperature above 30°C, appear to be quasi-insensitive to grafting density. This contrasts with preliminary results showing that cell adhesion strongly depends on brush density, and supports the idea that, as mentioned above, surface chemistry of brushes alone does not control cell-PNIPAM interactions.

3.2 Experiments

3.2.1 Surface Forces Apparatus

We have used a home-built surface forces apparatus (SFA) (Bureau 2007) to perform time-resolved measurements of force-thickness curves during compression/decompression of PNIPAM brushes in water, confined between two atomically smooth surfaces. Such measurements were performed at different temperatures, below and above the bulk collapse temperature of 32°C (Malham and Bureau 2010).

The experimental configuration involves two identical facing brushes grafted on curved mica sheets, as sketched in Figure 3.1. Normal forces are measured, with a resolution of about 10^{-6} N, by means of a flexure-hinge spring of stiffness 9,500 N/m equipped with a capacitive displacement sensor. The distance between the brush-bearing substrates is measured by white-light multiple-beam interferometry (MBI) (Tabor and Winterton 1969, Israelachvili and Tabor 1972): The fringes of equal chromatic order produced between the reflective backsides of the substrates are analyzed using the multilayer matrix method in order to deduce the thickness and the effective refractive index of the confined medium (Heuberger 2001, Bureau 2007).

Mica substrates were prepared as follows (Malham and Bureau 2010). A large mica sheet (~10 × 10 cm²) was cleaved down to a thickness of ~10 µm, and one side was coated with a 45 nm thick silver layer, in order to obtain a highly reflective surface suitable for multiple-beam interferometry. Samples of ~1 cm² were cut from this sheet and glued, silver side down, onto glass cylindrical lenses (radius of curvature R = 1 cm), using a UV-curing glue. A pair of glued mica sheets were then recleaved using adhesive tape, down to a thickness of 1–5 µm. The cylindrical lenses were mounted inside the SFA, with their axis crossed at a right angle, and the thickness of each substrate was determined by MBI, following a procedure described in Bureau and Arvengas (2008). Samples were then unmounted to undergo the surface treatments described below in order to obtain brushes of PNIPAM covalently grafted on the substrates. After grafting on both mica sheets, samples

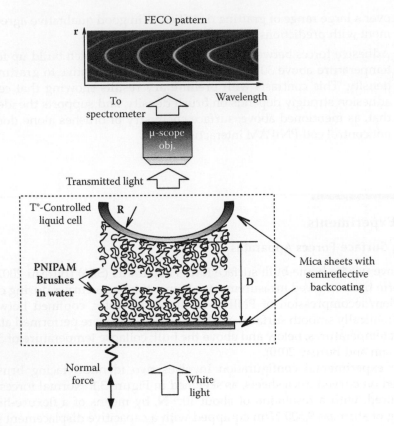

FIGURE 3.1
Sketch of the experimental configuration in the surface forces apparatus. PNIPAM brushes are grafted onto curved mica sheets. Interaction forces during brush compression/decompression are measured by means of a loading spring. The distance D between the mica surfaces is measured by multiple-beam interferometry: The Fabry–Perot cavity formed by the back-silvered mica sheets is shone with white light. The transmitted light is collected by a microscope objective and sent to a spectrometer for spectral analysis of the fringes of equal chromatic order, from which D is calculated. Experiments are performed over a temperature range of 20°C–40°C.

were installed back in the SFA, brushes were brought into contact in the absence of solvent, and the dry thickness, h_{dry}, of grafted PNIPAM was measured. The surfaces were then immersed in water and the temperature inside the SFA was set to a chosen value in the range 20°C–40°C, by means of a thermoregulation system described previously. The surfaces were approached, by driving the remote point of the force-measuring spring at a prescribed velocity V, until a normal load of ~10^{-2} N was reached, and then separated at the same driving velocity. The results presented below have been obtained with V = 4 nm/s, a velocity at which we observed hydrodynamic drag forces to be negligible compared to steric forces due to polymer brushes.

3.2.2 Brush Grafting

After thickness measurement of a pair of bare mica sheets, the following surface treatments, summarized in Figure 3.2, were performed in order to yield substrates grafted with a given surface density of initiator for atom transfer radical polymerization (ATRP). (1) The mica samples were transferred to a plasma reactor where they were exposed to a radio frequency plasma generated in water vapor at reduced pressure (Parker et al. 1989, Liberelle and Giasson 2008, Lego et al. 2009, Malham and Bureau 2009). This step produces silanol (Si-OH) groups at the mica surface, which then allows the covalent attachment of silane-based molecules. (2) Hydroxylated mica surfaces were immediately immersed in a solution of propyltrimethoxysilane (PTMS) in toluene (Smith and Chen 2008). This results in a fraction of the silanol groups at the mica surfaces being replaced by methyl-terminated molecules ($-Si-(CH_2)_2-CH_3$). This fraction was varied by adjusting the immersion time between 0 and 1 h. (3) PTMS-grafted samples were then immersed, at room temperature and for a duration ranging from a few seconds to 5 min, in an aqueous solution of aminopropyltriethoxysilane (APTES). This yields mica substrates exhibiting a mixed NH_2/CH_3 functionality at the surface, the ratio of amino to methyl groups being varied by adjusting immersion times in steps 2 and 3. (4) The silane-grafted mica samples were immersed in a solution of triethylamine in toluene to which 2-bromo-2-methylpropionyl bromide was then added. This step results in surface immobilization, on the amino-terminated sites, of the bromo-terminated initiator from which PNIPAM brushes are grown.

Brushes were grown by ATRP, following a protocol akin to that described in Kaholek et al. (2004), Plunkett et al. (2006), and Yim et al. (2006). A solution of N–isopropylacrylamide (NIPAM) dissolved in deionized water was deoxygenated by bubbling argon at room temperature for 30 min. A solution

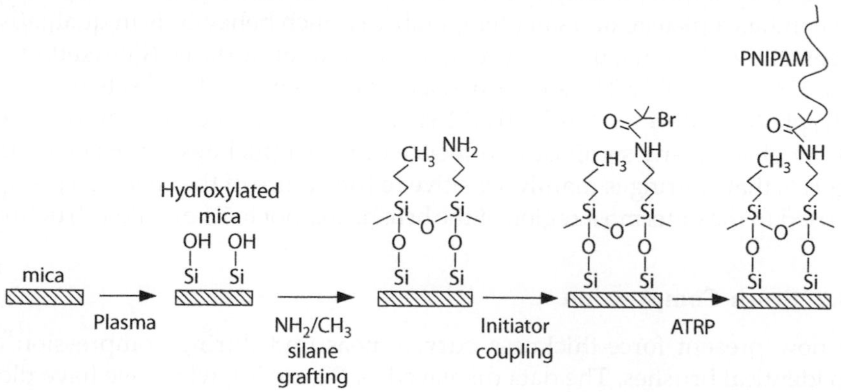

FIGURE 3.2
Scheme showing the successive grafting steps for ATRP growth of PNIPAM chains.

of copper(I) bromide in water was prepared and stirred while a ligand was added for copper complexation. This solution was then mixed to the NIPAM solution under argon bubbling. A pair of identically modified mica samples was then placed for 5 min in the aqueous solution of NIPAM and copper catalyst. Next, samples were immersed for 10 min in an aqueous solution of copper(II) bromide-ligand to quench the polymerization reaction. Finally, surfaces were rinsed repeatedly with water and dried carefully in an argon stream inside a laminar flow cabinet, where they were installed in the surface forces apparatus.

We have used the same polymerization time for each pair of mica substrates, which results in brushes formed of chains of comparable molecular weight (Mw ≈ 475 kg/mol), and of dry thickness varying between 10 and 215 nm, which translates into grafting densities ranging from 2.10^{-4} to 4.10^{-3} chain/Å^2 (Malham and Bureau 2010).

3.3 Results and Discussion

3.3.1 Wetting

We first present results regarding the temperature-dependent wetting properties of the PNIPAM brushes. We have measured the static contact angle (θ_s) of water, as a function of temperature, on flat mica substrates submitted, simultaneously with the SFA samples, to the sequence of chemical modifications described above. Experiments were performed on a custom-built instrument that allowed regulation of the temperature of the substrate and of the water droplets in the range 20°C–40°C. Figure 3.3 shows a typical evolution of θ_s with temperature. The contact angle is seen to stay constant for $T \leq 30$°C, increases by approximately 7°C–8°C, while T increases from 30°C to 34°C and exhibits a plateau at higher temperatures. Such behavior is in qualitative agreement with previous measurements on similar surfaces (Plunkett et al. 2006, Yim et al. 2006): The sharp increase of θ_s around 32°C reflects the more hydrophobic nature of the PNIPAM surface above its LCST. Moreover, we did not observe any significant influence of brush thickness on $\theta_s(T)$, which suggests that wetting is mainly sensitive to the nature of the chemical groups exposed in the outermost region of the brush, and not to its detailed structure.

3.3.2 Brush Compression

We now present force-thickness curves measured during compression of two identical brushes. The data displayed in Figure 3.4, where we have plotted the intersurface force F, normalized by the radius of curvature R, as a function of brush thickness, have been obtained with 50 nm thick brushes

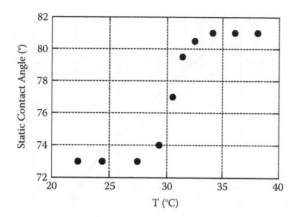

FIGURE 3.3
Static contact angle of water vs. temperature, measured on a PNIPAM brush of grafting density 0.0025 chain/Å². No influence of the grafting density was observed on the angle, within the measurement accuracy of ±1°.

(grafting density 0.0009 chain/Å²) immersed in water at $T = 23°C$. Thicker or thinner brushes displayed the same qualitative features. We observe that as the brushes are gradually approached, repulsive forces build up, due to steric interactions between grafted polymer chains. Repulsion forces are all the greater that brushes are more strongly compressed, up to a point where the polymer thickness reaches a value close to its dry state. Such force-thickness data obtained at low temperatures, i.e., in the situation of chains in good solvent, are valuable, from the standpoint of polymer brush physics, for they can be confronted to theoretical predictions on brush compressibility. We have compared our experimental data to predictions from the widely used framework proposed by Alexander and de Gennes (A-dG) to describe brushes in good solvent (Alexander 1977, de Gennes 1980). For two compressed brushes, A-dG predict the following form of the free energy per unit area, which, within the Derjaguin approximation, is proportional to the measured F/R (Israelachvili 1992, Heuberger et al. 2005, Dunlop et al. 2004):

$$\frac{F(D)}{R} = C\left[7\left(\frac{D}{2L}\right)^{-5/4} + 5\left(\frac{D}{2L}\right)^{7/4} - 12\right] \qquad (3.1)$$

with L the unperturbed thickness of one brush, D the total thickness of the brushes compressed under a force F, R the radius of curvature of the surfaces, and the prefactor $C \propto k_B T L/d^3$, where k_B is the Boltzmann constant and d the distance between grafting sites.

In Figure 3.4 we have plotted the theoretical compression curve $F(D)/R$ obtained using for L the brush thickness measured at the onset of repulsion

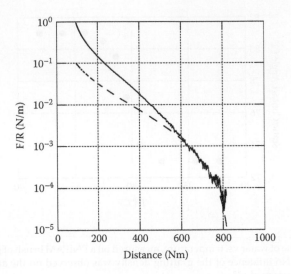

FIGURE 3.4

Continuous line: Repulsive force vs. thickness, measured at $T = 23°C$, upon compression of two facing brushes of grafting density 0.0009 chain/Å². The ratio F/R, with R the radius of curvature, is plotted for direct comparison with predictions from the Alexander-de Gennes model. Dashed line: Theoretical curve computed using an unperturbed thickness of $2L = 820$ nm and the above grafting density.

and for d the grafting density reported above. Although the qualitative shapes of the compression curves are similar, we find only a partial quantitative agreement between predictions and measurements: The low compression end of the $F(D)/R$ profile is reasonably well described by Equation 3.1, but the prediction is found to systematically underestimate the repulsive forces under strong compression. Such a discrepancy was observed over the whole range of grafting density explored in this study, the calculated $F(D)/R$ failing to reproduce the measurements for values of $D/2L$ lower than 0.6–0.8. This shows that the Alexander-de Gennes model, based on the assumption of semidilute chains in the brush, has only a limited range of validity when describing dense brushes under compression. Furthermore, one may evaluate the ratio of the dry thickness of the brush to the distance below, which A-dG predictions fail. We find this ratio to lie in the range 0.2–0.3, which yields an estimate of the monomer volume fraction above which the semidilute assumption breaks down.

We now come to the effect of temperature on $F(D)$ curves. It can be seen in Figure 3.5 that, as the temperature is increased, the onset of repulsive forces occurs at smaller intersurface separations (Figure 3.5a,b), and that hard wall repulsion, i.e., a situation of quasi-fully compressed brushes, occurs at lower normal forces. Such a temperature dependence of the repulsive range shows that the thickness of the brushes indeed decreases as the temperature is increased: grafted chains go from a low-temperature swollen state to a high-temperature collapsed state where most of the solvent is

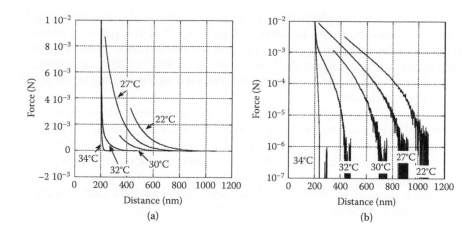

FIGURE 3.5

(a) Force (F)-distance (D) curves measured during approach of two brushes of total dry thickness 170 nm, at the temperatures indicated on the figure. (b) Same data as in (a), with forces displayed on a logarithmic scale.

expelled from the polymer layers. This result is fully consistent with the thermal response of grafted PNIPAM previously observed in experiments using atomic force microscopy (AFM) or SFA (Ishida and Biggs 2007, Kidoaki et al. 2001, Kaholek et al. 2004, Plunkett et al. 2006).

We can further characterize the effect of temperature on the range of repulsive forces as follows. We measure, as a function of temperature, the thickness h_0 reached under a normal force of 10^{-5} N, i.e., under low compression. The choice of such a criterion is arbitrary; it merely ensures that the applied force is well above the experimental resolution, and h_0 under such a finite normal force is therefore lower than the unperturbed brush thickness. We define the ratio $\alpha = h_0/h_{dry}$, which we will refer to as the swelling ratio in the following. Figure 3.5 presents the evolution of α with temperature, for different brush densities. We observe that, for all densities, α decreases gradually as T increases, displays a stronger temperature sensitivity in the range 30°C–35°C, and reaches, for $T > 35$°C, a constant value close to unity. Moreover, we note that the lower the brush density, the larger the value of α at low temperatures.

The temperature dependence of the swelling ratio clearly shows that the collapse of end-tethered chains, in the range of grafting densities explored here, occurs gradually rather than through a sharp coil-globule transition as observed for isolated chains in solution. Such behavior, observed for the first time by means of force measurements, is in good agreement with results obtained from neutron reflectivity (Balamurugan et al. 2003), surface plasmon resonance (Yim et al. 2006), or quartz crystal microbalance (Liu and Zhang 2005, Annaka et al. 2007). It is qualitatively consistent with theoretical predictions by Zhulina et al. (1991, Wijmans et al. 1992), who have shown that, due to repulsive interactions between densely grafted stretched chains, the collapse of a dense planar brush occurs gradually as the solvent strength is lowered. Besides, it can be seen in Figure 3.5 that the grafting density affects both the magnitude of swelling and the temperature sensitivity (the slope da/dT) of the brushes: As T increases from 30°C to 35°C, α exhibits a 6-fold decrease for the lowest density, whereas it drops by only a factor of 2.5 for the highest density.

Finally, in order to highlight the role of grafting density on brush swelling, we present in Figure 3.6b the low-temperature values of α (measured at 23°C) as a function of surface coverage. It can be clearly seen that the swelling ratio decreases in a nonlinear way as the grafting density increases. Such a non-linear dependence, observed over a large range of densities, provides a good qualitative support to predictions from self-consistent field calculations of brush swelling done by Mendez et al. (2005).

3.3.3 Adhesion

We now focus on adhesion between two identical PNIPAM brushes. Upon surface separation, we observe that adhesive forces build up for $T > 29°C–30°C$ and are of greater magnitude at higher temperatures, as illustrated in Figure 3.7a.

In Figure 3.7b, we have plotted the measured pull-off force as a function of temperature, for various brush densities. It can be seen that adhesion forces between contacting brushes are independent of their grafting density and increase with temperature above 30°C.

We propose the following picture to account for such an observation. The backbone of PNIPAM chains bears both hydrophobic (methyl (CH3)) and hydrophilic (amide (NH) and carbonyl (C=O)) groups. As already mentioned above, the temperature dependence of the water contact angle (θ_s) shown in

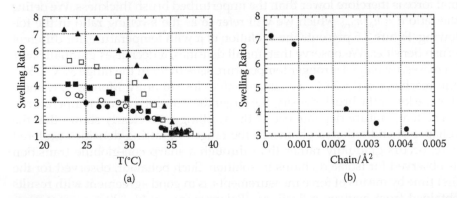

FIGURE 3.6
(a) Swelling ratio as a function of temperature, for brushes of density: (●) 0.0042, (○) 0.0033, (■) 0.0024, (□) 0.0014, and (▲) 0.0002 chain/Å². (b) Swelling ratio as a function of brush density, at $T = 23°C$. (Reprinted with permission from Malham, I.B. and Bureau, L. 2010. *Langmuir* 26:4762–4768, copyright (2010) American Chemical Society.)

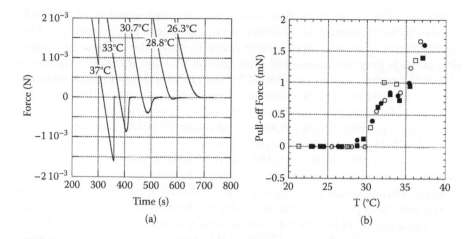

FIGURE 3.7
(a) Force as a function of time, during separation of two brushes, measured at the indicated temperatures. The curves have been horizontally shifted in order to the facilitate comparison. (b) Pull-off force as a function of temperature, for brushes of density: (●) 0.0014, (○) 0.0024, (■) 0.0033, and (□) 0.0042 chain/Å². (Reprinted with permission from Malham, I.B. and Bureau, L. 2010. *Langmuir* 26:4762–4768, copyright (2010). American Chemical Society.)

Figure 3.3 suggests that the outermost region of a PNIPAM brush is the seat of local molecular rearrangements, so that, at $T > 35°C$, mostly hydrophobic groups are exposed at the interface with water. Moreover, the temperature range over which θ_s increases corresponds to the range where brush thicknesses exhibit a steeper decrease with T. This supports the idea that, as T increases and chain dehydration and collapse take place, a growing fraction of methyl groups is exposed at the surface, until a maximum surface density of CH_3 is reached above 35°C. Akin to wetting, measurements of pull-off forces that do not depend on grafting density indicate that adhesion between brushes in contact is mainly controlled by the nature of the chemical groups exposed at the interface. However, we note that while the water contact angle levels off above 35°C, adhesion forces are found to increase steadily up to 38°C. This suggests that once the contact is established between two collapsed PNIPAM layers, a mechanism is at play that affects interbrush adhesion but not wetting. We believe that this mechanism corresponds to the formation of interchain hydrogen bonds, between –NH and –C=O groups, thus contributing to increase the pull-off force (Cheng et al. 2006).

3.4 Conclusion and Outlook

We have presented a systematic investigation of the effect of grafting density on the thermomechanical properties of PNIPAM brushes. By using plasma activation of mica surfaces that allows for covalent grafting of PNIPAM brushes by surface-initiated ATRP, we have been able to take advantage of the surface forces apparatus technique to probe in detail the collapse and adhesion of such polymer layers. We have shown that the thermal response of PNIPAM brushes is strongly affected by the grafting density: chain swelling is found to decrease at higher densities, a result that may have direct implications in applications relying on the magnitude of size variation of the polymer chains, as, e.g., microfluidics flow control using PNIPAM coatings (Lokuge et al. 2007). Moreover, the analysis of force-compression curves provides an insight into the range of validity of the classical Alexander-de Gennes framework when dealing with dense brushes. Finally, we have found that, in marked contrast to the swelling behavior, adhesive forces between brushes do not seem to depend on grafting density. Such an observation leaves largely open the question of how brush thickness may affect cell-substrate adhesive interactions in cell culture on PNIPAM (Mitzutani et al. 2008, Nagase et al. 2009). In relation to this, we conclude by showing, in Figure 3.8, preliminary results concerning cell proliferation on PNIPAM brushes of various densities (hence different thicknesses). Mouse fibroblasts (Swiss 3T3 cells) were cultured on oxidized silicon wafer grafted with PNIPAM layers of thickness 3, 5, and 37 nm, and cell morphology and proliferation was observed at 37°C after 48 h of culture. It can be seen that the cell did proliferate and spread on the two thinnest brushes, which indicates good cell-substrate adhesion at 37°C, while only a low density of weakly adhering and rounded cells was observed on the thick brush. Moreover, we have found that cells cultured on 3 and 5 nm thick brushes spontaneously detached from the surface after 30 min at room temperature. Such results are in good qualitative agreement with those of Nagase et al. (2009) and indicate that thinner (less dense) PNIPAM brushes indeed favor cell adhesion above the LCST. This suggests that the interaction between fibronectin and brushes is controlled by a mechanism that is sensitive to the grafting density. Protein insertion inside the brush is likely to be such a mechanism, as proposed in previous theoretical works (Halperin

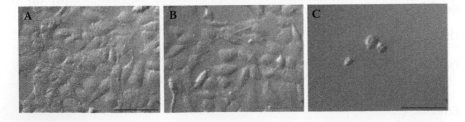

FIGURE 3.8
Swiss 3T3 fibroblast morphology 48 h postseeding on PNIPAM brushes of thickness (a) 3 nm, (b) 5 nm, and (c) 37 nm. Swiss 3T3 cells were seeded at 10^4 cells/cm^2 and cultured in Dulbecco's modified eagle medium (DMEM) supplemented with 10% fetal bovine serum, 2% L-glutamine, 1% penicillin (100 U/ml), and 1% streptomycin (100 U/ml) for 48 h. Cells were observed at 37°C using a Leitz Orthoplan microscope in bright field. Scale bar = 100 μm.

1999): the osmotic pressure penalty associated with protein adsorption within the polymer brush is indeed expected to be lower at smaller chain densities. Such a scenario, which remains to be experimentally verified, would allow one to rationalize the above observations and provide a guide for the optimal design of PNIPAM coatings in cell culture applications.

References

Akiyama, Y., Kikuchi, A., Yamato, M., Okano, T. 2004. Ultrathin poly(N-isopropylacrylamide) grafted layer on polystyrene surfaces for cell adhesion/detachment control. *Langmuir* 20:5506–5511.

Alexander, S. 1977. Adsorption of chain molecules with a polar head—a scaling description. *J. Phys. (Paris)* 38:983–987.

Annaka, M., Yahiro, C., Nagase, K., Kikuchi, A., Okano, T. 2007. Real-time observation of coil-to-globule transition in thermosensitive poly (N-isopropylacrylamide) brushes by quartz crystal microbalance. *Polymer* 48:5713–5720.

Balamurugan, S., Mendez, S., Balamurugan, S. S., O'Brien, M. J., Lopez, G. P. 2003. Thermal response of poly(N-isopropylacrylamide) brushes probed by surface plasmon resonance. *Langmuir* 19:2545–2549.

Bureau, L. 2007. Surface force apparatus for nanorheology under large shear strain. *Rev. Sci. Instrum.* 78:065110.

Bureau, L., Arvengas, A. 2008. Drainage of a nanoconfined simple fluid: Rate effects on squeeze-out dynamics. *Phys. Rev. E* 78:061501.

Chang, D. P., Dolbow, J. E., Zauscher, S. 2007. Switchable friction of stimulus-responsive hydrogels. *Langmuir* 23:250–257.

Cheng, H., Shen, L., Wu, C. 2006. LLS and FTIR studies on the hysteresis in association and dissociation of poly(N-isopropylacrylamide) chains in water. *Macromolecules* 39:2325–2329.

Cole, M.A., Voelcker, N.H., Thissen, H., Griesser, H.J. 2009. Stimuli-responsive interfaces and systems for the control of protein-surface and cell-surface interactions. *Biomaterials* 30:1827–1850.

de Gennes, P. G. 1980. Conformations of polymers attached to an interface. *Macromolecules* 13:1069–1075.

Dimitrov, D. I., Milchev, A., Binder, K. 2007. Polymer brushes in solvents of variable quality: Molecular dynamics simulations using explicit solvent. *J. Chem. Phys.* 127:084905.

Dunlop, I. E., Briscoe, W. H., Titmuss, S., Sakellariou, G., Hadjichristidis, N., Klein, J. 2004. Interactions between polymer brushes: Varying the number of end-attaching groups. *Macromol. Chem. Phys.* 205:2443–2450.

Ernst, O., Lieske, A., Jager, M., Lankenau, A., Duschl, C. 2007. Control of cell detachment in a microfluidic device using a thermo-responsive copolymer on a gold substrate. *Lab Chip* 7:1322–1329.

Freeman, P. I., Rowlinson, J. S. 1960. Lower critical points in polymer solutions. *Polymer* 1:20–26.

Gil, E. S., Hudson, S. M. 2004. Stimuli-responsive polymers and their bioconjugates. *Prog. Polym. Sci.* 29:1173–1222.

Goldstein, R. E. 1984. On the theory of lower critical solution points in hydrogen-bonded mixtures. *J. Chem. Phys.* 80:5340–5341.

Halperin, A. 1999. Polymer brushes that resist adsorption of model proteins: Design parameters. *Langmuir* 15:2525–2533.

Harmon, M. E., Tang, M., Frank, C. W. 2003. A microfluidic actuator based on thermo-responsive hydrogels. *Polymer* 44:4547–4556.

Heuberger, M. 2001. The extended surface forces apparatus. Part I. Fast spectral correlation interferometry. *Rev. Sci. Instrum.* 72:1700–1707.

Heuberger, M., Drobek, T., Spencer, N. D. 2005. Interaction forces and morphology of a protein-resistant poly(ethylene glycol) layer. *Biophys. J.* 88:495–504.

Hirschfelder, J., Stevensen, D., Eyring, H. 1937. A theory of liquid structure. *J. Chem. Phys.* 5:896–912.

Ikeuchi, K., Kouchiyama, M., Tomita, N., Uyama, Y. Ikada, Y. 1996. Friction control with a graft layer of a thermo-sensing polymer. *Wear* 199:197–201.

Ishida, N., Biggs, S. 2007. Direct observation of the phase transition for a poly(N-isopropylacryamide) layer grafted onto a solid surface by AFM and QCM-D. *Langmuir* 23:11083–11088.

Israelachvili, J. 1992. *Intermolecular and surface forces*. London: Academic Press.

Israelachvili, J. N., Tabor, D. 1972. Measurement of van der Waals dispersion forces in range 1.5 to 130 nm. *Proc. R. Soc. London A* 331:19–38.

Jones, D. M., Smith, J. R., Huck, W. T. S., Alexander, C. 2002. Variable adhesion of micropatterned thermoresponsive polymer brushes: AFM investigations of poly (N-isopropylacrylamide) brushes prepared by surface-initiated polymerizations. *Adv. Mater.* 14:1130–1134.

Kaholek, M., Lee, W. K., Ahn, S. J., Ma, H. W., Caster, K. C., LaMattina, B., Zauscher, S. 2004. Stimulus-responsive poly(N-isopropylacrylamide) brushes and nanopatterns prepared by surface-initiated polymerization. *Chem. Mater.* 16:3688–3696.

Kidoaki, S., Ohya, S., Nakayama, Y., Matsuda, T. 2001. Thermoresponsive structural change of a poly(N-isopropylacrylamide) graft layer measured with an atomic force microscope. *Langmuir* 17:2402–2407.

Lego, B., Francois, M., Skene, W. G., Giasson, S. 2009. Polymer brush covalently attached to OH-functionalized mica surface via surface-initiated ATRP: Control of grafting density and polymer chain length. *Langmuir* 25:5313–5321.

Li, L., Zhu, Y., Li. B., Gao, C. 2008. Fabrication of thermoresponsive polymer gradients for study of cell adhesion and detachment. *Langmuir* 24:13632–13639.

Liberelle, B., Giasson, S. 2008. Friction and normal interaction forces between irreversibly attached weakly charged polymer brushes. *Langmuir* 24:1550–1559.

Liu, G., Zhang, G. 2005. Collapse and swelling of thermally sensitive Poly(N-isopropylacrylamide) brushes monitored with a quartz crystal microbalance. *J. Phys. Chem. B* 109:743–747.

Lokuge, I., Wang, X., Bohn, P. W. 2007. Temperature-controlled flow switching in nano-capillary array membranes mediated by poly(N-isopropylacrylamide) polymer brushes grafted by atom transfer radical polymerization. *Langmuir* 23:305–311.

Luzinov, I., Minko, S., Tsukruk, V. V. 2004. Adaptive and responsive surfaces through controlled reorganization of interfacial polymer layers. *Prog. Polym. Sci.* 29:635–698.

Malham, I. B., Bureau, L. 2009. Growth and stability of a self-assembled monolayer on plasma-treated mica. *Langmuir* 25:5631–5636.

Malham, I. B., Bureau L. 2010. Density effects on collapse, compression and adhesion of thermoresponsive polymer brushes. *Langmuir* 26:4762–4768.

Mendez, S., Curro, J. G., McCoy, J. D., Lopez, G. P. 2005. Computational modeling of the temperature-induced structural changes of tethered poly(N-isopropylacrylamide) with self-consistent field theory. *Macromolecules* 38:174–181.

Mizutani, A., Kikuchi, A., Yamato, M., Kanazawa, H., Okano, T. 2008. Preparation of thermoresponsive polymer brush surfaces and their interaction with cells. *Biomaterials* 29:2073–2081.

Nagase, K., Kobayashi, J., Okano, T. 2009. Temperature-responsive intelligent interfaces for biomolecular separation and cell sheet engineering. *J. R. Soc. Interface* 6:S293–S309.

Ohashi, K., Yokoyama, T., Yamato, M., et al. 2007. Engineering functional two- and three-dimensional liver systems *in vivo* using hepatic tissue sheets. *Nature Med.* 13:880–885.

Parker, J. L., Cho, D. L., Claesson, P. M. 1989. Plasma modification of mica—forces between fluorocarbon surfaces in water and a nonpolar liquid. *J. Phys. Chem.* 93:6121–6125.

Plunkett, K. N., Zhu, X., Moore, J. S., Leckband, D. 2006. PNIPAM chain collapse depends on the molecular weight and grafting density. *Langmuir* 22:4259–4266.

Schild, H. G. 1992. Poly (n-isopropylacrylamide)—experiment, theory and application. *Prog. Polym. Sci.* 17:163–249.

Smith, A. E., Chen, W. 2008. How to prevent the loss of surface functionality derived from aminosilanes. *Langmuir* 24:12405–12409.

Tabor, D., Winterton, R. H. S. 1969. Direct measurement of normal and retarded van der Waals forces. *Proc. R. Soc. London A* 312:435–450.

Wijmans, C. M., Scheutjens, J. M. H. M., Zhulina, E. B. 1992. Self-consistent field-theories for polymer brushes—lattice calculations and an asymptotic analytical description. *Macromolecules* 25:2657–2665.

Wu, C., Wang, X. H. 1998. Globule-to-coil transition of a single homopolymer chain in solution. *Phys. Rev. Lett.* 80:4092–4094.

Wu, C., Zhou, S. Q. 1995. Laser-light scattering study of the phase-transition of poly(n-isopropylacrylamide) in water. 1. Single-chain. *Macromolecules* 28:8381–8387.

Xu, F. J., Zhong, S. P., Yung, L. Y. L., Kang, E. T., Neoh, K. G. 2004. Surface-active and stimuli-responsive polymer-Si(100) hybrids from surface-initiated atom transfer radical polymerization for control of cell adhesion. *Biomacromolecules* 5:2392–2403.

Yakushiji, T., Sakai, K., Kikuchi, A., Aoyagi, T., Sakurai, Y., Okano, T. 1998. Graft architectural effects on thermoresponsive wettability changes of poly(N-isopropylacrylamide)-modified surfaces. *Langmuir* 14: 4657–4662.

Yim, H., Kent, M. S., Mendez, S., Lopez, G. P., Satija, S., Seo, Y. 2006. Effects of grafting density and molecular weight on the temperature-dependent conformational change of poly(N-isopropylacrylamide) grafted chains in water. *Macromolecules* 39:3420–3426.

Zhang, K., Ma, J., Zhang, B., Zhao, S., Li, Y., Xu, Y., Yu, W., Wang, J. 2007. Synthesis of thermoresponsive silica nanoparticle/PNIPAM hybrids by aqueous surface-initiated atom transfer radical polymerization. *Mater. Lett.* 61:949–952.

Zhu, X., Yan, C., Winnik, F. M., Leckband, D. 2007. End-grafted low-molecular-weight PNIPAM does not collapse above the LCST. *Langmuir* 23:162–169.

Zhulina, E. B., Borisov, O. V., Pryamitsyn, V. A., Birshtein, T. M. 1991. Coil globule type transitions in polymers. 1. Collapse of layers of grafted polymer-chains. *Macromolecules* 24:140–149.

Halperin, A., Kröger, M., 2012. Theoretical considerations on mechanisms of harvesting cells cultured on thermoresponsive polymer brushes. Biomaterials 33, 4975–4987.

Kessel, S., Schmidt, S., Müller, R., et al., 2010. Thermoresponsive PEG-based polymer layers: surface characterization with AFM force measurements. Langmuir 26, 3462–3467.

Laloyaux, X., Mathy, B., Nysten, B., Jonas, A.M., 2010. Surface and bulk collapse transitions of thermoresponsive polymer brushes. Langmuir 26, 838–847.

Malham, I.B., Bureau, L., 2010. Density effects on collapse, compression, and adhesion of thermoresponsive polymer brushes. Langmuir 26, 4762–4768.

Mendez, S., Curro, J.G., McCoy, J.D., López, G.P., 2005. Computational modeling of the temperature-induced structural changes of tethered poly(N-isopropylacrylamide) with self-consistent field theory. Macromolecules 38, 174–181.

Mizutani, A., Kikuchi, A., Yamato, M., Kanazawa, H., Okano, T., 2008. Preparation of thermoresponsive polymer brush surfaces and their interaction with cells. Biomaterials 29, 2073–2081.

Nagase, K., Kobayashi, J., Okano, T., 2009. Temperature-responsive intelligent interfaces for biomolecular separation and cell sheet engineering. J. R. Soc. Interface 6, S293–S309.

Okano, T., Yamada, N., Sakai, H., et al., 1993. A novel recovery system for cultured cells using plasma-treated polystyrene dishes grafted with poly(N-isopropylacrylamide). J. Biomed. Mater. Res. 27, 1243–1251.

Park, S., Cho, D., Gleeson, F.M., 1989. Plasma modification of miscellaneous surfaces.

Plunkett, K.N., Zhu, X., Moore, J.S., Leckband, D., 2006. PNIPAM chain collapse depends on the molecular weight and grafting density. Langmuir 22, 4259–4266.

Safell, H.G., 1992. Poly(N-isopropylacrylamide): experiment, theory and application. Prog. Polym. Sci. 17, 163–249.

Schön, A.P., Chen, W., 2005. How to prevent the loss of surface functionality derived from amino silanes. Langmuir 24, 3960–13869.

Taunton, H.J., Toprakcioglu, C., et al., 1990. Interactions between surfaces bearing end-adsorbed chains in a good solvent. Macromolecules 23, 571–580.

Wu, C., Wang, X., 1998. Globule-to-coil transition of a single homopolymer chain in solution. Phys. Rev. Lett. 80, 4092–4094.

Wu, C., Zhou, S.Q., 1995. Laser light scattering study of the phase transition of poly(N-isopropylacrylamide) in water. 1. Single chain. Macromolecules 28, 8381–8387.

Xu, F.J., Zhong, S.P., Yung, L.Y.L., Kang, E.T., Neoh, K.G., 2004. Surface-active and stimuli-responsive polymer-Si(100) hybrids from surface-initiated atom transfer radical polymerization for control of cell adhesion. Biomacromolecules 5, 2392–2403.

Yamada, N., Okano, T., Sakai, H., Karikusa, F., Sawasaki, Y., Okano, T., 1990. Thermoresponsive polymeric surfaces: control of attachment and detachment of cultured cells. Makromol. Chem. Rapid Commun. 11, 571–576.

Yim, H., Kent, M.S., Mendez, S., López, G.P., Satija, S., Seo, Y., 2006. Effects of grafting density and molecular weight on the temperature-dependent conformational change of poly(N-isopropylacrylamide) grafted chains in water. Macromolecules 39, 3420–3426.

Zhang, Z., Xu, J., Zhou, H., Chen, Y., Li, Y., Xu, Y., Yu, W., Wang, J., 2007. Synthesis of thermoresponsive silica nanoparticle/PNIPAM hybrids by aqueous surface-initiated atom transfer radical polymerization. Mater. Lett. 61, 949–952.

Zhu, X., Yan, C., Winnik, F.M., Leckband, D., 2007. End-grafted low-molecular-weight PNIPAM does not collapse above the LCST. Langmuir 23, 162–169.

Zhulina, E.B., Borisov, O.V., Pryamitsyn, V.A., Birshtein, T.M., 1991. Coil-globule type transitions in polymers. 1. Collapse of layers of grafted polymer chains. Macromolecules 24, 140–149.

4

Ferrocene-Functionalized Polymer Brushes: Synthesis and Applications

Xu Li Qun and Kang En-Tang

Department of Chemical and Biomolecular Engineering, National University of Singapore, Kent Ridge, Singapore

Fu Guo Dong

School of Chemistry and Chemical Engineering, Southeast University, Nanjing, People's Republic of China

CONTENTS

4.1   Introduction

Since their discovery in the early 1950s, ferrocenes have been at the fore-front of chemical research (Amer et al. 2010). Ferrocenes are well known for their high electron density, unique redox and magnetic properties, and good electrochemical responses (Bildstein 2000; Bandoli and Dolmella 2000). Ferrocene and its derivatives have found applications in asymmetric catalysis (Arrayas et al. 2006) and electrochemistry (Kulbaba and Manners 2001), and as functional biomaterials (van Staveren and Metzler-Nolte 2004) and biomedical materials (Neuse 2005). In addition, the cyclopentadienyl rings of ferrocene can be readily functionalized, and their properties can be regulated or modified by different substituents (Horikoshi and Mochida 2010).

Ferrocene polymer is one of the most interesting metallopolymers because of its unique electronic and electrochemical properties, as well as chemical stability (Masson et al. 2008; Engtrakul and Sita 2001). Ferrocene polymers were prepared either by attaching the ferrocene to the polymer chain as a pendant group (side-chain polymer) or integrating the ferrocene group to the main chain as a composite structure unit (main-chain polymer). Earlier studies were focused on the synthesis of the side-chain polymers, such as poly(vinylferrocene), by conventional free radical, cationic, and anionic polymerization (Arimoto and Haven 1955; Deschenaux et al. 1996; Pittman et al. 1971; Rausch and Coleman 1958; Wright 1990). In the early 1990s, Manners and coworkers synthesized the high molecular weight main-chain ferrocene polymers by anionic ring-opening polymerization (Foucher et al. 1992). Since then, a variety of main-chain ferrocene polymers have been developed (Rulkens et al. 1994, 1996). Recent progress in controlled/living polymerization, such as atom transfer radical polymerization (ATRP) and ring-opening metathesis polymerization (ROMP), provides a novel approach for the preparation of ferrocene polymers with well-defined structure and controllable molecular weight (Albagli et al. 1992; Watson et al. 1999; Feng et al. 2009).

Tethering of polymer brushes on solid substrates is an effective approach to tailoring their surface properties, such as wettability, biocompatibility, corrosion resistance, and adhesion (Zhao and Brittain 2000; Barbey et al. 2009). Materials with surface-tethered functional polymer brushes have versatile applications, ranging from biotechnology to advanced microelectronics. In this regard, functional materials arising from surface-grafted ferrocene polymers will continue to be of great scientific interest. This chapter summarizes the recent development in the preparation and application of ferrocene polymer brushes.

4.2 Synthesis of Ferrocene Polymer Brushes

Ferrocene polymer brushes can be prepared via three main strategies: (A) physisorption, (B) grafting-to, and (C) grafting-from approaches (Figure 4.1). Physisorption is a reversible process. The grafting-to approach is carried out via chemical bond formation between the reactive groups of end-functionalized polymer chains and the reactive groups of the substrate.

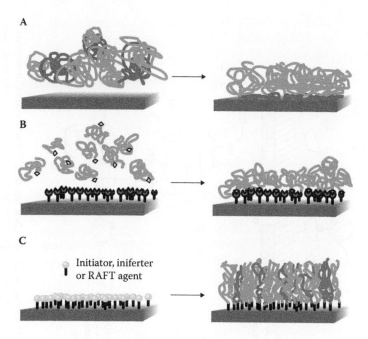

FIGURE 4.1
Synthetic strategies for the preparation of polymer brushes: (A) physisorption, (B) grafting-to approach via reaction of appropriately end-functionalized polymers with complementary functional groups on the substrate surface, (C) grafting-from approach via surface-initiated polymerization. (Reprinted with permission from Barbey et al. 2009. Polymer Brushes via Surface-Initiated Controlled Radical Polymerization: Synthesis, Characterization, Properties, and Applications. *Chemical Reviews* 109 (11):5437–5527, copyright (2009) American Chemical Society.)

The grafting-to approach is experimentally simple, but is limited by a low grafting density because of steric crowding of reactive sites by the adsorbed polymers. Grafting-from (also called surface-initiated polymerization) is a powerful method for the preparation of polymer brushes, involving polymerization of monomers from a substrate with surface-anchored initiator species or initiator sites (Barbey et al. 2009; Zhao and Brittain 2000).

4.2.1 Physisorption Method

Host-guest interaction-included physisorption can be used for the preparation of ferrocene polymer brushes. Dubacheva et al. (2010) prepared ferrocene polymer brushes on a gold surface via host-guest interaction between β-CD and the ferrocene group. Figure 4.2 shows the process for the preparation of physisorbed ferrocene polymer brushes on a gold surface. Initially, monolayers of HS-$(CH_2)_{11}$-EG$_4$-OH (EG = ethylene glycol) and HS-$(CH_2)_{11}$-EG$_6$-N$_3$ are immobilized on gold-coated quartz crystals via a thiol-gold reaction. OligoEG-terminated alkanethiol, HS-$(CH_2)_{11}$-EG$_4$-OH, is chosen as a diluting component to minimize nonspecific adsorption phenomena. Then, click chemistry

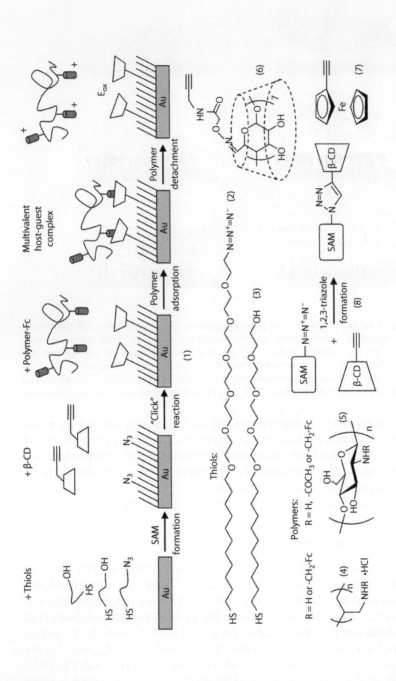

FIGURE 4.2

Schematic representation of self-assemble monolayer (SAM) formation, SAM-β-CD formation via the click reaction, the adsorption/desorption of ferrocene-modified polymers (1), and the chemical structures of HS-(CH₂)₁₁-EG₆-N₃ (2), HS-(CH₂)₁₁-EG₄-OH (3), ferrocene-derivated PAH (4), ferrocene-derivated CHI (5), alkyne-functionalized β-CD (6), ethynylferrocene (7), and triazole formation (8). (Reprinted with permission from Dubacheva et at. 2010. Electrochemically Controlled Adsorption of Fc-Functionalized Polymers on beta-CD-Modified Self-Assembled Monolayers. *Langmuir* 26 (17):13976–13986, copyright (2010) American Chemical Society.)

between the azide groups on a gold surface and alkyne-functionalized β-CD was carried out to introduce the host β-CD molecules on the gold substrate surface. Finally, chitosan polymers with ferrocene pendent groups were synthesized and adsorbed on the gold surface to produce the ferrocene polymer brushes via the host-guest interaction between the β-CD and ferrocene groups.

4.2.2 Grafting-from Strategy

With the development of controlled radical polymerization techniques in recent years, the grafting-from approach has attracted considerable attention in the preparation of polymer brushes on solid substrates. The tethered polymer brushes can be prepared by the immobilization of initiators and subsequent surface-initiated polymerization (Zhao and Brittain 2000).

4.2.2.1 Synthesis of Ferrocene-Functionalized Polymer Brushes by ATRP

ATRP, a recently developed controlled radical polymerization (CRP) technique, has been widely used to prepare macromolecules with controlled molecular weight and molecular architecture (Patten et al. 1996; Tsarevsky and Matyjaszewski 2007). ATRP allows the preparation of polymer brushes with a well-defined molecular structure on various substrates, such as polymers, silicon, titanium, and gold, via a surface-initiated process (Edmondson et al. 2004; Fristrup et al. 2009). Surface-initiated ATRP (SI-ATRP) has a number of attractive features, such as a high tolerance toward a large assortment of monomers and functional groups, and relatively low sensitivity toward residual traces of oxygen. Most of the standard SI-ATRP catalyst systems, as well as initiator coupling agents, are commercially available or can be readily synthesized with well-established procedures.

The synthesis of ferrocene polymer brushes on quartz and indium tin oxide (ITO) substrates via SI-ATRP was first reported in 2005 by Sakakiyama et al. The silane coupling agent, 2-(4-chlorosulfonylphenyl)-ethyltrichlorosilane, was immobilized on quartz and ITO surface to generate the initiator sites. SI-ATRP of ferrocenylmethyl methacrylate (FMMA) using CuBr-sparteine as the catalyst-ligand system gave rise to the well-defined ferrocene polymer brushes (Figure 4.3). The free initiator 4-toluenesulfonyl chloride (TsCl) was added into the polymerization system to control the catalytic balance of ATRP, and to estimate the molecular weight and molecular weight distribution of the grafted brushes. The gel permeation chromatography (GPC) results suggest that the process of polymerization is controllable and the prepared ferrocene polymer brushes are well defined. The ferrocene polymer brushes also have a higher graft density of about 0.10 chains/nm^2.

Ferrocene polymer brushes on the ITO surface were also prepared by Kim et al. (2010) by immobilizing a phosphonic acid initiator on the ITO surface, followed by SI-ATRP of a ferrocene-containing monomer (FcMA). Ferrocene polymer brushes of various sizes (M_n = 4,000 to 37,000 g/mol) and a low

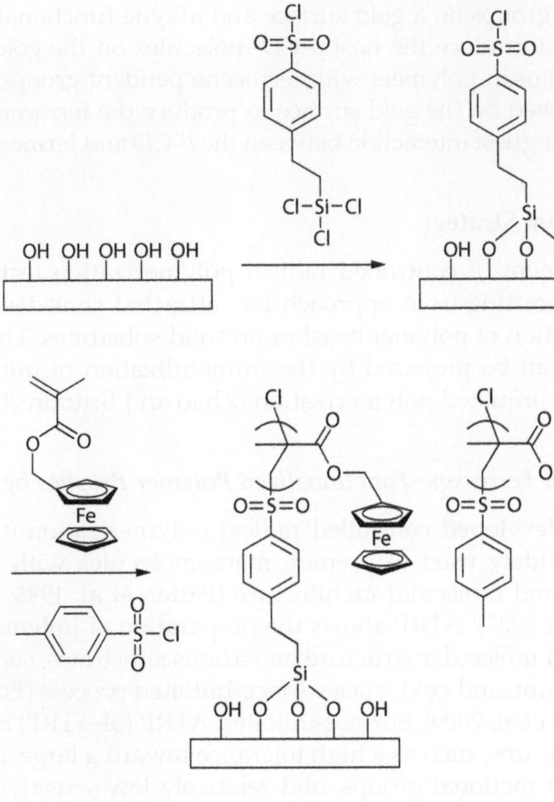

FIGURE 4.3
Procedures for the preparation of PFMMA graft polymer brushes on quartz and ITO substrates. (Reproduced from Sakakiyama et al. 2005. Fabrication and electrochemical properties of high-density graft films with ferrocene moieties on ITO substrates. *Chemistry Letters* 34 (10):1366–1367, with permission from The Chemical Society of Japan.)

polydispersity ($M_w/M_n < 1.3$) were obtained. To measure the thickness of the brush films, photopatterning of initiator-modified ITO substrates was conducted. Block copolymer brushes of methyl methacrylate (MMA) and FcMA on ITO were also prepared via SI-ATRP in order to investigate the effect of the copolymer sequence and distance to the ITO interface on electroactive properties of these ferrocene polymer brushes.

Ferrocene polymer brushes on ITO substrates were also prepared by Zhang et al. (2010) (Figure 4.4). Initially, the ITO surface was activated by oxygen plasma treatment. Then, trichlorosilane coupling agents, containing the sulfonyl halide ATRP initiator, were immobilized on ITO surface. Consecutive SI-ATRP of FMMA (inner block) and glycidyl methacrylate (GMA; outer block) gave rise to block copolymer brushes (PFMMA-*b*-PGMA). Glucose oxidase (GOD) was subsequently immobilized on the PFMMA-*b*-PGMA brushes via ring-opening reactions between the epoxide groups of GMA and the amine groups of GOD. Block copolymer brushes of GMA (inner block) and FMMA (outer block)

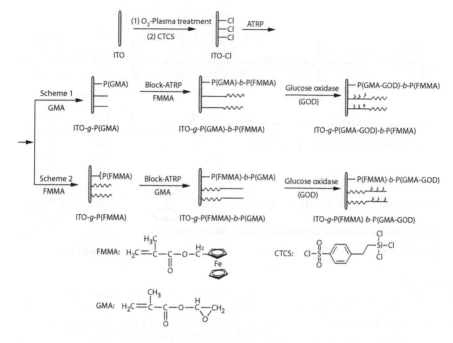

FIGURE 4.4
Schematic diagram illustrating the process of immobilization of trichlorosilane coupling agent on the ITO surface, surface-initiated ATRP of GMA from the ITO-Cl surface to produce the inner block, followed by block copolymer of FMMA and the coupling of GOD to the epoxide groups of the P(GMA) (scheme 1). In scheme 2, P(FMMA) is first grafted from the ITO-Cl surface, and the P(GMA) segment is then grafted as the outer block of the copolymer. (Reproduced from Zhang et al. 2010. Enzyme-mediated amperometric biosensors prepared via successive surface-initiated atom-transfer radical polymerization. *Biosensors & Bioelectronics* 25 (5):1102–1108, with permission from Elsevier.)

(PGMA-*b*-PFMMA) were also synthesized to investigate the effect of electron transfer mediator of ferrocene moieties on the sensitivity of the GOD.

Gold substrate with surface-grafted ferrocene polymer brushes, Au-*g*-PFTMA (PFTMA = poly(5-ferrocene-triazolyl methacrylate)), was prepared by combining SI-ATRP and click chemistry in a one-pot synthesis, as shown schematically in Figure 4.5 (Xu et al. 2010). Initially, cysteamine adsorption forms a uniform amino-terminated monolayer on the gold substrate. Then, N-hydroxysuccinimidyl bromoisobutyrate is used as a more active electrophile to react with the amine groups and to permanently attach the ATRP initiator on the gold surface. Finally, the simultaneous SI-ATRP of 2-azido-ethyl methacrylate (AzEMA) and click coupling of ethynyl ferrocene allow the preparation of gold substrates with surface-grafted ferrocene polymer brushes. A β-CD polymer is also synthesized via the reaction of β-CD with epichlorohydrin in NaOH solution. The redox-responsive surface is formed finally by exposing the gold substrate with ferrocene polymer brushes (guest) on a surface to the aqueous β-CD polymer (host) solution. A redox-controlled

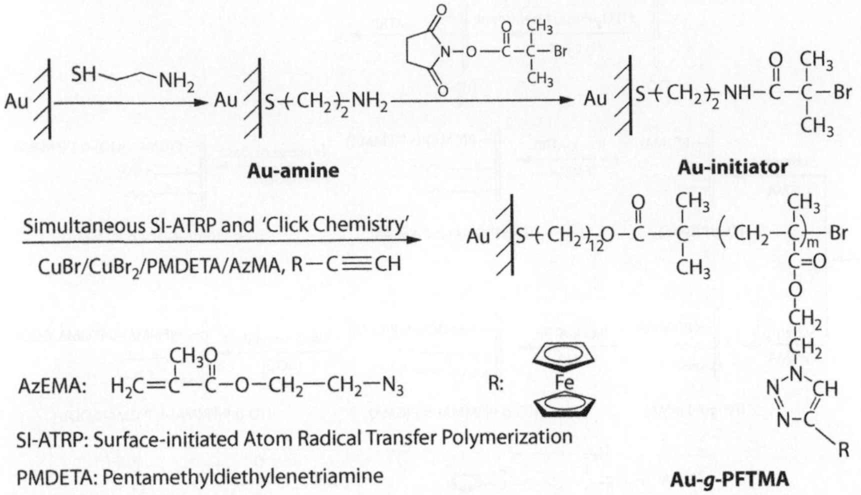

FIGURE 4.5
Synthesis of ferrocene-functionalized polymer brushes on a gold substrate by simultaneous SI-ATRP and click chemistry. (Reprinted with permission from Xu et al. 2010. One-Pot Preparation of "Ferrocene-Functionalized Polymer Brushes on Gold Substrates by Combined Surface-Initiated Atom Transfer Radical Polymerization and "Click Chemistry." *Langmuir* 26 (19):15376–15382, copyright (2010) American Chemical Society.)

"smart" surface with the β-CD polymer loaded on the ferrocene functional polymer brushes was realized by the host-guest interaction. Thus, reversible switching from the hydrophilic Au-*g*-PFTMA-capped-β-CD surface to the hydrophobic Au-*g*-PFTMA surface can be regulated through surface oxidation and reduction. The permanence and reversibility of redox-controlled surface properties of Au-*g*-PFTMA was demonstrated by several cycles of redox-controlled loading and disassociation. In comparison to the commonly used techniques, the preparation of ferrocene-containing polymer brushes by simultaneous click chemistry and SI-ATRP presents several advantages, including high click yield and reduced reaction steps.

Wu et al. (2009) also prepared the ferrocene polymer brushes on a Au electrode via a three-step synthesis: (1) immobilization of initiator-coupled thiolated DNA on a Au electrode through thiol-Au interaction, (2) activator-generated electron transfer for SI-ATRP (AGET SI-ATRP) of 2-hydroxyethyl methacrylate (HEMA) or GMA, and (3) coupling of amino-ferrocene ($FcNH_2$) to the functional groups of the side chains. For PGMA brushes, $FcNH_2$ can be coupled directly to the PGMA side chains via ring-opening reactions between the epoxide groups of GMA and the amine groups of $FcNH_2$. As for the poly(hydroxyethyl methacrylate) (PHEMA) brushes, 1,1'-carbonyldiimidazole (CDI) was used to transform the hydroxyl groups of PHEMA into the inidazolyl-carbamate groups for reaction with $FcNH_2$ to form the relatively stable *N*-alkyl carbamates (Figure 4.6).

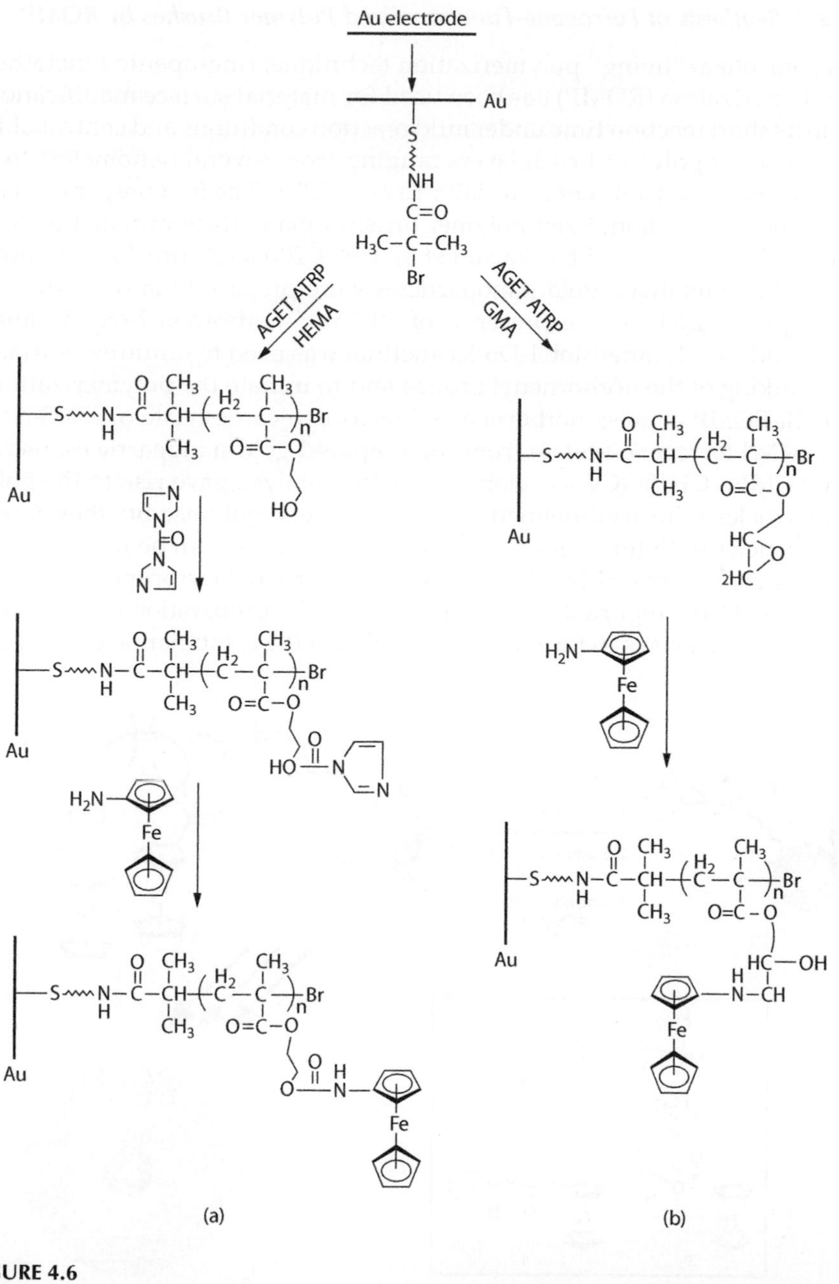

FIGURE 4.6
Schematic illustration of the two strategies used in the preparation of ferrocene-functionalized polymer brushes: (a) hydroxyl groups of the PHEMA brushes activated by CDI, followed by coupling with FcNH$_2$; and (b) epoxy groups of the PGMA brushes underwent direct ring-opening reaction with FcNH$_2$. (Reprinted with permission from Wu et al. 2009. Electrochemical Biosensing Using Amplification-by-Polymerization. *Analytical Chemistry* 81 (16):7015–7021, copyright (2009) American Chemical Society.)

4.2.2.2 Synthesis of Ferrocene-Functionalized Polymer Brushes by ROMP

As a controlled/"living" polymerization technique, ring-opening metathesis polymerization (ROMP) has been used for material surface modification due to its short reaction time under mild reaction conditions and controllable thicknesses of polymer brush layers ranging from several nanometers to a few micrometers (Faulkner et al. 2010; Ye et al. 2010). The first preparation of the ferrocene-functionalized polymer brushes via surface-initiated ROMP (SI-ROMP) was reported by Watson et al. (1999, 2000) (Figure 4.7). Initially, surface-functionalized gold nanoparticles were prepared via reduction of $HAuCl_4$ by $NaBH_4$ in the presence of 10-(*exo*-5-norbornen-2-oxy)decane-1-thiol and 1-dodecanethiol. 1-Dodecanethiol was used to minimize surface cross-linking of the norbornenyl groups and to initiate the polymerization. Then, SI-ROMP of *exo*-5-norbornen-2-yl ferrocene-carboxylate or *exo*-5-norbornen-2-yl ferrocene acetate from the prepared gold nanoparticles, using $(PCy_3)_2Cl_2Ru=CHPh$ (Cy = cyclohexyl) as the catalyst, gave rise to the gold nanoparticles with well-defined surface ferrocene polymer brushes. Gold nanoparticles with ferrocene block copolymer brushes can be feasibly prepared by consecutive SI-ROMP of another norbornenyl ferrocene monomer.

Liu et al. (2003) reported a novel approach to the preparation of ferrocene polymer brush arrays in the nanometer scale via the combination of dip-pen

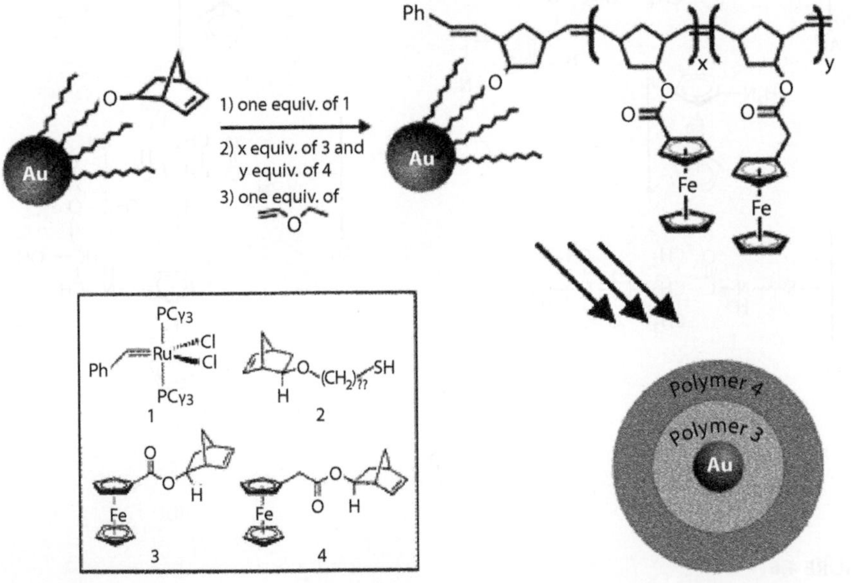

FIGURE 4.7
Graphical representation of the preparation of ferrocene-functionalized polymer brushes grafted on gold nanoparticles via SI-ROMP. (Reprinted with permission from Watson et at. 1999. Hybrid nanoparticles with block copolymer shell structures. *Journal of the American Chemical Society* 121 (2):462–463, copyright (1999) American Chemical Society.)

nanolithography (DPN) and SI-ROMP. Gold substrate was first immersed into a solution of 10-(*exo*-5-norbornen-2-oxy)decane-1-thiol to introduce a layer of functionalities on the surface. Then, the gold substrate was transferred into a solution of Grubbs' first-generation catalyst ([(PCy$_3$)$_2$Cl$_2$Ru=CHPh]) to generate the initiating sites on the surface. SI-ROMP of *exo*-5-norbornen-2-yl ferrocene-carboxylate or *exo*-5-norbornen-2-yl ferrocene-acetate produced the ferrocene polymer brush arrays (Figure 4.8).

Ferrocene-functionalized polymer brushes were grafted on the silicon surface in order to prevent the formation of silicon oxide (Figure 4.9). A silicon substrate modified with 5-(bicycloheptenyl)trichlorosilane was immersed in a solution of Grubbs' first-generation catalyst to introduce the ROMP initiator on the surface. Then, an atomic force microscopy (AFM) microcantilever was used to transport the ferrocene monomer to the catalyst-activated substrates. Thus, a wide range of ferrocene polymer brush arrays with varying numbers of repeating monomer units can be easily generated by deliberately controlling the contact time between the monomer-coated AFM tip and the substrate. Various ferrocene monomers can be polymerized in different spatially defined regions, resulting in different ferrocene-functionalized polymer brush arrays on the same substrate.

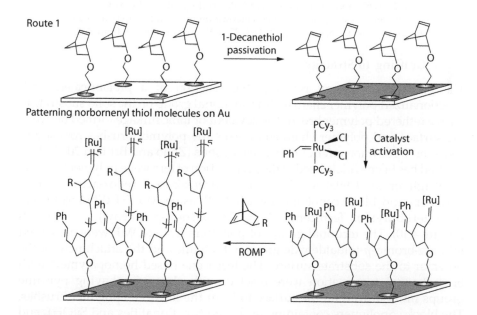

FIGURE 4.8
Graphical representation of the preparation of ferrocene-functionalized polymer brush arrays via SI-ROMP by DPN. (Reproduced from Liu et al. 2003. Surface and site-specific ring-opening metathesis polymerization initiated by dip-pen nanolithography. Angewandte Chemie-International Edition 42 (39):4785–4789 with permission from Wiley.)

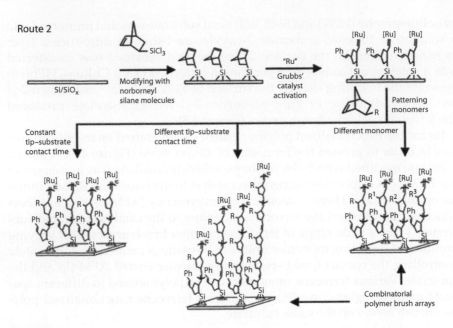

FIGURE 4.9
Graphical representation of the preparation of ferrocene-functionalized polymer brushes via ROMP, initiated by DPN and a monomer-coated AFM tip. (Reproduced from Liu et al. 2003. Surface and site-specific ring-opening metathesis polymerization initiated by dip-pen nanolithography. *Angewandte Chemie-International* Edition 42 (39):4785–4789 with permission from Wiley.)

4.2.3 Grafting-to Strategy

The grafting-to strategy refers to the reaction of a prefabricated, end-functionalized polymer chain with a functional group on the solid substrate to form a tethered polymer brush. The covalent or ionic linkage formed between the surface and polymer chain end makes the polymer brushes robust and resistant to common environmental conditions (Zhao and Brittain 2000). This method has been widely used in the preparation of ferrocene polymer brushes.

Albagli et al. (1993) synthesized ferrocene-based redox-active homo-polymers and block copolymers via ROMP, using $Mo(CHR)(NAr)(O\text{-}t\text{-}Bu)_2$ (R = tert-butyl or ferrocenyl, Ar = 2,6-dissopropylphenyl) as initiators (Figure 4.10). The terminal groups of polymers from ROMP were converted into a α-bromo-p-tolualdehyde group, which allowed the attachment of the polymer to the substrate surface. The ferrocene-based homopolymers with bromobenzyl termination were reacted with electrodes having pyridine groups immobilized on the surface to form the ferrocene polymer brushes. The block copolymers containing ferrocene functionalities and $Si(OH)_3$ end groups were also synthesized and used to decorate Pt, $In(Sn)O_3$, and n-Si electrodes via the formation of siloxane bonds.

Peter et al. (1999, 2001, 2004) prepared poly(ferrocenyldimethylsilane) (PFS) by anionic polymerization, followed by end-group modification with

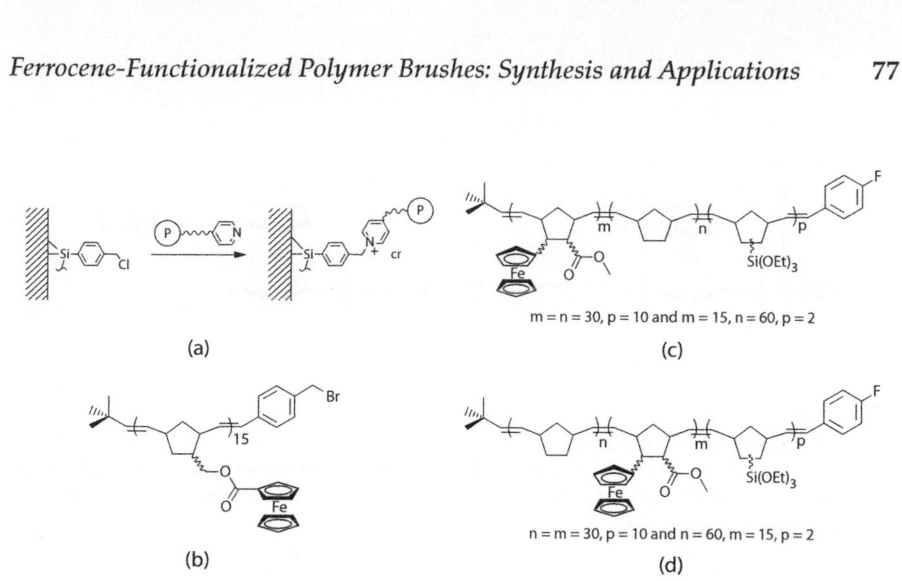
Si
Cl
P
N
Si
N
+
Cl⁻
P
(a)
Br
15
O
O
Fe
(b)
m
n
p
F
Fe
O
O
Si(OEt)3
m = n = 30, p = 10 and m = 15, n = 60, p = 2
(c)
n
m
p
F
Fe
O
O
Si(OEt)3
n = m = 30, p = 10 and n = 60, m = 15, p = 2
(d)

FIGURE 4.10
(a) Attachment of a polymer chain via nucleophilic attack of a terminal pyridine on a surface-confined benzyl chloride. (b) Ferrocene-based homopolymers terminated with bromobenzyl groups. (c,d) Errocene-based block copolymers containing $Si(OH)_3$. (Reprinted with permission from Albagli et al. 1993. Surface Attachment of Well-Defined Redox-Active Polymers and Block Polymers via Terminal Functional-Groups. *Journal of the American Chemical Society* 115 (16):7328–7334, copyright (1993) American Chemical Society.)

ethylene sulfide (Figure 4.11). The ferrocene polymer brushes, featuring a thiol end group, self-assembled on the gold surface via a gold-thiol interaction. Ferrocene polymer brushes on gold substrate were also prepared using thiol end-functionalized PFS from end-group modification of PFS by trimethylene sulfide (Peter et al. 2005).

Eder et al. (2001) prepared ferrocene polymer brushes on mesoporous and nonporous silica surfaces via ROMP of ferrocene monomers followed by a grafting-to approach. At first, norborn-2-ene-5-yltrichlorosilane was immobilized on the surface of mesoporous and nonporous silica from a mixture of chlorotrimethylsilane and dichlorodimethylsilane. Then, metathesis polymerization of ethynylferrocene, 2-[4-(ethynyl)phenyl ethenyl]-1′,2,2′,3,3′,4,4′,5-octamethylferrocene, and 2-[4-(ethynyl)phenyl ethenyl]-1′,2,2′,3,3′,4,4′,5-octamethylferrocene using the Schrock type catalyst $Mo(N\text{-}2,6\text{-}Me_2\text{-}C_6H_3)(CHCMe_2Ph)(OCMe(CF_3)_2)_2$ gave rise to the ferrocene polymers. Subsequent grafting of the prepared polymers onto the silica surface by the coupling reaction between the norborn-2-ene-5-yl groups on the silica support and the end group of ferrocene polymers produces the ferrocene polymer brushes (Figure 4.12).

Nagel et al. (2007) synthesized a thermoresponsive random copolymer of poly(*N*-isopropylacrylamide)-*co*-poly(vinylferrocene)-*co*-poly(methacrylic acid-2,3-epoxypropyl ester) by radical polymerization using 2,2′-azodiisobutyronitrile (AIBN) as the initiator. The copolymers were then attached to a cysteamine-modified gold electrode surface via reaction between the epoxy group of the copolymer and the amino group on the electrode surface

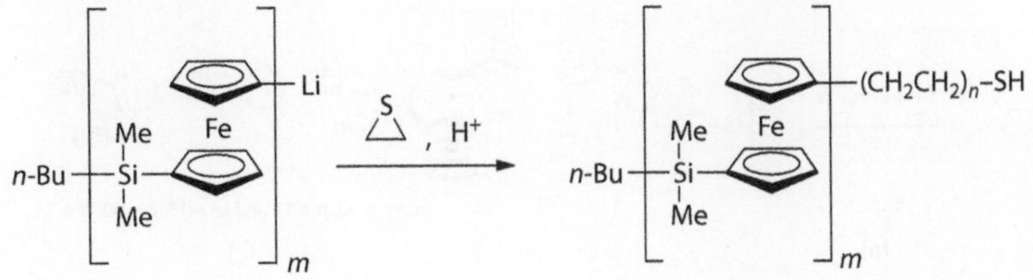

FIGURE 4.11
Graphical representation of the synthesis of thiol end-functionalized PFS by anionic polymerization. (Reproduced from Peter et al. 1999. Synthesis, characterization, and thin film formation of end-functionalized organometallic polymers. *Chemical Communications* (4):359–360 with permission from RSC.)

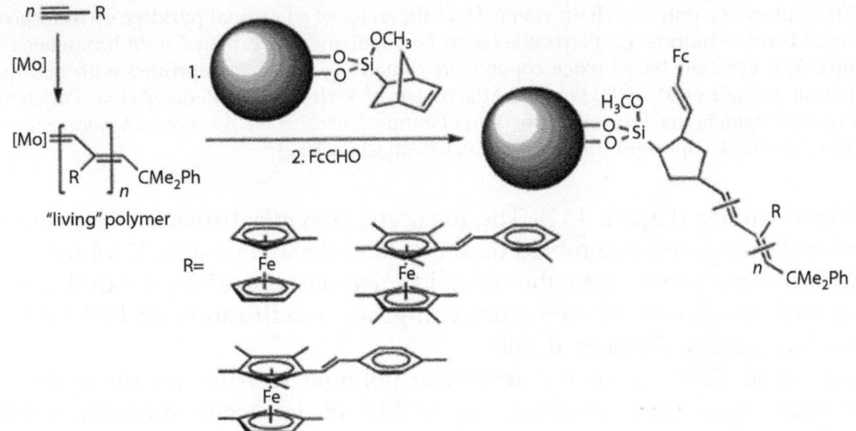

FIGURE 4.12
Synthesis of ferrocene-functionalized polymer brushes via ROMP and subsequent surface grafting. (Reprinted with permission from Eder et al. 2001. Alkyne metathesis graft polymerization: Synthesis of poly(ferricinium)-based silica supports for anion-exchange chromatography of oligonucleotides. *Macromolecules* 34 (13):4334–4341, copyright (2001) American Chemical Society.)

(Figure 4.13). Recently, Nagel et al. (2010) synthesized poly(ethylene glycol) (PEG)-based copolymers bearing both an electrochemically active ferrocene group and thiol anchoring functionality. These copolymers are prepared from radical polymerization of a mixture of triethylene glycol monoethyl ether methacrylate, diethylene glycol monoethyl ether methacrylate, vinylferrocene, and 6-(pyridine-2-yl disulfanyl)hexyl methacrylate (or dithiodi(2-propanoyloxy)ethyl methacrylate). The copolymers were attached to the gold surfaces after the cleavage of the disulfide group by tris(2-carboxyethyl) phosphine hydrochloride (TCEP HCl).

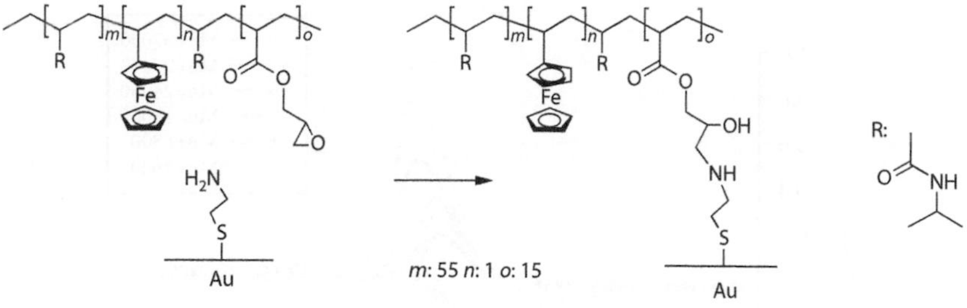

FIGURE 4.13
Attachment of a polymer bearing additional epoxy groups on an amine-modified gold electrode. (Reprinted with permission from Nagel et al. 2007. Enzyme activity control by responsive redoxpolymers. *Langmuir* 23 (12):6807–6811, copyright (2007) American Chemical Society.)

4.2.4 Estimation of Surface Coverage and Grafting Density of Polymer Brushes

To date, characterization of electroactive polymer-modified electrodes using coulometric analysis of the voltammetric peaks has been widely employed, since coulometric analysis is a useful technique elucidating the surface parameters (Blaedel and Mabbott 1981). The quantity of charge consumed for a given reaction can be calculated from the integral of the current-voltage (CV) curves. If the number of electrons transferred per molecule is known, the surface coverage can be estimated from Faraday's law:

$$Q = nFA\Gamma \tag{4.1}$$

where Γ represents the surface coverage in mol/cm^2, n represents the number of electrons transferred per molecule, F is Faraday's constant, and A is the electrode area. The grafting density (σ) of the polymer brushes can be calculated from the surface coverage derived from coulometric analysis, or from

$$\sigma = (\Gamma N_a)/DP \tag{4.2}$$

where Γ is surface coverage, N_a is Avogadro's number, and DP is the number average degree polymerization of the grafted polymer brushes.

A series of ferrocene polymer brushes with various number average molecular weights (M_n, ranging from 4,600 to 37,000 g/mol) were synthesized by SI-ATRP and used to estimate the grafting density from the CV measurements (Kim et al. 2010). The CV curves of the ferrocene polymer brushes with various M_n are shown in Figure 4.14. With the increase in M_n, the CV curves exhibit a progressive increase in peak current and charge density Q (calculated from integration of the current-voltage plots along the baseline from the voltammetry). Based on the results of surface coverage, a grafting density of the ferrocene brushes with a value of 0.6 ± 0.2 chain/nm^2 was obtained.

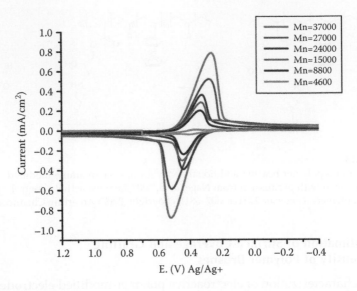

FIGURE 4.14

Stacked CV of ferrocene-functionalized polymer brushes of varying M_n grafted on ITO in acetonitrile with tetrabutylammoniumhexafluorophosphate as the electrolyte at a scan rate of 100 mV/s. (Reprinted with permission from Kim et al 2010. Ferrocene Functional Polymer Brushes on Indium Tin Oxide via Surface-Initiated Atom Transfer Radical Polymerization. *Langmuir* 26 (3):2083–2092, copyright (2010) American Chemical Society.)

4.3 Applications of the Ferrocene Functionalized Polymer Brushes

4.3.1 Tailoring Surface Wetting Property

The surface property of materials with ferrocene polymer brushes can be regulated by the reversible host-guest interaction between the ferrocene group and β-CD under redox switching. A smart surface resulting from reversible coupling reaction between the ferrocene-functionalized polymer brushes and the β-CD polymer is shown schematically in Figure 4.15 (Xu et al. 2010). Simultaneous SI-ATRP and "click chemistry" of 2-azidoethyl methacrylate (AzEMA) and ethynyl ferrocene allow the preparation of gold substrates with surface-grafted ferrocene functional polymer brushes. A β-CD polymer is also synthesized via the reaction of β-CD with epichlorohydrin in NaOH solution. The loading of β-CD polymer on the ferrocene-functionalized polymer brushes can be achieved by immersing of the ferrocene-functionalized polymer brush-grafted substrate in the aqueous solution of β-CD polymer. The static water contact angle of the gold substrate with surface-grafted ferrocene polymer brushes is about 98°. The loading of β-CD polymer affords the gold substrate a hydrophilic surface with a static water contact angle of

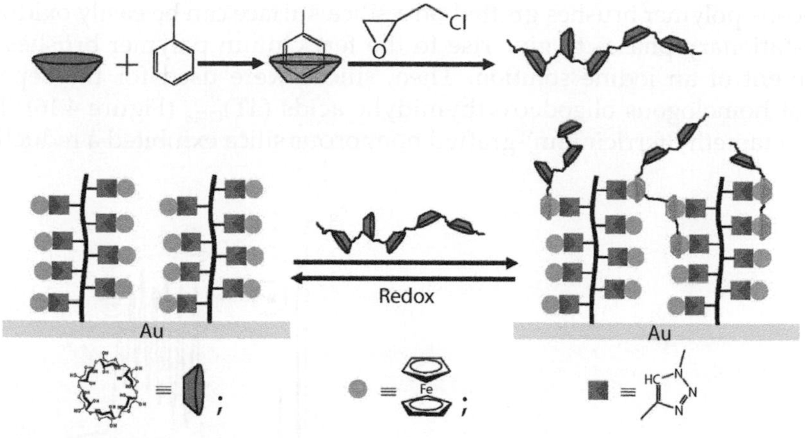

FIGURE 4.15 (See color insert.)
Redox-controlled reversible loading of the β-CD polymer on the gold substrates with surface-grafted and ferrocene-functionalized polymer brushes. (Reprinted with permission from Xu et al. 2010. One-Pot Preparation of "Ferrocene-Functionalized Polymer Brushes on Gold Substrates by Combined Surface-Initiated Atom Transfer Radical Polymerization and "Click Chemistry." *Langmuir* 26 (19):15376–15382, copyright (2010) American Chemical Society.)

about 26°. Redox-controlled disassociation of the β-CD polymer from the ferrocene-functionalized polymer brushes was achieved by applying a voltage of around 1 V for 60 min to oxidize the ferrocence unit. Upon oxidative transformation of the ferrocene groups to ferrocenium moieties, positively charged ferrocenium ions do not bound effectively with CD. The static water contact angle of the β-CD-capped surface returns to >90° after unloading of the β-CD polymer.

4.3.2 Anion-Exchange Chromatography

Oligonucleotides synthesized by standard solid-phase synthesis are widely used in hybridization experiments, either as primers in the polymerase chain reaction or as adapters for the construction of deletions, insertions, and other biologically relevant mutations. However, failure and partially deprotected nucleotide sequences are usually generated by the standard solid-phase synthesis. In general, oligonucleotides are usually separated and purified by high-performance liquid chromatography (HPLC) techniques, such as ion pair reversed-phase, mixed-mode, size exclusion, and anion-exchange HPLC (Eder et al. 2001). Ferrocene polymer brushes were synthesized and used for anion-exchange chromatography (Eder et al. 2001). Three different ethynyl-substituted ferrocene- and octamethylferrocenes were prepared via alkyne metathesis polymerization using the Schrock-type catalyst $Mo(N\text{-}2,6\text{-}Me_2\text{-}C_6H_3)(CHCMe_2Ph)(OCMe(CF_3)_2)_2$ and grafted on mesoporous or nonporous silica (Nucleosil 300–5 or Micra). The ferrocene polymer brushes grafted on a silica surface can be easily oxidized into stationary phases to give rise to the ferricinium polymer brushes by treatment of an iodine solution. Then, silicas were used for the separation of homologous oligodeoxythymidylic acids $(dT)_{12-18}$ (Figure 4.16). The poly(octamethylferricinium)-grafted nonporous silica exhibited a reduction

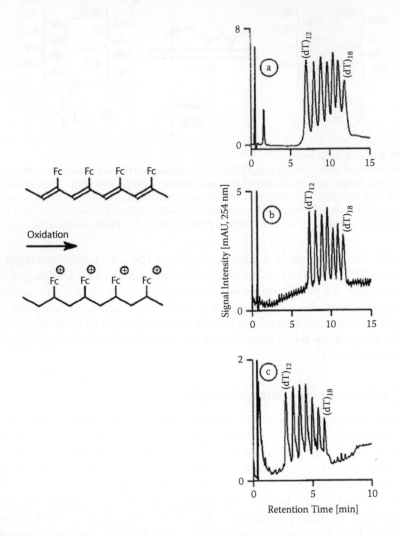

FIGURE 4.16
Schematic illustration of the oxidation of poly(ferrocene) to the corresponding poly(ferricinium). Chromatographic separation of $(dT)_{12-18}$ on (a) poly(ferricinium)-grafted mesoporous Nucleosil 300–5, (b) poly(octamethylferricinium)-grafted mesoporous Nucleosil 300–5, and (c) poly(octamethylferricinium)-grafted nonporous Micra. (Reprinted with permission from Eder et al. 2001. Alkyne metathesis graft polymerization: Synthesis of poly(ferricinium)-based silica supports for anion-exchange chromatography of oligonucleotides. *Macromolecules* 34 (13):4334–4341, copyright (2001) American Chemical Society.)

in analysis time and slightly improved peak resolution. Separation was accomplished within less than 7 min.

4.3.3 Amperometric Biosensors

The conventional oxidase-based amperometric biosensors were fabricated via the immobilization of oxidase emzymes on the electrode surface. However, the electron transfer efficiency of the oxidase-based amperometric biosensors is very poor in the absence of a mediator (Amer et al. 2010). The detection sensitivity of the amperometric biosensors can be significantly enhanced via the attachment of electroactive ferrocene mediators in the matrices for the modification of enzymes and antibodies (Nagel et al. 2007). An amperometric glucose biosensor, consisting of ITO electrodes with ferrocene brushes on the surface, was fabricated via consecutive SI-ATRP of FMMA and GMA (Zhang et al. 2010). Glucose oxidase (GOD) was subsequently immobilized on the modified ITO electrode surface via coupling reactions between the epoxide groups of GMA and the amine groups of GOD. With the introduction of redox P(FMMA) block as the electron transfer mediator, the enzyme-mediated ITO electrode exhibits high sensitivity, as revealed by cyclic voltammetry measurement. The sensitivities of the ITO-*g*-P(GMA–GOD)-*b*-P(FMMA) and ITO-*g*-P(FMMA)-*b*-P(GMA–GOD) electrodes are about 3.6 µA/(mM cm^2) (in the linear concentration range 0–5 mM of glucose) and 10.9 µA/(mM cm^2) (in the linear concentration range of 0–17 mM of glucose), respectively. For both biosensors, the steady-state response time and the detection limits are estimated to be less than 20s and 0.4 ± 0.1 mM of glucose concentration, respectively. Furthermore, the spatial effect of the redox mediator on the electrode surface is revealed by the fact that the block copolymer brush-functionalized ITO electrode with P(FMMA) as the inner (first) block is more sensitive to glucose than that with P(GMA) as the inner block.

A new thermoresponsive poly-*N*-isopropylacrylamide (PNIPAM)-ferrocene polymer was synthesized by Nagel et al. (2007), and was used to modify a gold electrode via self-assembly to create a thin hydrophilic film. The ferrocene moieties in the grafted polymer chains enable electrical communication between the pyrrolinoquinoline quinone (PQQ) cofactor of glucose dehydrogenase (sGDH) and the electrode surface for sensitive detection. The thermoresponsive property of poly(*N*-isopropylacrylamide) segments can effect the mediated electron transfer between enzyme and electrode surface. As illustrated in Figure 4.17, poly(*N*-isopropylacrylamide) is an aggregated state at high temperature, and the brush-like structure has disappeared. Therefore, the enzyme molecules are not able to penetrate the polymer network and the mediated electron transfer is decreased. The highest efficiency of mediated electron transfer for the immobilized thermoresponsive ferrocene polymers is observed at 24°C. The efficiency is twice as high as that of its soluble counterpart (without the anchoring sites to the gold electrode). A steady-state

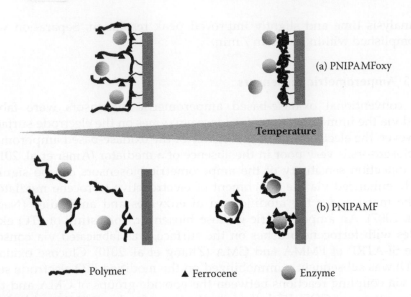

FIGURE 4.17
Proposed scheme of the temperature effect of (a) thermoresponsive ferrocene-containing polymers grafted with gold electrode and (b) free thermoresponsive ferrocene-containing polymers without the anchoring sites. (Reprinted with permission from Eder et al. 2001. Alkyne metathesis graft polymerization: Synthesis of poly(ferricinium)-based silica supports for anion-exchange chromatography of oligonucleotides. *Macromolecules* 34 (13):4334–4341, copyright (2001) American Chemical Society.)

electrooxidation current density of 4.5 µA/cm^2 is observed in the presence of 10 nM sGDH and 5 mM glucose.

4.3.4 DNA and Protein Sensors

The demand for disease screening and testing at their early stage has increased in recent years. The increase in demand has inevitably attracted great efforts toward exploring novel means to enhance detection sensitivity of biological species at an extremely low level of expression (Wu et al. 2009). Various strategies have been employed to enhance ultrasensitive detection, such as integration of enzyme-assisted signal amplification processes, incorporation of nanomaterials to increase the unloading of electrochemical tags, and introduction of multiple redox species per binding event in polymeric systems (Tuncagil et al. 2009; Chen et al. 2007; Wang et al. 2009; Tang and Ren 2008; Zhang et al. 2008). Among the above-mentioned successful strategies, biomolecule sensors based on polymeric materials have been reported to have a remarkable advantage over conventional molecule-based sensors, due to their amplification properties by a collective system response.

Wu et al. (2009) used the ferrocene polymer brushes as the electrochemical biosensors for DNA and protein detection. Figure 4.18 shows the

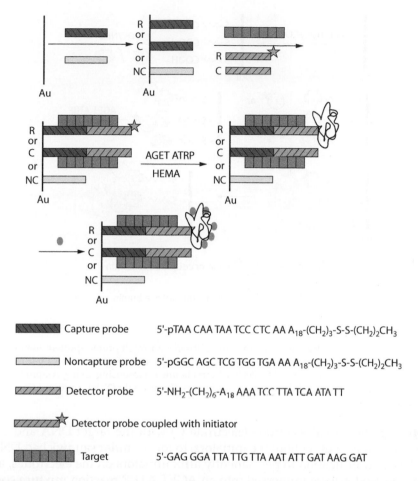

Capture probe — 5'-pTAA CAA TAA TCC CTC AA A_{18}-$(CH_2)_3$-S-S-$(CH_2)_2CH_3$

Noncapture probe — 5'-pGGC AGC TCG TGG TGA AA A_{18}-$(CH_2)_3$-S-S-$(CH_2)_2CH_3$

Detector probe — 5'-NH_2-$(CH_2)_6$-A_{18} AAA TCC TTA TCA ATA TT

Detector probe coupled with initiator

Target — 5'-GAG GGA TTA TTG TTA AAT ATT GAT AAG GAT

FIGURE 4.18
Fabrication of ferrocene-functionalized polymer brushes through DNA hybridization and AGET SI-ATRP for the electrochemical detection of DNA. (Reprinted with permission from Nagel et al. 2007. Enzyme activity control by responsive redoxpolymers. *Langmuir* 23 (12):6807–6811, copyright (2007) American Chemical Society.)

experimental procedure for DNA detection through hybridization and subsequent polymerization-assisted signal amplification. The DNA capture probe of the complementary sequence to the target DNA was first immobilized on an Au electrode (i.e., electrode R) through thiol-Au interactions. A noncomplementary (NC) capture DNA probe to the target DNA was immobilized on a separate electrode (i.e., electrode NC) in a similar fashion. Both electrodes were then immersed into the hybridization buffer containing the target DNA sequence and the initiator-labeled detection probe to form sandwiched DNA duplexes on the R electrode, but not on the electrode NC. An additional control experiment was carried out to eliminate any possible side reaction between $FcNH_2$ and DNA by incubating a DNA

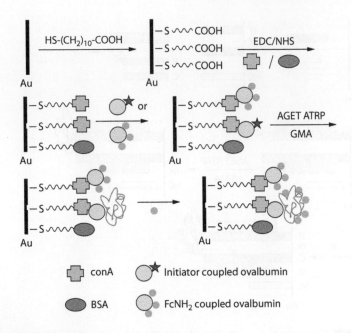

FIGURE 4.19 (See color insert.)
Fabrication of ferrocene-functionalized polymer brushes through protein-protein interaction and AGET SI-ATRP for the electrochemical detection of protein. (Reprinted with permission from Wu, Y. F., S. Q. Liu, and L. He. 2009. Electrochemical Biosensing Using Amplification-by-Polymerization. *Analytical Chemistry* 81 (16):7015–7021, copyright (2009) American Chemical Society.)

capture probe-coated electrode (electrode C) with the target DNA and the detection probe, but without the initiators, to form similar sandwiched DNA duplexes. After ligation to permanently affix initiators on the electrodes, all three electrodes were immersed into an AGET ATRP reaction mixture containing HEMA as the monomer. A layer of PHEMA film was formed on electrode R that allowed further FcNH$_2$ coupling. Neither polymer growth nor much FcNH$_2$ coupling was observed on electrodes C and NC. Quantitative analysis of the target DNA concentration shows that the measured electrocatalytic current is also proportional to the logarithm of DNA concentration. The calculated limit of detection (LOD) for target DNA detection is ~15 pM or ~30 amol of materials in solution, at the same level as other amplified electrochemical sensing results.

Ferrocene polymer brushes have also found applications in protein assay. Figure 4.19 illustrates the major steps undertaken in the detection of ovalbumin, a main energy storage protein that binds to concanavalin A (con A) through one of its four glycosylation sites. HS-(CH$_2$)$_{10}$-COOH (mercaptoundecanoic acid [MUA]) was first immobilized on an Au electrode where the carboxylic group at the distal end of MUA was used for coupling the primary amines in con A and bovine serum albumin (BSA), respectively (as in the electrodes R and C). Meanwhile, ovalbumin was derivatized with the ATRP initiators. Both electrodes R and C were incubated with initiator-coupled ovalbumin for 30 min. In addition, in a separate experiment, ovalbumin was preconjugated with $FcNH_2$ at a 1:100 molar ratio to reach its maximal loading of ferrocene. The ferrocene-conjugated ovalbumin was incubated with a con A-coated electrode (i.e., electrode D) to provide a baseline for sensitivity comparison to the conventional electrochemical detection methods where the redox tags are directly derivatized to protein targets. All three electrodes and the glass slides were immersed in an AGET ATRP reaction mixture containing GMA as the monomer. The pendant epoxide groups on PGMA were used for direct coupling of $FcNH_2$. A positive redox peak was observed from the con A-coated electrode R, whereas no catalytic wave was observed from the BSA-coated electrode C. Quantitative measurements of ovalbumin show that the measured current was proportional to the logarithm of ovalbumin concentrations in the range of 0.1–500 ng/ml, with a correlation coefficient of 0.999. The calculated limit of detection of ovalbumin is ~0.07 ng/ml, comparable to the reported conventional enzyme-linked immunosorbent assay (ELISA) assays in which the readout signal is amplified with an enzymatic reaction.

4.4 Conclusions

A comprehensive summary of the syntheses and applications of the ferrocene polymer brushes has been presented. Ferrocene-functionalized polymer brushes covalently bonded or physisorbed on solid substrate surfaces can be readily prepared via well-known surface modification techniques, such as surface-initiated polymerizations, reaction of the end-functional groups of polymer chains with substrate surfaces, and host-guest interaction. These electroactive polymer brushes have found applications in biosensors, anion-exchange chromatography, and redox-controlled modification of surface wetting properties.

References

Albagli, D., G. C. Bazan, R. R. Schrock, and M. S. Wrighton. 1993. Surface attachment of well-defined redox-active polymers and block polymers via terminal functional-groups. *Journal of the American Chemical Society* 115 (16):7328–7334.
Albagli, D., G. Bazan, M. S. Wrighton, and R. R. Schrock. 1992. Well-defined redox-active polymers and block copolymers prepared by living ring-opening metathesis polymerization. *Journal of the American Chemical Society* 114 (11):4150–4158.

Amer, W. A., L. Wang, A. M. Amin, L. A. Ma, and H. J. Yu. 2010. Recent progress in the synthesis and applications of some ferrocene derivatives and ferrocene-based polymers. *Journal of Inorganic and Organometallic Polymers and Materials* 20 (4):605–615.

Arimoto, F. S., and A. C. Haven. 1955. Derivatives of dicyclopentadienyliron. *Journal of the American Chemical Society* 77 (23):6295–6297.

Arrayas, R. G., J. Adrio, and J. C. Carretero. 2006. Recent applications of chiral ferrocene ligands in asymmetric catalysis. *Angewandte Chemie—International Edition* 45 (46):7674–7715.

Bandoli, G., and A. Dolmella. 2000. Ligating ability of 1,1'-bis(diphenylphosphino) ferrocene: a structural survey (1994–1998). *Coordination Chemistry Reviews* 209:161–196.

Barbey, R., L. Lavanant, D. Paripovic, N. Schuwer, C. Sugnaux, S. Tugulu, and H. A. Klok. 2009. Polymer brushes via surface-initiated controlled radical polymerization: synthesis, characterization, properties, and applications. *Chemical Reviews* 109 (11):5437–5527.

Bildstein, B. 2000. Cationic and neutral cumulene sp-carbon chains with ferrocenyl termini. *Coordination Chemistry Reviews* 206:369–394.

Blaedel, W. J., and G. A. Mabbott. 1981. Surface voltammetry and surface coulometry at solid electrodes. *Analytical Chemistry* 53 (14):2270–2274.

Chen, Z. P., Z. F. Peng, Y. Luo, B. Qu, J. H. Jiang, X. B. Zhang, G. L. Shen, and R. Q. Yu. 2007. Successively amplified electrochemical immunoassay based on biocatalytic deposition of silver nanoparticles and silver enhancement. *Biosensors and Bioelectronics* 23 (4):485–491.

Deschenaux, R., V. Izvolenski, F. Turpin, D. Guillon, and B. Heinrich. 1996. Ferrocene-containing thermotropic side-chain liquid-crystalline polymethacrylates. *Chemical Communications* (3):439–440.

Dubacheva, G. V., A. Van der Heyden, P. Dumy, O. Kaftan, R. Auzely-Velty, L. Coche-Guerente, and P. Labbe. 2010. Electrochemically controlled adsorption of Fc-functionalized polymers on beta-CD-modified self-assembled monolayers. *Langmuir* 26 (17):13976–13986.

Eder, K., E. Reichel, H. Schottenberger, C. G. Huber, and M. R. Buchmeiser. 2001. Alkyne metathesis graft polymerization: synthesis of poly(ferricinium)-based silica supports for anion-exchange chromatography of oligonucleotides. *Macromolecules* 34 (13):4334–4341.

Edmondson, S., V. L. Osborne, and W. T. S. Huck. 2004. Polymer brushes via surface-initiated polymerizations. *Chemical Society Reviews* 33 (1):14–22.

Engtrakul, C., and L. R. Sita. 2001. Ferrocene-based nanoelectronics: 2,5-diethynyl-pyridine as a reversible switching element. *Nano Letters* 1 (10):541–549.

Faulkner, C. J., R. E. Fischer, and G. K. Jennings. 2010. Surface-initiated polymerization of 5-(perfluoro-n-alkyl)norbornenes from gold substrates. *Macromolecules* 43 (3):1203–1209.

Feng, C., Z. Shen, D. Yang, Y. G. Li, J. H. Hu, G. L. Lu, and X. Y. Huang. 2009. Synthesis of well-defined amphiphilic graft copolymer bearing poly(2-acryloyloxyethyl ferrocenecarboxylate) side chains via successive SET-LRP and ATRP. *Journal of Polymer Science Part A—Polymer Chemistry* 47 (17):4346–4357.

Foucher, D. A., B. Z. Tang, and I. Manners. 1992. Ring-opening polymerization of strained, ring-tilted ferrocenophanes—a route to high-molecular-weight poly(ferrocenylsilanes). *Journal of the American Chemical Society* 114 (15):6246–6248.

Fristrup, C. J., K. Jankova, and S. Hvilsted. 2009. Surface-initiated atom transfer radical polymerization—a technique to develop biofunctional coatings. *Soft Matter* 5 (23):4623–4634.

Horikoshi, R., and T. Mochida. 2010. Ferrocene-containing coordination polymers: ligand design and assembled structures. *European Journal of Inorganic Chemistry* (34):5355–5371.

Kim, B. Y., E. L. Ratcliff, N. R. Armstrong, T. Kowalewski, and J. Pyun. 2010. Ferrocene functional polymer brushes on indium tin oxide via surface-initiated atom transfer radical polymerization. *Langmuir* 26 (3):2083–2092.

Kulbaba, K., and I. Manners. 2001. Polyferrocenylsilanes: metal-containing polymers for materials science, self-assembly and nanostructure applications. *Macromolecular Rapid Communications* 22 (10):711–724.

Liu, X. G., S. W. Guo, and C. A. Mirkin. 2003. Surface and site-specific ring-opening metathesis polymerization initiated by dip-pen nanolithography. *Angewandte Chemie—International Edition* 42 (39):4785–4789.

Masson, G., A. J. Lough, and I. Manners. 2008. Soluble poly(ferrocenylenevinylene) with t-butyl substituents on the cyclopentadienyl ligands via ring-opening metathesis polymerization. *Macromolecules* 41 (3):539–547.

Nagel, B., N. Gajovic-Eichelmann, F. W. Scheller, and M. Katterle. 2010. Ionic topochemical tuned biosensor interface. *Langmuir* 26 (11):9088–9093.

Nagel, B., A. Warsinke, and M. Katterle. 2007. Enzyme activity control by responsive redoxpolymers. *Langmuir* 23 (12):6807–6811.

Neuse, E. W. 2005. Macromolecular ferrocene compounds as cancer drug models. *Journal of Inorganic and Organometallic Polymers and Materials* 15 (1):3–32.

Patten, T. E., J. H. Xia, T. Abernathy, and K. Matyjaszewski. 1996. Polymers with very low polydispersities from atom transfer radical polymerization. *Science* 272 (5263):866–868.

Peter, M., M. A. Hempenius, E. S. Kooij, T. A. Jenkins, S. J. Roser, W. Knoll, and G. J. Vancso. 2004. Electrochemically induced morphology and volume changes in surface-grafted poly(ferrocenyldimethylsilane) monolayers. *Langmuir* 20 (3):891–897.

Peter, M., M. A. Hempenius, R. G. H. Lammertink, and J. G. Vancso. 2001. Electrochemical AFM on surface grafted poly(ferrocenylsilanes). *Macromolecular Symposia* 167:285–296.

Peter, M., R. G. H. Lammertink, M. A. Hempenius, and G. J. Vancso. 2005. Electrochemistry of surface-grafted stimulus-responsive monolayers of poly(ferrocenyldimethylsilane) on gold. *Langmuir* 21 (11):5115–5123.

Peter, M., R. G. H. Lammertink, M. A. Hempenius, M. van Os, M. W. J. Beulen, D. N. Reinhoudt, W. Knoll, and G. J. Vancso. 1999. Synthesis, characterization and thin film formation of end-functionalized organometallic polymers. *Chemical Communications* (4):359–360.

Pittman, C. U., O. E. Ayers, S. P. Mcmanus, J. E. Sheats, and C. E. Whitten. 1971. Organometallic polymers. 9. Polyesters of 1,1'-bis(chlorocarbonyl)cobalticinium hexafluorophosphate. *Macromolecules* 4 (3):360.

Rausch, M. D., and L. E. Coleman. 1958. Derivatives of ferrocene. 4. Ferrocene-containing unsaturated ketones. *Journal of Organic Chemistry* 23 (1):107–108.

Rulkens, R., A. J. Lough, and I. Manners. 1994. Anionic ring-opening oligomerization and polymerization of silicon-bridged [1]ferrocenophanes—characterization of short-chain models for poly(ferrocenylsilane) high polymers. *Journal of the American Chemical Society* 116 (2):797–798.

Rulkens, R., A. J. Lough, I. Manners, S. R. Lovelace, C. Grant, and W. E. Geiger. 1996. Linear oligo(ferrocenyldimethylsilanes) with between two and nine ferrocene units: electrochemical and structural models for poly(ferrocenylsilane) high polymers. *Journal of the American Chemical Society* 118 (50):12683–12695.

Sakakiyama, T., H. Ohkita, M. Ohoka, S. Ito, Y. Tsujii, and T. Fukudal. 2005. Fabrication and electrochemical properties of high-density graft films with ferrocene moieties on ITO substrates. *Chemistry Letters* 34 (10):1366–1367.

Tang, D. P., and J. J. Ren. 2008. *In situ* amplified electrochemical immunoassay for carcinoembryonic antigen using horseradish peroxidase-encapsulated nanogold hollow microspheres as labels. *Analytical Chemistry* 80 (21):8064–8070.

Tsarevsky, N. V., and K. Matyjaszewski. 2007. "Green" atom transfer radical polymerization: from process design to preparation of well-defined environmentally friendly polymeric materials. *Chemical Reviews* 107 (6):2270–2299.

Tuncagil, S., D. Odaci, E. Yidiz, S. Timur, and L. Toppare. 2009. Design of a microbial sensor using conducting polymer of 4-(2,5-di(thiophen-2-yl)-1H-pyrrole-1-l) benzenamine. *Sensors and Actuators B—Chemical* 137 (1):42–47.

van Staveren, D. R., and N. Metzler-Nolte. 2004. Bioorganometallic chemistry of ferrocene. *Chemical Reviews* 104 (12):5931–5985.

Wang, J., W. Y. Meng, X. F. Zheng, S. L. Liu, and G. X. Li. 2009. Combination of aptamer with gold nanoparticles for electrochemical signal amplification: application to sensitive detection of platelet-derived growth factor. *Biosensors and Bioelectronics* 24 (6):1598–1602.

Watson, K. J., J. Zhu, S. T. Nguyen, and C. A. Mirkin. 1999. Hybrid nanoparticles with block copolymer shell structures. *Journal of the American Chemical Society* 121 (2):462–463.

Watson, K. J., J. Zhu, S. T. Nguyen, and C. A. Mirkin. 2000. Redox-active polymer-nanoparticle hybrid materials. *Pure and Applied Chemistry* 72 (1–2):67–72.

Wright, M. E. 1990. 1,1'-Bis(tri-normal-butylstannyl)ferrocene—selective transmetalation applied to the synthesis of new ferrocenyl ligands. *Organometallics* 9 (3):853–856.

Wu, Y. F., S. Q. Liu, and L. He. 2009. Electrochemical biosensing using amplification-by-polymerization. *Analytical Chemistry* 81 (16):7015–7021.

Xu, L. Q., D. Wan, H. F. Gong, K. G. Neoh, E. T. Kang, and G. D. Fu. 2010. One-pot preparation of ferrocene-functionalized polymer brushes on gold substrates by combined surface-initiated atom transfer radical polymerization and "click chemistry." *Langmuir* 26 (19):15376–15382.

Ye, Q. A., X. L. Wang, S. B. Li, and F. Zhou. 2010. Surface-initiated ring-opening metathesis polymerization of pentadecafluorooctyl-5-norbornene-2-carboxylate from variable substrates modified with sticky biomimic initiator. *Macromolecules* 43 (13):5554–5560.

Zhang, S. S., H. Zhong, and C. F. Ding. 2008. Ultrasensitive flow injection chemiluminescence detection of DNA hybridization using signal DNA probe modified with Au and CuS nanoparticles. *Analytical Chemistry* 80 (19):7206–7212.

Zhang, Z. B., S. J. Yuan, X. L. Zhu, K. G. Neoh, and E. T. Kang. 2010. Enzyme-mediated amperometric biosensors prepared via successive surface-initiated atom-transfer radical polymerization.

Zhao, B., and W. J. Brittain. 2000. Polymer brushes: surface-immobilized macromolecules. *Progress in Polymer Science* 25 (5):677–710.

5

Preparation and Characterization of Nonfouling Polymer Brushes on Poly(Ethylene Terephthalate) Film Surfaces

Hong Tan, Jiehua Li, and Qiang Fu

College of Polymer Science and Engineering, State Key Lab Polymer and Materials Engineering, Sichuan University, Chengdu, China

CONTENTS

5.1 Introduction

Poly(ethylene terephthalate) (PET) is a typical aromatic polyester with excellent properties, such as mechanical strength, permeability to gases, transparency, chemical resistance, and moderate biocompatibility [1]. This polymer is widely applied in the textile industry, packaging, high-strength fibers, filtration membranes, automobile parts, biomedical field, and others. However, the unsatisfactory biocompatibility and functionality of PET have limited its use in some industrial and medical fields, especially in biomedical device and filtration membrane applications. For example, Dacron [poly(ethylene terephthalate)] has been successfully used to fabricate large-diameter (>6 mm) vascular substitutes in wide clinical applications [2], but PET for small-diameter (≤6 mm) applications such as coronary artery bypass grafting has been extensively unsuccessful mainly due to early graft occlusion [3,4]. Moreover, the infection is common to this kind of cardiovascular implants, which can lead to significant morbidity and mortality [5]. These are because many biological molecules have a tendency to physically adsorb onto a solid substrate without specific receptor recognition interactions (non-specific adsorption), while solid substrate contacts with biological systems [6]. Clearly, the surface structures and properties play a major role in nonspecific adsorption of material surfaces, as far as applications to biomaterials are concerned. Many biological nonspecific adsorptions are triggered by chemical structure, topography, and flexibility of the materials near the surface, such as protein, cell, and bacterial adsorption, subsequently causing blood coagulation, complement activation, inflammation, biodegradation, infection, and biofilm [7–10], resulting in implantable device failure because of the biofouling [11–14]. Thus, nonfouling surfaces of materials is of great importance for biomedical implants, textile fibers, filtration membranes, food packaging, and so on, due to their functionalities being drastically reduced by this type of contamination. To date, numerous strategies focusing on surface modification without affecting their bulk properties have been developed to inhibit or prevent nonspecific adsorption on PET surfaces.

Grafting polymer on solid surfaces is being extensively studied to control surface properties such as wettability, friction, adsorption, antifouling ability, and biocompatibility of substrates [6,15,16]. As polymer chains tethered to a surface or interface with a sufficiently high grafting density, the grafting polymers on solid surfaces are described as polymer brushes [17,18]. Polymer brushes can increase the spatial density of functional groups on a surface, thereby avoiding polymer chain overlap. Thus, polymer brushes have recently attracted considerable attention, applied to solid surface modifications of different shapes, including flat substrates, particles, fiber, and porous or tube-like structures, and of various materials, including metal, inorganic materials, and polymers [19]. Also, polymer brushes have extensively been studied to graft on PET films and microporous membranes for enhancing antifouling ability and biocompatibility.

5.2 Preparation of Reactive Functional Groups on PET Surfaces

For preparing polymer brushes on PET surfaces, reactive functional groups, such as a hydroxyl, carboxyl, or amine groups, are often deliberately introduced onto PET surfaces to act as an initiation or conjugation site. Currently, various techniques have been used to introduce reactive functional groups on PET surfaces (Table 5.1). Carboxyl groups can be introduced onto PET surfaces through hydrolysis/oxidation [20–24] and plasma deposition [33–37]. Amino groups are usually produced by aminolysis reactions with multifunctional amines, such as ethylenediamine (EDA), triethylenetetraamine (TTETA), diethylenetriamine (DETA), etc. [24–28]. Fadeev and McCarthy reported that reactions of PET film with 3-aminopropyltrialkoxysilanes at

TABLE 5.1

The Reactive Functional Groups on PET Surfaces Prepared by Different Methods

Methods	Conditions	Functional Groups	Reference
Hydrolysis/oxidation	1. NaOH, 2 N, 70°C; 2. CH₃COOH, H₂O rinsing; 3. KMnO₄/H₂SO₄, 60°C; 4. HCl, H₂O, THF rinsing	Carboxyl groups	20–24
Aminolysis	Ethylenediamine, 40 or 60°C	Amino groups	24–27
	Triethylenetetraamine (TTETA) and diethylenetriamine (DETA), 85°C	Amino groups	28
Amidation reaction	1. 3-Aminopropyltrimethoxysilane (APTMS) or 3-aminopropyltriethoxysilane (APTES), 55°C; 2. Hydrolysis	Silanol	29
Reduction	LiAlH₄, THF solution, room temperature 4.5 h, 0.1 M HCl	Hydroxyl groups	30
UV irradiation	UV light-induced surface aminolysis reaction (USAR): 1. A sandwich assembly is designed to form a thin layer of dimethylformamide (middle) between a PP film (top) and PET film (bottom); 2. Irradiation with UV light forms the top side at room temperature	Amino groups	31
	The new photo-cross-linker with three hydroxyl groups (4-benzoyl-N-(2-hydroxy-1,1-bishydroxymethyl-ethyl)-benzamide (BHAM)) solution were deposited on the PET surface via spin coating; UV irradiated	Hydroxyl groups	32
Plasma deposition	O₂ plasma; in air	Carboxyl groups	33
	Ar plasma; in air	Carboxyl groups	34–37, 53, 55
	Ammonia plasma	Amino groups	38

the film solution interface and subsequent hydrolysis render silanol (Si-OH) functionality attached to the PET surface [29]. Those silanol surfaces could be further modified to introduce alkyl, perfluoroalkyl, bromoalkyl, and aminoalkyl functionality to the PET film surfaces. Yang and coworkers developed an impressive and simple method, using a UV light-induced surface aminolysis reaction (USAR) for preparing a tertiary amine-terminated PET surface [31]. They designed a sandwich assembly, a thin layer of dimethylformamide (DMF; middle) between a polypropylene (PP) film (top) and PET film (bottom), which was irradiated with UV light from the top side at room temperature for a certain time. Under UV irradiation, the PET film surface underwent aminolysis to obtain amine functional groups on the irradiated region. Li et al. synthesized a new photo-cross-linker with three hydroxyl groups and functionalized PET film with hydroxyl groups by UV irradiation, which can be applied on surfaces of all hydrocarbon polymers [32]. PET has also been surface modified by other techniques, including plasma [38], corona discharge [39], and laser treatment [40].

5.3 Preparation of Polymer Brushes on PET Surfaces

Polymer brushes are classically prepared from physical adsorption or covalent grafting. Physical adsorption of polymer brushes on surfaces is often unstable due to no covalent bonding under some special conditions, such as high shear forces. Covalent grafting polymer brushes onto surfaces can be accomplished by the grafting to or grafting from method [16–18]. For the grafting to approach, the polymer chain is covalently grafted to surfaces, via chemical reaction between reactive groups on the surface and functional end groups of polymer chain. The grafting from method (also called surface-initiated polymerization (SIP)) means the polymer chains grow from initiators bound onto the surface. The grafting from is a more widely used approach than grafting to, since it can control the functionality, density, and thickness of the polymer brushes.

5.3.1 Graft to Polymer Brushes on PET Surfaces

Since Ikada in the early 1980s proposed that a diffused hydrophilic biomaterial surface would be good biocompatible [41], many hydrophilic polymers, including poly(ethylene glycol) (PEG), polysaccharides, polyamides, phosphobetaine-based polymers, sulfobetaine, and carboxybetaine polymers, as low-fouling or nonfouling materials were introduced to various solid substrates for enhancing the antifouling ability using the grafting to method. Among them, PEG and its derivatives have been the most widely used antifouling materials to suppress undesirable protein adsorption, platelet adhesion, and cell adhesion, because of their outstanding properties, such as nontoxicity, nonimmunogenesis, nonantigenicity, good biocompatibility, and solubility [16,42,43]. It is essential for the antifouling ability that PEG units have the large excluded volume and the mobility or flexibility of highly hydrated chains in water. The PEG layer could provide an interfacial barrier to prevent the protein from interacting with the underlying substrate. More importantly, the PEG chain length and graft density are key factors for minimizing nonspecific adsorption [43,44].

A number of studies have focused on the PEG covalently grafted to PET surfaces with different molecular weight and density for the prevention of protein adsorption. Cohn and Stern grafted PEG chains onto the PET surface via a hexamethylene diisocyanate (HDI) spacer to obtain a hydrophilic surface [38]. Fukai et al. investigated protein adsorption to covalently grafted PEGs of various chain lengths (Mw = 400, 1,000, 3,400, 5,000, and 20,000) onto PET film substrates [45]. Graft density consistently decreased as PEG chain length increased due to the higher steric requirement of higher molecular weight PEG molecules. Their results demonstrated that both longer PEG chains and higher chain density, especially the former, were more resistant to protein adsorption. However, Wang et al. found that the hemocompatibility of PET grafted with PEG with molecular weights of 200, 1,000, 6,000, and 10,000, respectively, was closely related to the PEG chain length, and the best hemocompatibility was achieved when the grafted molecular weight of PEG

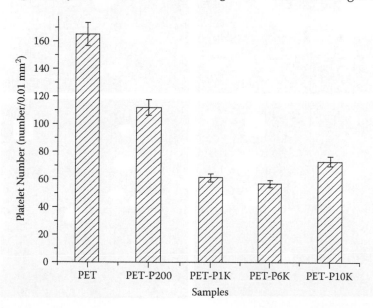

FIGURE 5.1.
Statistical data of platelet adhesion on the untreated PET and PEG-grafted PET after incubation for 15 min. (Reproduced from Wang, J., et al., *Surf. Coat. Tech.*, 196, 307–311, 2005. With permission from Elsevier.)

was 6,000 (Figures 5.1 and 5.2) [46]. Additionally, Kingshott et al. suggested that the graft density should be determined by the diameter of the PEG coils rather than the pinning group density [43]. The longer chains might be better able to cover transient gaps between chains and thereby prevent protein reaching the substrate polymer, if the chains adopted a configuration partway between brush and mushroom.

It is worth noting that the grafting to method is indeed experimentally simpler compared with the graft from method, but it has some limitations. It is difficult to form a stretched, dense polymer brush conformation on the substrate because of steric crowding of the surface reactive sites by the already adsorbed polymer chains. Furthermore, the thickness of the graft polymer brushes is limited by the prefixed molecular weight of the polymer in solution.

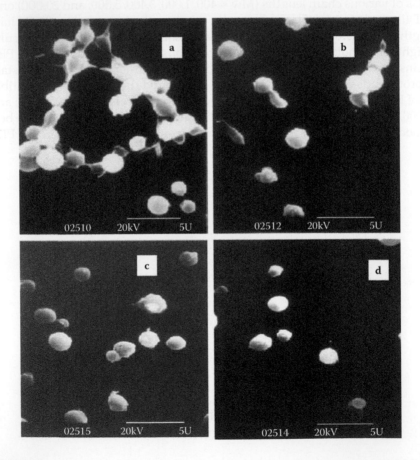

FIGURE 5.2
Typical SEM photographs of in vitro platelet adhesion tests: (a) control PET, (b) PET-P200, (c) PET-P1K, and (d) PET-P6K. The cultivation time is 120 min. (Reproduced from Wang, J., et al., *Surf. Coat. Tech.*, 196, 307–311, 2005. With permission from Elsevier.)

5.3.2 Graft from Polymer Brushes on PET Surfaces

Various grafting from techniques have been employed to grow polymer brushes on solid surfaces, including surface-initiated radical polymerizations, living ring-opening polymerization (ROP), anionic polymerization, cationic polymerization, nitroxide-mediated polymerization (NMP), reversible addition fragmentation chain transfer (RAFT) polymerization, atom transfer radical polymerization (ATRP), and so on [18]. However, UV-induced surface radical polymerization and surface-initiated ATRP have become the most popular routes for grafting from polymer brushes on PET surfaces.

5.3.2.1 UV-Induced Surface Radical Polymerization

UV-induced surface radical polymerization has been widely applied to improve the surface properties of polymers as a simple and versatile approach. To improve the biocompatibility and antibacterial adhesion of PET, graft polymer brushes via UV-induced surface radical polymerization have been extensively investigated. Usually, UV-induced polymerization is performed in the presence of a photoinitiator or photosensitizer, such as benzophenone (BP), isopropylthioxanthone (ITX), xanthone (XAN), anthraquinone (AQ), benzoyl peroxide (BPO), or 2,2′-azo-bis-isobutyronitrile (AIBN) [47]. Fortunately, since PET is one of the substrates containing carbonyl or ester groups, photografting polymerization on the surfaces can smoothly proceed even without any photoinitiator. For example, after UV irradiated with a high-pressure mercury lamp, PET films immersed in a 10 wt.% acrylamide aqueous solution containing an appropriate concentration of periodate ($NaIO_4$) gained a highly hydrophilic surface [48].

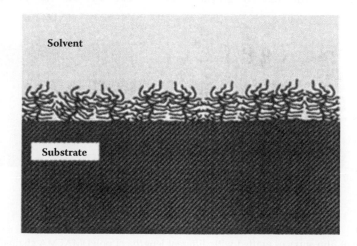

FIGURE 5.3
Schematic representation of the structure of the graft layer. (From Uchida, E., et al., *Langmuir*, 10, 481–485, 1994.)

In 1994 Uchida et al. grafted poly(ethylene glycol) methacrylate onto the surfaces of PET film by a simultaneous UV irradiation-grafted polymerization to improve wettability, antistatic property, and adhesion, and presumed that the model for the grafted surface in direct contact with a solvent of grafted chains was just like brushes (Figure 5.3) [49]. Subsequently, Uchida and Ikada also confirmed that the topography of water-soluble polymer

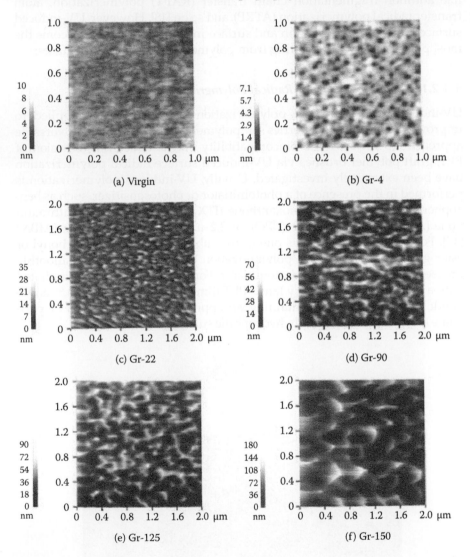

FIGURE 5.4
AFM image of film surfaces grafted with different lengths of polyDMAEMA chains in (a)-(f) (5 nN load, 2 μm/s scanning speed, and 2 × 2 μm² scan area except for (a) and (b) (1 × 2 μm²)). Brighter areas indicate higher levels. (Reprinted with permission from Uchida, E., and Ikada, Y. *Macromolecules*, 30, 5464–5469, 1997, copyright (1997) American Chemical Society.)

chains immobilized on a rigid polymer surface underwater was compared with that predicted by brush theories using atomic force microscopy (AFM) [50]. They immobilized different chain lengths of 2-(dimethylamino)ethyl methacrylate (DMAEMA) polymer onto the PET film using the UV-induced graft polymerization method, and observed the topography using AFM (Figure 5.4). They found that the graft polymer chains with a certain length stretched out in water to form a brush-like structure, and the models for this structure also presented, as shown in Figure 5.5.

In addition, poly(N-isopropylacrylamide) (PNIPAAm) is a well-known thermoresponsive polymer with a lower critical solution temperature (LCST) of 32°C in aqueous solution [20]. The chain of PNIPAAm is a random coil structure (hydrophilic state) below the LCST and a collapsed globular structure (hydrophobic state) above the LCST due to rapid, reversible chain dehydration

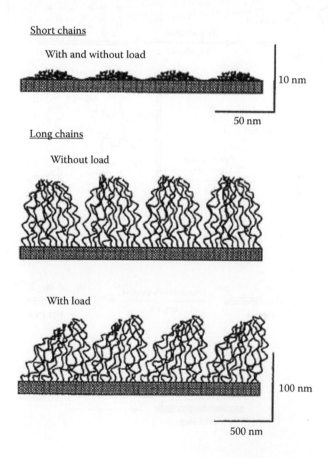

FIGURE 5.5
Topographical models for short and long graft chains underwater. (Reprinted with permission from Uchida, E., and Ikada, Y. *Macromolecules*, 30, 5464–5469, 1997, copyright (1997) American Chemical Society.)

and aggregation [16]. Thus, PNIPAAm brushes have been used in the preparation of stimuli-responsive PET films [51] or track-etched membrane surfaces [20,52] for controlling cell adhesion and permeability of membranes.

Besides that PET films were directly grafted to improve their nonfouling abilities using various hydrophilic monomers, these surfaces were further

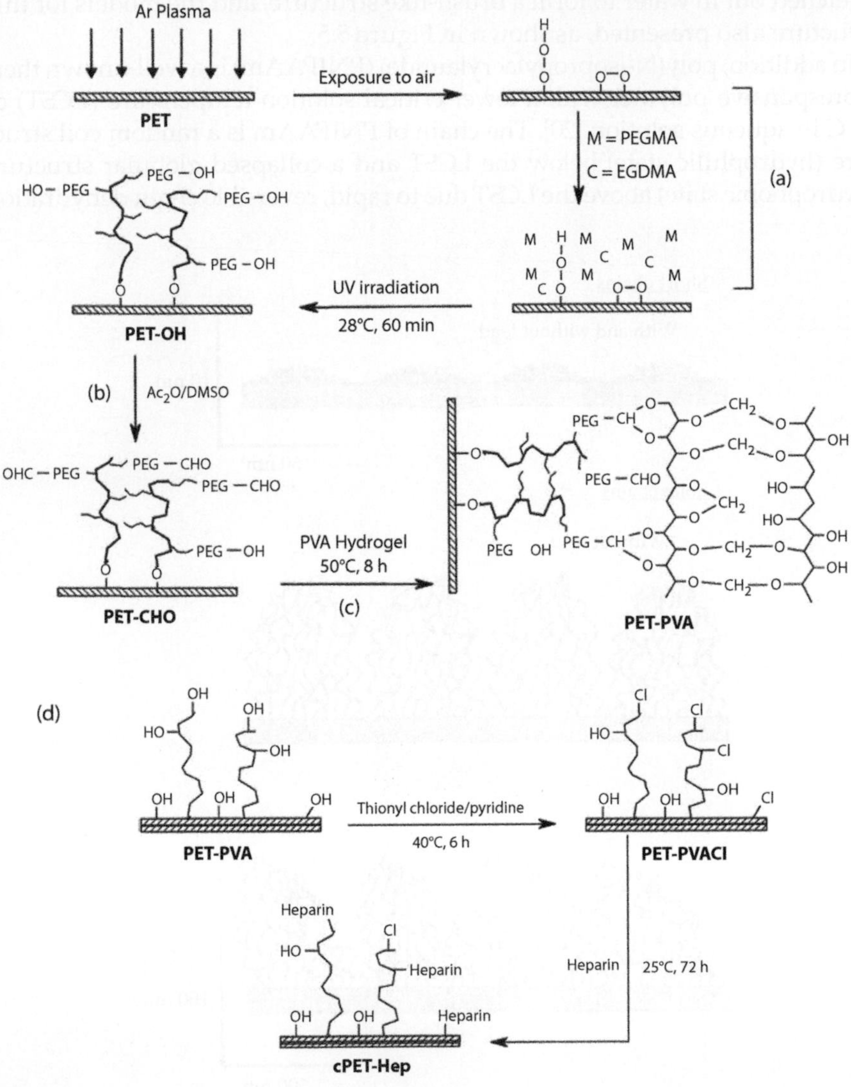

FIGURE 5.6
Surface immobilization of PET with PVA: (a) Graft copolymerization of PEGMA on plasma-pretreated PET, (b) oxidation of PET-OH, (c) PVA hydrogel immobilization on PET-CHO, and (d) covalent immobilization of heparin on PET-PVA. (Reproduced from Li, Y. L., et al., *Polymer*, 45, 8779–8789, 2004. With permission from Elsevier.)

modified to endow their better functionality. PET films were first modified by UV-induced graft polymerization with functional monomers, such as acrylic acid (AAc) or poly(ethylene glycol) monomethacrylate (PEGMA), and then coupled to another biomolecule by a functional group of polymer brushes for improving PET biocompatibility and functionality. For example, Li et al. grafted PEGMA onto PET films using UV-induced graft polymerization as the first step, and then oxidized hydroxyl end groups to produce aldehyde groups, which subsequently reacted with the hydroxyl groups of the poly(vinyl alcohol) (PVA) hydrogel [34]. Finally, heparin was immobilized on the PVA-layered PET using covalent bonding, to further improve the biocompatibility of PET, as indicated in Figure 5.6. This approach is effective in further enhancing the biocompatibility of the film, as indicated by the significantly prolonged plasma recalcification time (PRT) and very low platelet adhesion (Figures 5.7 and 5.8).

On the other hand, cationic polymer brushes grafted on PET surfaces could prevent microbial infection in industrial and medical fields, such as alkyl pyridinium or quaternary ammonium moieties. Cen et al. grafted copoly-merized with 4-vinylpyridine (4VP) and subsequently quaternized with hexylbromide [53,54], and N-hexyl-N'-(4-vinylbenzyl)-4,4'-bipyridinium bromide chloride (HVV) [55] onto PET films for enhancing antibacterial properties. The PET films were pretreated by argon plasma to form surface

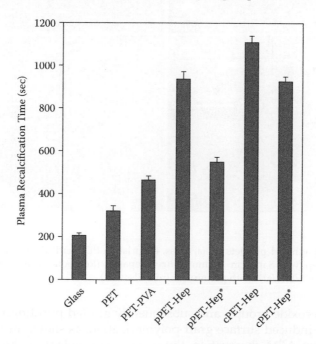

FIGURE 5.7
PRT on glass and various surface-modified PET films. The films indicated with * are those that have been immersed in distilled water for 7 days. (Reproduced from Li, Y. L., et al., *Polymer*, 45, 8779–8789, 2004. With permission from Elsevier.)

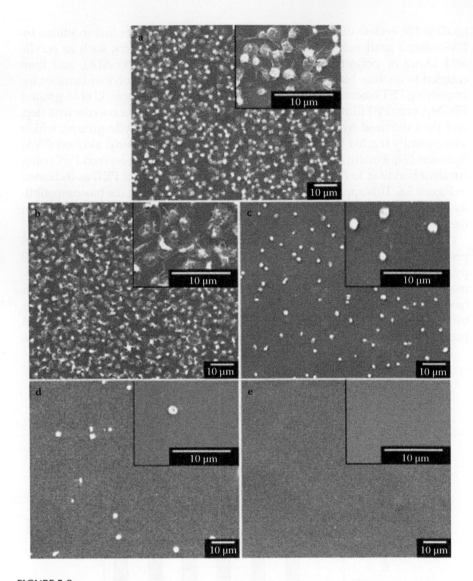

FIGURE 5.8
SEM images of adhered platelets on (a) glass, (b) pristine PET, (c) PET-PVA, (d) pPET-Hep, and (e) cPET-Hep (with 8 mg/cm² of immobilized heparin). (Reproduced from Li, Y. L., et al., *Polymer*, 45, 8779–8789, 2004. With permission from Elsevier.)

oxide and peroxide groups, and subsequently grafted pyridinium polymer through UV-induced surface graft polymerization, as shown in Figures 5.9 and 5.10. From AFM observation, the surface morphology of the PET films after UV-induced graft copolymerization with 4VP using 10 vol.% monomer and subsequent quaternization is uniform (Figure 5.12a), which ensures the homogeneity of the surface antibacterial properties. The bacterial killing

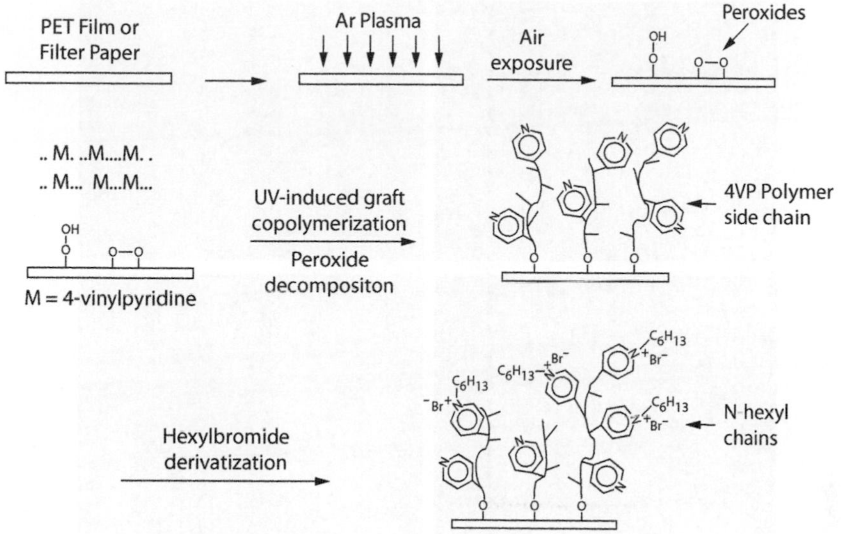

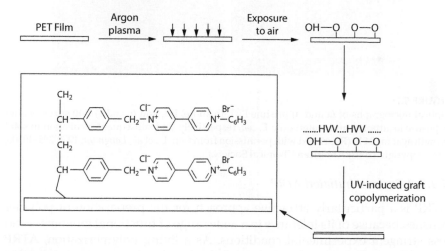

FIGURE 5.9
Schematic illustration of the surface functionalization of PET films with 4VP. (From Cen, L., et al., *Langmuir*, 19, 10295–10303, 2003.)

FIGURE 5.10
Schematic illustration of the surface functionalization of PET films with HVV. (Reproduced from Shi, Z., et al., *Biomaterials*, 26, 501–508, 2005. With permission from Elsevier.)

efficiency is dependent on the surface pyridinium concentration (Figure 5.11), and a surface concentration of 15 nmol/cm^2 (P-10) on PET has been shown to be highly effective. The P-10 film could rapidly kill *E. coli* bacteria on contact, and the antibacterial properties could be preserved for long periods of time (Figure 5.12b) [54].

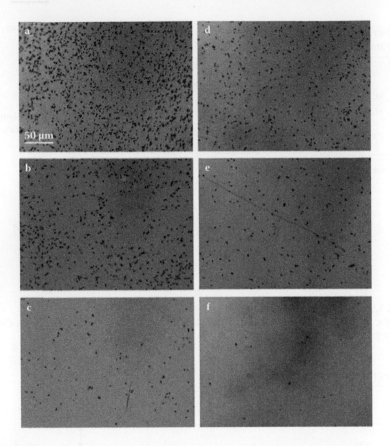

FIGURE 5.11
Optical micrographs of (a and d) pristine PET, (b and e) P-1, and (c and f) P-10 surfaces after exposure to airborne and waterborne *E. coli*, respectively, and subsequent incubation in solid growth agar for 24 h. (Reprinted with permission from Cen, L., et al., *Langmuir*, 19, 10295–10303, 2003, copyright (2003) American Chemical Society.)

5.3.2.2 Surface-Initiated ATRP

ATRP is a particularly attractive approach for the preparation of polymer brushes, because of its tolerance to a wide range of functional monomers and less stringent experimental conditions. As a living polymerization, ATRP allows for control of molecular weight and molecular weight distribution and affords the opportunity to prepare block copolymers. For the preparation of functional polymer brushes onto PET surfaces via surface-initiated ATRP, the presence of alkyl halide initiators on the surfaces is indispensable. ATRP initiators are often introduced onto PET surfaces via reaction of the reactive functional groups (as described in Section 5.2) with functional alkyl halides, such as bromoisobutyryl bromide or 3-trichlorosilyl-propyl-2-bromo-2-methyl propanoate. Farhan and Huck modified PET films by plasma oxidation to create hydroxyl groups on their surfaces, followed by

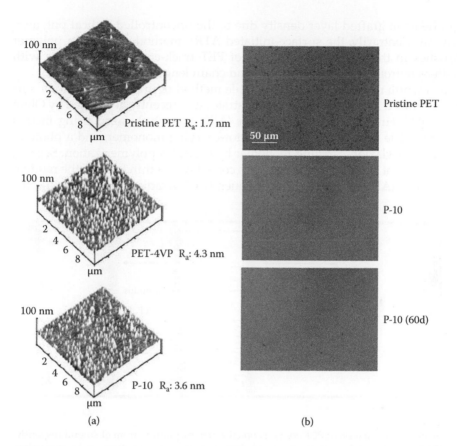

FIGURE 5.12

(a) AFM images of the pristine PET film, the PET film after UV-induced graft copolymerization with 4VP using 10 vol.% monomer in isopropanol, and P-10 film. (b) Optical micrographs of pristine PET, P-10, and P-10 (60 d) surfaces after exposure to waterborne *E. coli*, and subsequent incubation in solid growth agar for 24 h. P-10 (60 d) is the P-10 film that has been stored in air for 60 days before the antibacterial assay. (Reproduced from Cen, L., et al., *Surf. Interface Anal.*, 36, 716–719, 2004. With permission from Wiley-VCH Verlag GmbH & Co.)

the reaction with an ATRP initiator-carrying trichlorosilane [56]. Friebe and Ulbricht carboxylated the PET surfaces by oxidative hydrolysis beforehand via a two-step chemical reaction to immobilize ATRP initiators on PET films: first condensation with ethanolamine, then esterification of the hydroxyl groups with α-bromoisobutyrylbromide [57]. After the immobilization of ATRP initiators on the PET, surface-initiated ATRP can be carried out in the presence of a copper halide/nitrogen-based ligand catalyst system.

Friebe and Ulbricht functionalized PET track-etched membrane pores by grafting from of stimuli-responsive PNIPAAm via surface-initiated ATRP, photografting-from method as comparison [57]. It was found that the photografting-from method led to lower grafting density and larger gradients in grafted layer density due to the uncontrolled radical polymerization. Contrarily, the surface-initiated ATRP method can obtain polymer brushes in the isocylindrical pores of PET track-etched membranes with better-controlled grafting densities and chain lengths (Figure 5.13).

It is worth mentioning that a versatile method of initiator fixation for surface-initiated ATRP on polymeric substrates was recently developed by Ohno et al. [58]. First, a random copolymer (AIP) was synthesized using methyl methacrylate (MMA), an ATRP initiator-carrying monomer, and a photoreactive phenylazide-carrying monomer by radical copolymerization. Second, the surface of polymeric substrate was coated with a thin layer of the random copolymer (AIP) obtained, and subsequently UV irradiated to immobilize the

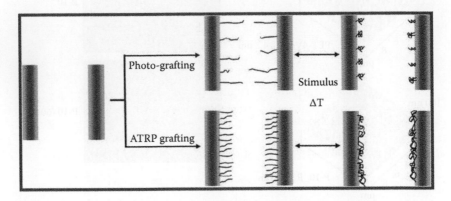

FIGURE 5.13
Schematic representation of PET pore functionalization by grafting from of stimuli-responsive PNIPAAm via two different strategies and change of effective pore diameter by increasing/decreasing the temperature around the LCST (32°C). (From Friebe, A., and Ulbricht, M., *Langmuir*, 23, 10316–10322, 2007.)

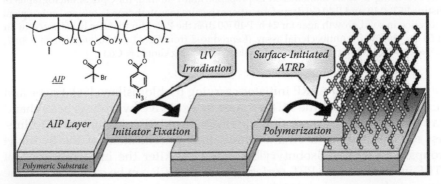

FIGURE 5.14
Chemical structure of phenyl azide and atom transfer radical polymerization (ATRP) initiator-carrying random copolymer (AIP) and schematic representation for the immobilization of AIP on polymeric substrate followed by surface-initiated ATRP. (From Ohno, K., et al., *Macromolecules*, 43, 5569–5574, 2010.)

copolymer by utilizing the photoreactivity of the phenylazido groups of the copolymer, as shown in Figure 5.14. Then the hydrophilic poly(ethylene glycol) methacrylate (PEGMA) initiated polymerization on the initiator-immobilized PET films, yielding polymeric substrates grafted with well-defined, concentrated poly(poly(ethylene glycol) methyl ether methacrylate) (PPEGMA) brushes. The PPEGMA brushes could effectively suppress the nonspecific adsorption of proteins. Figure 5.15 shows the confocal laser scanning microscopy (CLSM) image of the bovine serum albumin (BSA)-treated substrate; the fluorescence emission was hardly detected on the AIP-immobilized regions, indicating that PPEGMA brushes have been generated.

Additionally, Li et al. also reported a facile approach to chemically fix ATRP initiators onto surfaces of various hydrocarbon polymers (Figure 5.16) [32]. The hydroxyl groups were immobilized onto the surfaces of polymeric substrate via UV irradiation using the photoreactivity of a new photo-cross-linker containing three hydroxyl groups, then ATRP initiators were prepared through the esterification of the tethered hydroxyl groups with bromoisobutyryl bromide (BIBB). Subsequently, poly(ethylene glycol) methyl ether acrylate (PEGMA) polymer brushes with various main chain lengths were grafted onto PET via surface-initiated ATRP to improve hemocompatibility of polymer-based biomaterials. It was found that different aggregation structures of polymer brushes onto PET films grafted of various chain length of poly(PEGMA) could be obtained, which were observed by AFM

FIGURE 5.15
Confocal laser scanning microscopy images of patterned poly(poly(ethylene glycol) methyl ether methacrylate) (PPEGMA) brushes on (a) poly(ethylene terephthalate) (PET) after the treatment with fluorescein isothiocyanate (FITC)-labeled bovine serum albumin (BSA). (Reprinted with permission from Ohno, K., et al., *Macromolecules*, 43, 5569–5574, 2010, copyright (2010) American Chemical Society.)

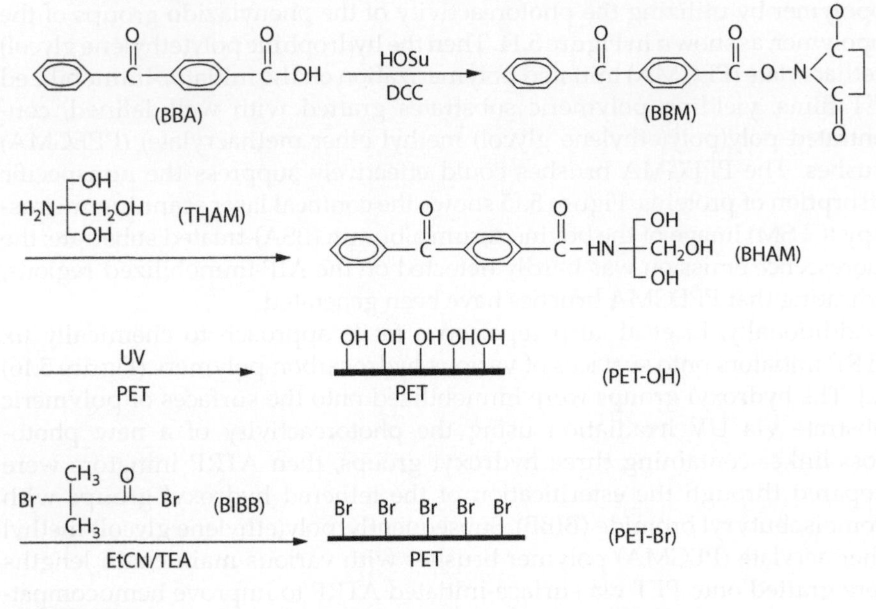

FIGURE 5.16
The synthesis route of 4-benzoyl-N-(2-hydroxy-1,1- bishydroxymethyl-ethyl)-benzamide (BHAM) and immobilization of ATRP initiators on the PET films. (Reproduced from Li, J. H., et al., *Colloid. Surface. B*, 78, 343–350, 2010. With permission from Elsevier.)

(Figure 5.17). The poly(PEGMA) brushes with proper chain length could form real brush structures on PET films due to their strong intermolecular interactions, which have been higher surface tension and lower root mean square (RMS). However, the longer polymer chains onto PET films may twist together, not form brushes, as shown in Figure 5.18. The aggregation structures of polymer brushes have great influence on the protein adsorption, and the real poly(PEGMA) brush structure on the surface of the PET could well suppress protein adsorption (Figure 5.19), which should be a nonfouling surface to meet the clinical applications.

Markedly, surface-initiated ATRP provides the opportunity to use a variety of functional monomers, to tailor the composition and thickness of the brushes, and to prepare multiple functional surfaces. For example, Zhang et al. prepared poly(methyl methacrylate) (PMMA), poly(acrylamide) (PAAM), and their diblock copolymer (PMMA/PAAM) on the surface of PET film by surface-initiated ATRP [59]. The results indicated that the surface properties of PET film were greatly improved by grafted polymer, the surface of PET film modified by PAAM was hydrophilic, and the surface of PET film modified by diblock copolymer was amphiphilic. This kind of modified PET film may be applied as biocompatible materials, amphiphilic functional film, or conductive film.

Also, the surfaces of PET with both pH and thermal stimuli sensitivity, consisting of pH-responsive poly(acrylic acid) (PAA) and thermoresponsive

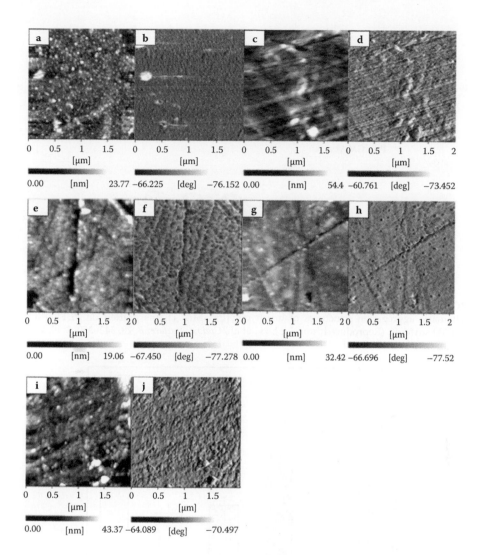

FIGURE 5.17
Topography and phase images of PET (a and b), PET10 (c and d), PET50 (e and f), PET100 (g and h), and PET200 (i and j) at 2 μm scan sizes. (Reproduced from Li, J. H., et al., *Colloid. Surface. B*, 78, 343–350, 2010. With permission from Elsevier.)

PNIPAAm blocks, have been prepared via surface-initiated ATRP onto the pore surfaces of track-etched PET membranes with pore diameters of 790 and 1,900 nm (Figure 5.20) [60]. The functional porous membrane will be relevant for the development of new functional (smart) modules, such as lab-on-a-chip systems for bio- and medicine technologies.

In conclusion, polymer brushes prepared on material surfaces, as a powerful tool and approach to surface modification, have greatly improved antifouling ability, biocompatibility, and functionality of PET. The graft to

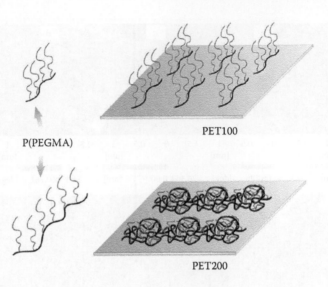

FIGURE 5.18
Schematic aggregation structures of brush poly(PEGMA) chains grafted on PET100 and PET200 surfaces. (Reproduced from Li, J. H., et al., *Colloid. Surface. B*, 78, 343–350, 2010. With permission from Elsevier.)

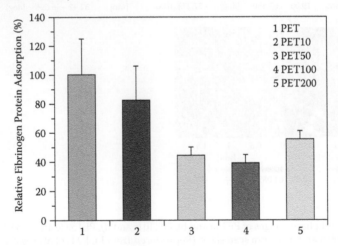

FIGURE 5.19
Relative fibrinogen protein adsorption on various film surfaces. (Reproduced from Li, J. H., et al., *Colloid. Surface. B*, 78, 343–350, 2010. With permission from Elsevier.)

and graft from approaches were generally employed to fabricate polymer brushes on PET surfaces. Existing research results suggested that surface-initiated ATRP was a particularly attractive approach for the preparation of polymer brushes on PET surfaces due to its tolerance to a wide range of functional monomers and less stringent experimental conditions, and well-defined polymer brushes obtained.

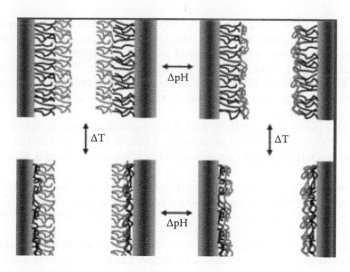

FIGURE 5.20
Combination of two stimuli-responsive polymers with low polydispersity as grafted diblock copolymer brushes in cylindrical pores leading to four different effective pore diameters as a function of the combination of the two stimuli (temperature change, ΔT, around the lower critical solution temperature of the first polymer block/here: PNIPAAm; pH change, ΔpH, around the pKa of the second polymer block/here: PAA) (From Friebe, A., and Ulbricht, M., *Macromolecules*, 42, 1838–1848, 2009.)

References

1. Sánchez-Arrieta, N., Martínez de Ilarduya, A., Alla, A., Muñoz-Guerra, S. 2005. Poly(ethylene terephthalate) copolymers containing 1,4-cyclohexane dicarboxylate units. *Eur. Polym. J.* 41: 1493–1501.
2. Edward, A., Carson, R.T., Szycher, H., Bowald, S. 1998. *In vitro* and *in vivo* biodurability of a compliant microporous vascular graft. *J. Biomater. Appl.* 13: 23–45.
3. Clowes, A.W. 1993. Intimal hyperplasia and graft failure. *Cardiovasc. Pathol.* 2 (Suppl): 179S–186S.
4. Thompson, M. M., Budd, J. S., Eady, S. L., James, R. F. L., Bell, P. R. F. 1994. Effect of pulsatile shear stress on endothelial attachment to native vascular surfaces. *Br. J. Surg.* 81: 1121–1127.
5. Wang, J., Huang, N., Yang, P., Leng, Y. X., Sun, H., Liu, Z. Y., Chu, P. K. 2004. The effects of amorphous carbon films deposited on polyethylene terephthalate on bacterial adhesion. *Biomaterials* 25: 3063–3070.
6. Senaratne, W., Andruzzi, L., Ober, C. K. 2005. Self-assembled monolayers and polymer brushes in biotechnology: Current applications and future perspectives. *Biomacromolecules* 6: 2427–2448.
7. Patel, J. D., Ebert, M., Stokes, K., Ward, R., Anderson, J. M. 2003. Inhibition of bacterial and leukocyte adhesion under shear stress conditions by material surface chemistry. *J. Biomater. Sci. Polym. Ed.* 14: 279–295.

8. Fu, J. H., Ji, J., Fan, D. Z., Shen, J. C. 2006. Construction of antibacterial multi-layer films containing nanosilver via layer-by-layer assembly of heparin and chitosan-silver ions complex. *J. Biomed. Mater. Res.* 79A: 665–674.

9. Brunstedt, M. R., Ziats, N. P., Schubert, M., Hiltner, P. A., Anderson, J. M., Lodoen, G. A., Payet C. R. 1993. Protein adsorption onto poly(ether urethane ureas) containing methacrol 2138F: A surface-active amphiphilic additive. *J. Biomed. Mater. Res.* 27: 255–267.

10. Zhao, K., Deng, Y., Chen, G. Q. 2003. Effects of surface morphology on the biocompatibility of polyhydroxyalkanoates. *Biochem. Eng. J.* 16: 115–123.

11. Ratner, B. D. 1993. The blood compatibility catastrophe. *J. Biomed. Mater. Res.* 27: 283–288.

12. Bruinsma, G. M., van derMei, H. C., Busscher, H. J. 2001. Bacterial adhesion to surface hydrophilic and hydrophobic contact lenses. *Biomaterials* 22: 3217–3224.

13. Bryers, J. D. 1994. Biofilms and the technological implications of microbial cell adhesion. *Colloids Surf. B Biointerfaces* 2: 9–23.

14. Chen, Y., and Thayumanavan, S. 2009. Amphiphilicity in homopolymer surfaces reduces nonspecific protein adsorption. *Langmuir* 25: 13795–13799.

15. Kim, J. H., Park, K., Nam, H. Y., Lee, S., Kim, K., Kwon, I. C. 2007. Polymers for bioimaging. *Prog. Polym. Sci.* 32: 1031–1053.

16. Xu, F. J., Neoh, K. G., Kang, E. T. 2009. Bioactive surfaces and biomaterials via atom transfer radical polymerization. *Prog. Polym. Sci.* 34: 719–761.

17. Zhao, B., Brittain, W. J. 2000. Polymer brushes: Surface-immobilized macromolecules. *Prog. Polym. Sci.* 25: 677–710.

18. Edmondson S., Osborne, V. L., Huck, W. T. S. 2004. Polymer brushes via surface-initiated polymerizations. *Chem. Soc. Rev.* 33: 14–22.

19. Barbey, R., Lavanant, L., Paripovic, D., Schuwer, N., Sugnaux, C., Tugulu, S., Klok, H. A. 2009. Polymer brushes via surface-initiated controlled radical polymerisation: Synthesis, characterisation, properties, and applications. *Chem. Rev.* 109: 5437–5527.

20. Geismann, C., Yaroshchuk, A., Ulbricht, M. 2007. Permeability and lectrokinetic characterization of poly(ethylene terephthalate) capillary pore membranes with grafted temperature-responsive polymers. *Langmuir* 23: 76–83.

21. Geismann, C., Ulbricht, M. 2005. Photoreactive functionalization of poly(ethylene terephthalate) track-etched pore surfaces with "smart" polymer systems. *Macromol. Chem. Phys.* 206: 268–281.

22. Roux, S., Demoustier-Champagne, S. 2003. Surface initiated polymerization from poly(ethylene terephtalate). *J. Polym. Sci. A Polym. Chem.* 41: 1347–1359.

23. Zhang, H., Shouro, D., Itoh, K., Takata, T., Jiang, Y. 2008. Grafting polymer from poly(ethylene terephthalate) films by surface-initiated ATRP. *J. Appl. Polym. Sci.* 108: 351–357.

24. Muthuvijayan, V., Gu, J., Lewis, R. S. 2009. Analysis of functionalized polyethylene terephthalate with immobilized NTPDase and cysteine. *Acta Biomater.* 5(9): 3382–3393.

25. Gappa-Fahlenkamp, H., Lewis, R. S. 2005. Improved hemocompatibility of poly(ethylene terephthalate) modified with various thiol-conatining groups. *Biomaterials* 26: 3479–3485.

26. Bhat, V. T., James, N. R., Jayakrishnan, A. 2008. A photochemical method for immobilization of dextran onto poly(ethylene terephthalate) surfaces. *Polym. Int.* 57: 124–132.

27. Liu, Y., Chen, J. R., Yang, Y., Wu, F. 2008. Improved blood compatibility of poly(ethylene terephthalate) films modified with L-arginine. *J. Biomater. Sci. Polymer Ed.* 19: 497–507.
28. Nissen, K. E., Stuart, B. H., Stevens, M. G., Baker, A. T. 2008. Characterization of aminated poly(ethylene terephthalate) surfaces for biomedical applications. *J. Appl. Polym. Sci.* 107: 2394–2403.
29. Fadeev, A. Y., McCarthy, T. J. 1998. Surface modification of poly(ethylene terephthalate) to prepare surfaces with silica-like reactivity. *Langmuir* 14: 5586–5593.
30. Chen, W., McCarthy, T. J. 1998. Chemical surface modification of poly(ethylene terephthalate). *Macromolecules* 31: 3648–3655.
31. Yang, P., Zhang, X. X., Yang, B., Zhao, H. C., Chen, J. C., Yang, W. T. 2005. Facile preparation of a patterned, aminated polymer surface by UV-light-induced surface aminolysis. *Adv. Funct. Mater.* 15: 1415–1425.
32. Li, J. H., Tan, D. S., Zhang, X. Q., Tan, H., Ding, M. M., Wan, C. X., Fu, Q. 2010. Preparation and characterization of nonfouling polymer brushes on poly(ethylene terephthalate) film surfaces. *Colloid. Surface. B* 78: 343–350.
33. Kim, Y. J., Kang, I. K., Huh, M. W., Yoon. S. C. 2000. Surface characterization and *in vitro* blood compatibility of poly(ethylene terephthalate) immobilized with insulin and/or heparin using plasma glow discharge. *Biomaterials* 21: 121–130.
34. Li, Y. L., Neoh, K. G., Kang, E.T. 2004. Poly(vinyl alcohol) hydrogel fixation on poly(ethylene terephthalate) surface for biomedical application. *Polymer* 45: 8779–8789.
35. Bisson, I., Kosinski, M., Ruault, S., Gupta, B., Hilborn, J., Wurm, F., Frey, P. 2002. Acrylic acid grafting and collagen immobilization on poly(ethylene terephthalate) surfaces for adherence and growth of human bladder smooth muscle cells. *Biomaterials* 23: 3149–3158.
36. Singh, N., Bridges, A. W., Garcia, A. J., Lyon, L. A. 2007. Covalent tethering of functional microgel films onto poly(ethylene terephthalate) surfaces. *Biomacromolecules* 8: 3271–3275.
37. Ying, L., Yin, C., Zhou, R. X., Leong, K. W., Mao, H. Q., Kang, E. T., Neoh, K. G. 2003. Immobilization of galactose ligands on acrylic acid graft-copolymerized poly(ethylene terephthalate) film and its application to hepatocyte culture. *Biomacromolecules* 4: 157–165.
38. Cohn, D., Stern, T. 2000. Sequential surface derivatization of PET films. *Macromolecules* 33: 137–142.
39. Strobel, M., Lyons, C. S., Strobel, J. M., Kapaun, R. S. 1992. Analysis of air-corona-treated polypropylene and poly(ethylene-terephthalate) films by contact angle measurements and x-ray photoelectron spectroscopy. *J. Adhes. Sci. Technol.* 6: 429–523.
40. Arenolz, E., Heitz, J., Wagner, M., Baeuerle, D., Hibst, H., Hagemeyer, A. 1993. Laser-induced surface modification and structure formation of polymers. *Appl. Surf. Sci.* 69: 16–19.
41. Ikada, Y. 1984. Blood-compatible surfaces. *Adv. Polym. Sci.* 57: 103–140.
42. Kingshott, P., Griesser, H. J. 1999. Surfaces that resist bioadhesion. *Curr. Opinions. Sol. State. Mater. Sci.* 4: 403–412.
43. Kingshott, P., Thissen, H., Griesser, H. J. 2002. Effects of cloud-point grafting, chain length, and density of PEG layers on competitive adsorption of ocular proteins. *Biomaterials* 23: 2043–2056.

44. Ma, H., Hyun, J., Stiller, P., Chilkoti, A. 2004. "Non-fouling" oligo(ethyleneglycol)-functionalized polymer brushes synthesized by surface-initiated atom transfer radical polymerization. *Adv. Mater.* 16: 338–341.

45. Fukai, R., Dakwa, P. H. R., Chen, W. 2004. Strategies toward biocompatible artificial implants: Grafting of functionalized poly(ethylene glycol)s to poly(ethylene terephthalate) surfaces. *J. Polym. Sci. A Polym. Chem.* 42: 5389–5400.

46. Wang, J., Pan, C. J., Huang, N., Sun, H., Yang, P., Leng, Y. X., Chen, J. Y., Wan, G. J., Chu, P. K. 2005. Surface characterization and blood compatibility of poly(ethylene terephthalate) modified by plasma surface grafting. *Surf. Coat. Tech.* 196: 307–311.

47. Deng, J., Wang, L., Liu, L., Yang, W. 2009. Developments and new applications of UV-induced surface graft polymerizations. *Prog. Polym. Sci.* 34: 156–193.

48. Uchina, E., Uyama, Y., Ikada, Y. 1990. A novel method for graft polymerization onto poly(ethylene terephthalate) film surface by UV irradiation without degassing. *J. Appl. Polym. Sci.* 41: 677–687.

49. Uchida, E., Uyama, Y., Ikada, Y. 1994. Grafting of water-soluble chains onto a polymer surface. *Langmuir* 10: 481–485.

50. Uchida, E., Ikada, Y. 1997. Topography of polymer chains grafted on a polymer surface underwater. *Macromolecules* 30: 5464–5469.

51. Curti, P. S., de Moura, M. R., Veiga, W., Radovanovic, E., Rubira, A. F., Muniz, E. C. 2005. Characterization of PNIPAAm photografted on PET and PS surfaces. *Appl. Surf. Sci.* 245: 223–233.

52. Yang, B., Yang, W. T. 2003. Thermo-sensitive switching membranes regulated by pore-covering polymer brushes. *J. Membr. Sci.* 218: 247–255.

53. Cen, L., Neoh, K. G., Kang, E. T. 2003. Surface functionalization technique for conferring antibacterial properties to polymeric and cellulosic surfaces. *Langmuir* 19: 10295–10303.

54. Cen, L., Neoh, K. G., Ying, L., Kang, E. T. 2004. Surface modification of polymeric films and membranes to achieve antibacterial properties. *Surf. Interface Anal.* 36: 716–719.

55. Shi, Z., Neoh, K. G., Kang, E. T. 2005. Antibacterial activity of polymeric substrate with surface grafted viologen moieties. *Biomaterials* 26: 501–508.

56. Farhan, T., Huck, W. T. S. 2004. Synthesis of patterned polymer brushes from flexible polymeric films. *Eur. Polym. J.* 40: 1599–1604.

57. Friebe, A., and Ulbricht, M. 2007. Controlled pore functionalization of poly(ethylene terephthalate) track-etched membranes via surface-initiated atom transfer radical polymerization. *Langmuir* 23: 10316–10322.

58. Ohno, K., Kayama, Y., Ladmiral, V., Fukuda, T., Tsujii, Y. 2010. A versatile method of initiator fixation for surface-initiated living radical polymerization on polymeric substrates. *Macromolecules* 43: 5569–5574.

59. Zhang, H., Shouro, D., Itoh, K., Takata, T., Jiang, Y. 2008. Grafting polymer from poly(ethylene terephthalate) films by surface-initiated ATRP. *J. Appl. Polym. Sci.* 108: 351–357.

60. Friebe, A., and Ulbricht, M. 2009. Cylindrical pores responding to two different stimuli via surface-initiated atom transfer radical polymerization for synthesis of grafted diblock copolymers. *Macromolecules* 42: 1838–1848.

6

Formation of Polymer Brushes Inside Cylindrical Pores

Alexandros G. Koutsioubas

Laboratoire Léon Brillouin, CEA/CNRS UMR 12,
Gif-sur-Yvette, France

CONTENTS

6.1 Introduction

In the vast literature that covers the subject of polymer brushes, one may find numerous studies concerning brush formation and properties on flat surfaces. Since the introduction of the concept, polymer brushes have been studied traditionally on flat macroscopic surfaces due to the applicability of well-established experimental techniques and also because of the relative simplicity of the theoretical description on a flat geometry. However, the properties of polymer brushes grown on curved interfaces (especially on the surface of micro- and nanoparticles) have also received interest because of early-suggested applications in the stabilization of colloid dispersions [1].

As it is well established, polymer brushes on flat substrates can be treated to a first approximation as a linear string of blobs of equal size [2]. On the other hand, in the case of a brush grafted on a curved surface (convex or concave), the theoretical analysis of the system is somewhat more complicated than the case of flat surfaces, and that is because the surface curvature leads to the modification of the available space for each grafted chain. Following the classic theoretical analysis of Daoud and Cotton for star polymers [3], brushes on curved surfaces can be envisioned as an array of concentric shells of blobs that have variable size as we move away from the grafting surface. Using this somewhat simplified scaling picture, an initial insight may be gained for these systems [4,5].

The convex or concave nature of the grafting surface is intuitively expected to induce opposite effects on the brush structure. For positive curvatures (convex geometry) the deformation energy is lowered with respect to the flat brush due to the increased space availability per grafted chain. On the contrary, negative curvatures (concave geometry) lead to increased chain confinement and overlap. The intensity of these effects depends strongly on the ratio between the curvature of the grafting surface and the characteristic lengths (height L, interanchor distance s) of the brush. So we may understand that as the curvature increases, deviations from flat brush behavior should become more pronounced.

In this chapter, we will focus on the case of concave geometry and more specifically on brushes formed on the inner surface of cylindrical pores (Figure 6.1) that have a characteristic diameter in the nanometer to micrometer range, i.e., close to the characteristic length of polymer brushes. The increased interest in the field of nanotechnology during the last decades has led to the development of many new methods for the easy fabrication of highly aligned porous membranes like porous anodic alumina (PAA) and porous silicon (p-Si). These relatively new types of membranes add up to previously developed structures such as track-etched polymeric membranes. The widespread availability of these materials makes the experimental realization of confined brush systems much more attainable.

Probably the first notion of polymer brushes grafted inside pores was given by Sevick [6] in the mid 1990s in relation to the potential fabrication of pressure-sensitive microvalves. Essentially it was proposed that adsorbed polymer brushes on the inner surface of pores might sense the solvent flow through the pores due to a pressure difference and then respond by swelling. This response that actively changes the pore cross section is an

FIGURE 6.1
Simulation snapshots of a polymer brush inside a cylindrical pore. (Reprinted with permission from Koutsioubas et al., *J. Chem. Phys.*, 131, 044901. Copyright 2009, American Institute of Physics.)

important feature related to flow regulation applications. However, a more straightforward way for the active manipulation of the pore/brush system was proposed and demonstrated by Ito et al. [7] and concerns the use of signal-responsive polymers (usually charged polymers/polyelectrolytes) that may extend or collapse in response to variations of pH, ionic strength, temperature, redox reaction, and photoirradiation.

Going beyond the case of multipore membranes, a more recent area of research that is related to single nanochannels decorated with polymer brushes has emerged [8]. Single nanoscale fluidic channels have attracted a lot of interest lately due to their applications in the control of ionic species flow and also due to their usefulness in single molecule experiments of biomacromolecular translocation. By the incorporation of signal-responsive polymer chains inside these single channels, the resulting structures bear unique properties of sensing, switching, and regulating ionic species in aqueous solutions. Nanochannels that can be triggered from an on to an off state by a change of pH conditions, essentially mimic the function of many biological channels that display pH-dependent ion transport through biological membranes in living cells.

These interesting applications that we have briefly summarized require the fabrication of polymer brushes in nanoporous membranes or inside single nanochannels. However, it is not straightforward that well-established fabrication techniques for brushes on flat surfaces also apply in the case of a confined geometry. In general, there are two main approaches for the fabrication of polymer brushes: the grafting to (self-assembly from solution) and the grafting from (polymerization from surface-bound initiators) methods. Both of these methods may be applied in the fabrication of polymer brushes inside pores; however, as it has been shown, due to the special geometry of the pores, there are several limitations associated with each technique that have to be taken into account.

The need for the precise control of brush properties inside the pores also brings about the necessity of characterization methods and experiments that can give direct or indirect information concerning the detailed structure of the brush (grafting density, thickness, uniformity, etc.). It should be noted that as for the fabrication methods, the special geometry of the system renders some of the traditional characterization methods nonapplicable for the study of the present system. So special care should be devoted in the use of appropriate experimental techniques that may give valuable information.

In the coming sections, we will first review the general theoretical predictions for polymer brushes that are confined in concave geometries, and then we will pass to the fabrication aspects of brush formation inside porous membranes. Finally, characterization methods will be discussed together with potential practical applications of the pore/brush systems.

6.2 Theoretical Considerations

Every attempt to describe the properties of polymer brushes grafted on the inner surface of cylinders takes as a starting point the theoretical treatment of the flat polymer brush. Concavity is then included in the model by proper assumptions, and obviously at the limit of very small curvature the flat brush behavior has to be recovered. It is natural to think that depending on the relative values of pore diameter D or radius R, grafting density σ and chain size N (number of monomers of size α), different chain conformation regimes should appear inside the pore. As we will see, four different general regimes may be distinguished, while two of them are directly relevant to brush structure.

For a system of linear chains grafted on a plane surface under good solvent conditions, we may identify two main regimes. When the grafted chains are isolated from each other (small σ or N) they assume the form of mushrooms of size equal to the radius of gyration of the chains in solution $R_g = \alpha N^{3/5}$. For larger chains and denser grafting, the chains start to overlap and thus form a semidilute polymer brush where the screening length ξ (or blob radius) is of the order $\sigma^{-1/2}$. Each blob contributes $k_B T$ to the system's free energy and also contains $n = \sigma^{-5/6}$ monomers. By considering that each chain in the semidilute brush regime may be represented as a string of blobs and by using simple arguments, we may calculate the extension of the brush layer normal to the surface as

$$L = \alpha N \sigma^{1/3} \tag{6.1}$$

This line of thought represents the foundations of the classic blob model of Alexander [9] and de Gennes [2], which gives a valid approximation of the layer extension. However, the model is not so successful in the prediction of the monomer density $\phi(z)$ as a function of the distance from the grafting surface z. In the Alexander–de Gennes model, monomer density is considered constant and all chain ends are positioned at the outer extremities of the layer, a fact that largely underestimates the conformational entropy of the grafted chains. Later, self-consistent mean field (SCF) [10] approaches to the problem (that where also backed up by computer simulation results) showed that the monomer volume fraction profile of a brush in a good solvent decreases monotonically away from the surface and has an approximately parabolic form ($\phi(z) = \phi(0)[1 - (z/L)^2]$), while chain ends populate the entire brush layer even close to the grafting surface [11]. Neutron reflectivity experimental studies [12,13] have verified the SCF theory predictions for adequately dense brushes.

Now, in the case of chain grafting in a long cylindrical tube, if the radius of the cylinder is very large ($R \gg L$) and the grafting density small ($\sigma \ll (\alpha/R_g)^2$), then essentially we fall in the regime of isolated noninteracting mushrooms [14] of size $R_g = \alpha N^{3/5}$. In the other extreme case, where the tube radius is smaller than the chain's radius of gyration ($R_g \gg R$), the chains form the so-called cigars that are strings of blobs (with $\xi = D$) oriented parallel to the tube axis. Depending on the grafting density in the cigar regime, different scaling relations for the extension of the chains parallel to the tube axis are expected [15,16].

A more interesting regime (at least for the scope of this chapter) occurs when the interanchor distance is comparable to or somewhat larger than the radius of gyration of the chains ($\sigma > (\alpha/R_g)^2$). In this regime chains are stretched due to excluded volume interactions and form a brush whose structure is essentially the same as in the planar geometry. All the predictions that have been mentioned for plane brushes apply, and the thickness of the brush layer will be given to a good approximation by Equation 6.1. The curvature of the cylindrical surface does not appreciably affect the brush structure as long as the brush extension is smaller that the radius of the cylinder ($R < \alpha N\sigma^{1/3}$). But as we move to smaller radii that are comparable to brush thickness ($R \cong \alpha N\sigma^{1/3}$), there is a crossover between a brush and a confined (or compressed) brush state. In this regime of the compressed brush the interactions between neighboring chains are important and the density of monomers ϕ tends to become constant inside the cylinder, with values typical of a polymer semidilute solution.

For the intermediate regime of noncompressed concave brushes that we have mentioned, there have been two initial scaling descriptions based on (1) the Daoud and Cotton model of star polymers [3] and (2) Sevick [6]. The Daoud and Cotton model is particularly relevant to convex polymer brushes, because star polymers that have a large number of arms behave as a set of linear chains grafted on a spherical surface, which is the center of the star structure and is characterized by a high monomer concentration. The calculation of the chain free energy is based on the assumption that the blob size is not constant (contrary to chains grafted on a plane), but gets larger as we move away from the center (Figure 6.2). For a concave geometry we may argue that due to the geometrical similarity of the problem, the same analysis applies; i.e., the blob size is decreasing as we move away from the grafting surface (inner surface). Using this inverse Daoud and Cotton model for a concave brush, we get the following expression for the brush thickness:

$$L_{D-C} = R\left(1 - \left(1 - \frac{\alpha N\sigma^{1/3}}{R}\right)^{3/5}\right) \tag{6.2}$$

while for the monomer concentration as a function of the radial distance to the surface, we have

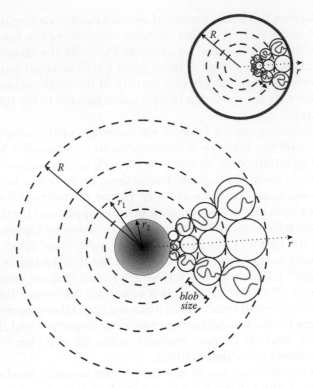

FIGURE 6.2
Schematic representation of the Daoud and Cotton model for star polymers (main figure) that can be adapted for the study of convex and concave brushes (top right).

$$\phi(r) = \frac{\sigma^{2/3}}{(1 - r/R)^{4/3}} \tag{6.3}$$

Quite surprisingly the model predicts that the monomer density is larger close to the cylinder axis than near the grafting surface.

The other theoretical approach by Sevick [6] is essentially a variation of the Alexander–de Gennes model that takes into account the cylindrical geometry. The result for the brush thickness is given by

$$L_{Sevick} = N^{3/4} (s/\alpha)^{-1/2} (R/\alpha)^{1/4} \tag{6.4}$$

Note that Equation 6.4 predicts that for a given interanchor distance s, brush thickness L is an increasing function of cylinder radius R, which is a counterintuitive result if we recall the simple argument that smaller curvatures lead to less confined chains that should also be less extended.

In a later theoretical work, Manghi et al. [5] have criticized* the use of both these models for the description of brushes in a concave geometry, and they have provided an SCF description of the system (in spherical geometry) by calculating both brush monomer and free-end density for the concave brush and compressed brush regime. They have shown that for a very large radius (R), compared to the brush height of the equivalent plane brush (L_{flat}), the concave brush extension (L) is approximately given by the relation

$$L \cong L_{flat}\left(1+\frac{L_{flat}}{4R}\right) \tag{6.5}$$

This gives a small correction to the estimation of the concave brush height, compared to Equation 6.1. Note that since brush extension is a decreasing function of R, this behavior is in contradiction with the result from Sevick's approach (Equation 6.4).

In addition, the SCF model by Manghi et al. [5] predicts that in the concave brush regime the monomer and free-end radial distributions have essentially the same form as for brushes on flat surfaces, i.e., a parabolic decay of the monomer density as a function of the radial distance to the surface (contrary to the prediction of the inverse Daoud and Cotton model; Equation 6.3) and a broad free-end distribution with a maximum near the brush extremities. Furthermore, for the compressed brush regime where the entire space is occupied by monomers, $\phi(r)$ has a truncated parabolic profile with a finite monomer concentration at the center ($r = 0$). The free-end distribution is also altered in this regime, being a monotonically increasing function of the radial distance from the surface with a large number of free ends situated at the edge of the distribution. As the brush gets highly confined, we expect that the monomer density would be the same everywhere inside the tube and free-end distribution will have a clear maximum at $r = 0$.

In a series of papers Dimitrov et al. [15,16] have addressed the problem of polymer brushes in cylindrical pores using coarse-grained molecular dynamics simulations of a bead-spring polymer model, and they have also provided a detailed scaling analysis of the various regimes for grafted chains inside a pore. Their simulation results concerning monomer and free-end distribution reproduce the general behavior that SCF theory predicts. Also, it is very interesting that for very small curvatures, the simulations reproduce the result of Equation 6.5. From an experimental point of view, the predictions of SCF theory and computer simulations concerning noncompressed concave brushes are of great interest, since very dense brushes inside nanopores are

---

* The main line of criticism for the inverse Daoud and Cotton model concerns the fact that the concentric blob structure does not represent the most stable state of the concave brush. On the other hand, the assumption that all chain ends are positioned at the end of the layer in the Sevick model is obviously not accurate and tends to become completely unjustified as we move closer to the confined brush regime.

difficult to achieve in the laboratory. Consequently, the described theoretical framework provides a fairly adequate description when neutral polymer brushes in a good solvent are considered, and may serve as a guide for the fabrication and characterization of polymer brushes inside cylindrical pores.

6.3 Fabrication Aspects

In the past few years there has been a real surge of new techniques for the preparation of porous materials that are characterized by well-defined cylindrical pores of sizes from a few micrometers, down to the nanometer range. Most notably, porous anodic alumina (PAA) [17] and porous silicon (p-Si) [18,19] that are prepared by electrochemical anodization, and track-etched polymer membranes (polycarbonate, polyimide, polyethylene terephtalate, etc.), represent the most well-known cases of porous membranes that are candidates for filtration applications and also for their use as templates in nanotechnology (nanowire fabrication [20]). The pore diameter range of these membranes is comparable to the typical thickness of polymer brushes that are usually prepared in the laboratory.

For the fabrication of polymer brushes on flat surfaces or in other surface geometries there exist two general methods: grafting to and grafting from. End-functionalized chains may be grafted on a surface from solution (grafting to), or in other words, self-assemble on the surface by end attachment of their one end. An alternative route is to proceed by polymerization of the chains from appropriate initiators on the surface (grafting from). In general, the grafting to method suffers from low final grafting densities, but control of chain size distribution (polydispersity) may be very accurate. On the contrary, by the grafting from method we may fabricate very dense brush layers but with broader polydispersity. Both of these techniques have been used for the preparation of brushes inside porous media. As it is expected, the special geometry of the porous medium poses some challenges during the fabrication procedure and also on the characterization of the resulting structures.

We will start our discussion with the grafting to approach, which may be characterized as the simplest. Depending on the nature of the surface of the porous membrane that we want to decorate with a brush, we must choose an end-functionalized polymer chain (or a highly asymmetric block copolymer of an amphipathic nature) that may strongly adsorb from its functionalized part on the inner surface. Then in principle we may use the same method as for the case of a flat surface, i.e., dip the membrane in a solution that contains these polymers and then wait for the full formation of the brush. However, it has been shown [21] that self-assembly inside the pores is prohibited when the ratio of pore diameter to chain radius of gyration is less than about 10 ($D/R_g > 10$). Despite that the characteristic size of the chains in this regime is much smaller than pore diameter, it appears that initial adsorption near the pore entrance creates a barrier for later incoming chains, thus restricting their approach in the pore interior (pore plugging) and preventing uniform pore filling by grafted chains. This effect has been observed for the self-assembly of polystyrene-b-polyethyleneoxide (PS-PEO) block copolymers inside porous alumina membranes and also in dynamic Monte Carlo simulations of brush formation inside cylindrical pores [22].

Except from the effect of pore plugging, the special geometry of the system induces a large change in the kinetics of adsorption. From the initial stage of adsorption kinetics, the chain diffusion rate toward the surface may be calculated. In fact, it is found that the overall diffusion rate is four orders of magnitude smaller than adsorption of the same polymers on flat surfaces. This dramatic reduction is a clear consequence of geometric chain confinement inside the pores. Under these conditions, full brush formation may last for many days, depending also on the thickness of the porous membrane.

As long as $D/R_g > 10$ and a uniform brush gets formed all along the pore length, the measured adsorbed amounts are identical to those measured on flat substrates and also follow the same scaling law as a function of chain molecular weight ($\Gamma \sim M_w^{-1/5}$). This indicates that the Alexander–de Gennes scaling applies to self-assembled concave brushes. In fact, by making some calculations (based on the measured adsorption) about the brush thickness and interanchor distance, we find that we are clearly in the nonconfined concave brush regime ($R > L$, $s > R_g$), where according to SCF theory [5] and simulations [15], the overall properties of the brush are approximately the same as for a flat brush.

Also, in the grafting to approach we should include methods that require the pretreatment of the pore surface in order for the end-functionalized polymer to graft. For example, amine end-functionalized polystyrene chains were grafted on epoxy-derivatized walls of porous alumina membranes, disulfide end-functionalized polypeptides (polyglutamic acid) have been grafted on gold-coated track-etched membranes [23], and poly(methyl methacrylate) (PMMA) chains have been grafted on glow discharge treated polycarbonate membranes [24].

Now let's turn to the second general method for preparing polymer brushes, the grafting from approach where polymer chains are prepared in situ. This method is based on the initial decoration of a surface with appropriate initiators and the subsequent polymerization directly on the surface. Depending on the surface coverage of the initiator molecules, very dense polymer brushes may be produced while final chain length may be controlled by the reaction time. The grafting from method has been successfully implemented for the fabrication of homogeneous layers of end-tethered homopolymers and block copolymers on flat surfaces [25], with a reasonable polydispersity.

There have been various experimental efforts for the fabrication of concave polymer brushes inside cylindrical pores by grafting from techniques. The most notable examples include

- A series of publications of Ito et al. [7,26] concerning signal-responsive brushes that were polymerized from the inner surface of porous membranes
- The preparation of poly(N-isopropylacrylamide) (PNIPAAm) on poly (ethylene terephthalate) (PET) track-etched membranes by photograft copolymerization [27]
- The radiografting of poly(acrylic acid) chains in poly(vinyliden fluoride) (b-PVDF) track-etched nanoporous membranes [28]
- The use of surface-initiated atom transfer radical polymerization (ATRP) for the preparation of PNIPAAm-decorated track-etched PET [29] and porous silicon membranes [30]
- The ATRP fabrication of polyacrylonitrile (PAN), poly (2-(dimethyl-amino)ethyl methacrylate), and polystyrene end-grafted chains on the inner walls of ordered mesoporous silica [31]
- The fabrication of a polyvinylpyridine (PVP) brush inside a single PET nanopore by surface-initiated free radical polymerization [8]

It is evident that ATRP is the most widely used technique for the decoration of various different membranes by end-grafted polymers. ATRP has been extensively explored for surface functionalization in a planar geometry, but in the case of porous membranes there are much less reports. In principle the confinement effect inside narrow pores should affect the characteristics of the grown polymers (maximum attainable molecular weight, chain polydispersity), an effect that we also discussed for the self-assembly of flexible polymers inside cylindrical pores. In a very interesting experimental study, Gorman et al. [32] have explored the effects of surface geometry on the molecular weight and polydispersity of PMMA brushes grown by ATRP in porous alumina (200 nm pore diameter) and porous silicon (50 nm diameter) membranes. It was found that confinement effects imposed inside porous alumina and porous silicon membranes give rise to substantially reduced polymerization rates and also to higher chain polydispersity. This trend is understood by considering that as the polymerization inside the pore evolves, the effective pore diameter decreases, thus influencing the transport of monomers and catalyst along the pore.

However, in a later work Kruk et al. [31] reported the successful use of ATRP for the controlled preparation of brushes inside quite narrow pores (~15 nm diameter) of ordered mesoporous silica with a low polydispersity index (<1.1), as long as the grafted chains do not fill the pore space completely. It is very probable that system-specific issues affect properties of the resulting concave brushes, something that has to be studied further.

We have to note that in the overwhelming part of reports related to concave brushes, the chains that are chemically grown are polyelectrolytes and not neutral polymers. That means that the general theoretical framework concerning the equilibrium properties of concave brushes consisting of neutral polymers is not directly applicable in the case of charged molecules (as polyelectrolytes) since additional electrostatic interactions arise between the monomers inside the brush layer. Only for high salt concentrations that are associated with strong screening of the electrostatic interactions will the polyelectrolyte brushes resemble the properties of their neutral counterparts. As we will see in Section 6.5, the interest in polyelectrolytes stems from their ability to respond to external stimuli, creating a vast number of potential applications.

6.4 Characterization Methods

Among the various experimental techniques that have been used in the past for the characterization of polymer brushes on flat surfaces, surface-sensitive techniques such as neutron reflectivity, ellipsometry, and in general reflection techniques have given valuable information concerning the structure of end-attached polymer layers in a planar geometry [12,33,34]. However, these techniques cannot be applied for the study of concave polymer brushes since the brush layer is distributed along the pore in the bulk of the membrane. In the following we will briefly describe some of the techniques that have been proposed in the literature for the experimental investigation of polymer brushes inside porous matrices.

Probably the most used experimental technique is the one that involves solvent permeation measurements through brush-coated membranes [7,27,35]. The rate of solvent permeation through a membrane of cylindrical pores depends on the pore size and may be related to their radius by the Hagen–Poiseuille equation:

$$\frac{V}{\Delta t} = \frac{\pi \Delta P R^4}{8nL} \tag{6.6}$$

where V is the volume of the permeate relating to a single cylindrical pore, Δt is the time of permeation, ΔP is the transmembrane pressure, R the pore radius, n is the solvent viscosity, and L is the thickness of the membrane. The calculated radius from Equation 6.6 represents the hydrodynamic radius of the pore. Obviously a polymer layer on the pore surface would decrease the hydrodynamic pore radius, so by comparison with measurements of unmodified membranes, we may estimate the thickness of the grafted layer.

This technique may be used for all types of membranes and the obtained results stay valid as long as the cylindrical pores are uniformly covered with polymer chains along their axis. Investigations of brush uniformity along the pore axis have been reported by the use of imaging techniques such as scanning electron microscopy (SEM) of the cross-sectional morphology of dry brush-coated membranes [36] and confocal laser scanning microscopy (CLSM) of concave PAA brushes that were labeled by appropriate fluorescent molecules [28].

A simple way for the determination of the polymer mass per unit area is through gravimetric analysis. Usually the porous membranes that are used for the preparation of concave brushes have a very large surface area, so the formation of a polymer brush inside the pores produces an easily measurable change of the sample's weight. Precise knowledge of the membrane porosity and pore size permits the straightforward calculation of the polymer mass per unit area, which for the case of self-assembled brushes is the adsorbed amount Γ.

Another useful technique for the qualitative and quantitative characterization of brushes inside porous membranes is transmission mode Fourier transform infrared (FTIR) spectroscopy [22,29]. If the membrane material does not appreciably adsorb light near the characteristic frequency of the C-H stretching peak, which is the footprint of carbon hydrogen bonds in polymer chains, then the measured FTIR absorbance peak can be used (after proper calibration) for the calculation of the amount of polymer inside the pore. As in the case of gravimetric analysis, transmission FTIR measurements are performed on dried solvent-free samples where the grafted chains are collapsed on the inner pore surface.

Electrokinetic phenomena associated with the flow of electrolyte solutions through porous materials [37] is also another alternative for characterization of pore-brush systems [38]. Especially the streaming current/potential originating from the pressure-driven flow of an electrolyte can be used for the measurement of the zeta potential of the pore surface [27]. The presence of a brush inside the pore is expected to alter the properties of the interfacial double layer, including the value of the zeta potential, thus providing additional information about the grafted layer.

Essentially all these experimental methods apply when we deal with high-porosity multipore membranes, which are the majority of the cases. However, polymer brush decoration of single nanometric pores on macroscopic membranes has been examined lately [8]. A quite interesting method for the characterization of this system is the measurement of the ionic current that flows through the channel in an electrolyte solution under different electric potential differences. The current-voltage (I-V) response of the system can be correlated to the conductance of the polymer-decorated channel and consequently to the effective free cross section of the pore-brush structure.

Additionally, all the previously described experimental techniques focus on the study of the bulk properties of the membrane-confined brushes.

However, it is also of interest to examine the properties of the brush near the entrance of the pores up to a depth of several nanometers. For such a scope, surface analytical techniques may be implemented for the determination of near-surface composition. Ito et al. [24] have visualized in situ, by atomic force microscopy (AFM), the entrance of brush-coated pores and found a correlation between the observed pore opening and its permeation properties. Also x-ray photoelectron spectroscopy (XPS) was performed [22] on layer-by-layer sputtered dry brush-coated membranes, in order to determine the in-depth distribution of polymer chains.

The discussion of experimental techniques that we have presented is not exhaustive and only covers their key features. There are also many other experimental methods that may give direct or indirect information concerning concave brushes. We may briefly mention techniques such as small-angle neutron scattering (SANS) [39], surface plasmon resonance (SPR) [40], and optical waveguide spectroscopy (OWS) [41] that have not yet been exploited for the characterization of polymer brushes in cylindrical pores but have given insights concerning adsorbed molecular layers inside ordered porous membranes (a relatively congenital problem to the one we are considering here). Depending on the specific nature of the membranes and grafted polymer molecules, the experimentalist may adapt well-established techniques to the problem at hand.

6.5 Applications

In Section 6.1, it was mentioned that probably one of the first studies of a polymer brush inside cylindrical pores was motivated by their potential applicability as a new type of responsive microvalves [42]. The basic idea was that under the flow of solvent through a channel that is coated by end-tethered chains, the shear swelling of the layer would produce a dynamic response that would tend to maintain a constant flow rate over a range of applied pressure gradients [43]. In general, the externally controlled manipulation of the brush layer extension inside a nanochannel is related to many applications and drives the interest on the subject. The large variety of different polymers that may get end-grafted inside cylindrical pores and their diverse responses to various stimuli, like pH, temperature, ionic strength, light irradiation, etc., create many opportunities for the design and fabrication of the so-called smart nanochannels, which are capable of switching/adaptive behavior upon the change of environmental conditions.

Before the introduction of brush-decorated porous membranes, the best candidates for signal-responsive permeation were hydrogel membranes that contract or expand in response to a change of an external parameter [44].

However, in a series of publications Ito et al. [7,23,24,26] have shown that polymer brush-decorated membranes offer a much faster response; while depending on the mechanical properties of the porous membrane, they can be more mechanically robust. In these publications it was also demonstrated that different types of external stimuli, such as variation of pH, temperature, solvent quality, and ultraviolet irradiation, might be used for the control of solvent permeation through the membranes.

Depending on the nature of the grafted chains, the membrane-brush system responds to a specific type of stimulus. Known types of stimuli-responsive polymers are PAA, which has a transition between the collapsed and swollen state around pK_α, and PNIPAAm, which has a pronounced tendency to swell and deswell as a function of temperature. Essentially numerous combinations can be envisioned for the fabrication of concave brushes with diverse properties, including multiblock grafted polymers [45] or mixed brushes that are sensitive to multiple stimuli [36].

Except from the tuned permeability that concave brushes may present, there are also other interesting possibilities involving (1) the trapping of molecules inside membranes and their controlled release upon collapse of the brush layer, a function of obvious relation to drug delivery; (2) the development of size-selective composite membranes for dissolved polymers in solution (nanofiltration) [35]; and (3) the fabrication of pore-brush systems that are sensitive to the presence of specific molecules by the appropriate modification/functionalization of the grafted chains [46].

Due to the importance of chemical gates in biology (for example, cellular membrane ion channels), biomimetic artificial brush-coated channels that are able to control and manipulate the transport of ionic species receive increasing attention. These nanofluidic elements are candidates for the construction of ionic circuits capable of sensing, regulating, or separating chemical species. Especially with the current ability to fabricate and study single nanometric channels, we may acquire information about the behavior of the pore-brush system at the single pore level. It has been demonstrated that abiotic PVP brushes inside PET single nanopores display the pH-dependent ionic current switching behavior that is typical in many biological channels. These successes also trigger theoretical studies [47] that address the properties of these somewhat complex structures where channel and brush dimensions are comparable to the range of electrostatic interactions in aqueous solutions.

It is evident that the continuing breakthroughs in nanofabrication techniques will permit in the near future the reproducible fabrication of complex pore-brush structures that will be specifically tailored for single or multiple tasks. The rich chemistry and physics of polymer brushes provide many advantages and would certainly lead to their incorporation in a plethora of applications.

References

1. T. A. Witten and P. A. Pincus. Colloid stabilization by long grafted polymers. *Macromolecules*, 19(10):2509–2513, 1986.
2. P. G. de Gennes. Conformations of polymers attached to an interface. *Macromolecules*, 13(5):1069–1075, 1980.
3. M. Daoud and J. P. Cotton. Star shaped polymers: a model for the conformation and its concentration dependence. *J. Phys. France*, 43:531–538, 1982.
4. C. Ligoure and L. Leibler. Decoration of rough surfaces by chain grafting. *Macromolecules*, 23(23):5044–5046, 1990.
5. M. Manghi, M. Aubouy, C. Gay, and C. Ligoure. Inwardly curved polymer brushes: concave is not like convex. *Eur. Phys. J. E* 5:519–530, 2001.
6. E. M. Sevick. Shear swelling of polymer brushes grafted onto convex and concave surfaces. *Macromolecules*, 29(21):6952–6958, 1996.
7. Y. Ito and Y. S. Park. Signal-responsive gating of porous membranes by polymer brushes. *Polym. Adv. Technol.*, 11(3):136–144, 2000.
8. B. Yameen, M. Ali, R. Neumann, W. Ensinger, W. Knoll, and O. Azzaroni. Synthetic proton-gated ion channels via single solid-state nanochannels modified with responsive polymer brushes. *Nano Lett.*, 9(7):2788–2793, 2009.
9. S. Alexander. Adsorption of chain molecules with a polar head: a scaling description. *J. Phys. France*, 38(8):983–987, 1977.
10. S. T. Milner, T. A. Witten, and M. E. Cates. Theory of the grafted polymer brush. *Macromolecules*, 21(8):2610–2619, 1988.
11. N. Spiliopoulos, A. G. Koutsioubas, D. L. Anastassopoulos, A. A. Vradis, C. Toprakcioglu, A. Menelle, G. Mountrichas, and S. Pispas. Neutron reflectivity study of free-end distribution in polymer brushes. *Macromolecules*, 42:6209, 2009.
12. J. B. Field, C. Toprakcioglu, R. C. Ball, H. B. Stanley, L. Dai, W. Barford, J. Penfold, G. Smith, and W. Hamilton. Determination of end-adsorbed polymer density profiles by neutron reflectometry. *Macromolecules*, 25(1):434–439, 1992.
13. A. Karim, S. K. Satija, J. F. Douglas, J. F. Ankner, and L. J. Fetters. Neutron reflectivity study of the density profile of a model end-grafted polymer brush: influence of solvent quality. *Phys. Rev. Lett.*, 73(25):3407–3410, 1994.
14. R. Wang, P. Virnau, and K. Binder. Conformational properties of polymer mushrooms under spherical and cylindrical confinement. *Macromol. Theory Simulations*, 19(5):258–268, 2010.
15. D. I. Dimitrov, A. Milchev, and K. Binder. Polymer brushes in cylindrical pores: simulation versus scaling theory. *J. Chem. Phys.*, 125(3):034905, 2006.
16. D. Dimitrov, A. Milchev, and K. Binder. Polymer brushes on flat and curved substrates: scaling concepts and computer simulations. *Macromol. Symp.*, 252(1):47–57, 2007.
17. H. Masuda and K. Fukuda. Ordered metal nanohole arrays made by a two-step replication of honeycomb structures of anodic alumina. *Science*, 268(5216):1466–1468, 1995.
18. J. Schilling, R. B. Wehrspohn, A. Birner, F. Müller, R. Hillebrand, U. Gösele, S. W. Leonard, J. P. Mondia, F. Genereux, H. M. van Driel, P. Kramper,

V. Sandoghdar, and K. Busch. A model system for two-dimensional and three-dimensional photonic crystals: macroporous silicon. *J. Optics A Pure Appl. Optics*, 3(6):S121, 2001.

19. S. Matthias, F. Müller, C. Jamois, R. B. Wehrspohn, and U. Gösele. Large-area three-dimensional structuring by electrochemical etching and lithography. *Adv. Mater.*, 16(23–24):2166–2170, 2004.

20. Z. L. Xiao, C. Y. Han, U. Welp, H. H. Wang, W. K. Kwok, G. A. Willing, J. M. Hiller, R. E. Cook, D. J. Miller, and G. W. Crabtree. Fabrication of alumina nanotubes and nanowires by etching porous alumina membranes. *Nano Lett.*, 2(11):1293–1297, 2002.

21. A. G. Koutsioubas, N. Spiliopoulos, D. L. Anastassopoulos, A. A. Vradis, and C. Toprakcioglu. Formation of polymer brushes inside cylindrical pores: a computer simulation study. *J. Chem. Phys.*, 131:044901, 2009.

22. S. Karagiovanaki, A. G. Koutsioubas, N. Spiliopoulos, D. L. Anastassopoulos, A. A. Vradis, C. Toprakcioglu, and A. E. Siokou. Adsorption of block copolymers in nanoporous alumina. *J. Polym. Sci. B Polym. Phys.*, 48(14):1676–1682, 2010.

23. Y. Ito. Signal-responsive gating by a polyelectrolyte pelage on a nanoporous membrane. *Nanotechnology*, 9(3):205, 1998.

24. Y. Ito, Y. S. Park, and Y. Imanishi. Visualization of critical pH-controlled gating of a porous membrane grafted with polyelectrolyte brushes. *J. Am. Chem. Soc.*, 119(11):2739–2740, 1997.

25. B. Zhao and W. J. Brittain. Polymer brushes: surface-immobilized macromolecules. *Progr. Polym. Sci.*, 25(5):677–710, 2000.

26. Y. S. Park, Y. Ito, and Y. Imanishi. Photocontrolled gating by polymer brushes grafted on porous glass filter. *Macromolecules*, 31(8):2606–2610, 1998.

27. C. Geismann, A. Yaroshchuk, and M. Ulbricht. Permeability and electro-kinetic characterization of poly(ethylene terephthalate) capillary pore membranes with grafted temperature-responsive polymers. *Langmuir*, 23(1):76–83, 2007.

28. O. Cuscito, M.-C. Clochard, S. Esnouf, N. Betz, and D. Lairez. Nanoporous [beta]-pvdf membranes with selectively functionalized pores. *Nucl. Instrum. Methods Phys. Res. B Beam Interact. Mater. Atoms*, 265(1):309–313, 2007.

29. A. Friebe and M. Ulbricht. Controlled pore functionalization of poly(ethylene terephthalate) track-etched membranes via surface-initiated atom transfer radical polymerization. *Langmuir*, 23(20):10316–10322, 2007.

30. Q. Fu, G. V. Rama Rao, S. B. Basame, D. J. Keller, K. Artyushkova, J. E. Fulghum, and G. P. López. Reversible control of free energy and topography of nanostructured surfaces. *J. Am. Chem. Soc.*, 126(29):8904–8905, 2004.

31. M. Kruk, B. Dufour, E. B. Celer, T. Kowalewski, M. Jaroniec, and K. Matyjaszewski. Grafting monodisperse polymer chains from concave surfaces of ordered mesoporous silicas. *Macromolecules*, 41(22):8584–8591, 2008.

32. C. B. Gorman, R. J. Petrie, and J. Genzer. Effect of substrate geometry on polymer molecular weight and polydispersity during surface-initiated polymerization. *Macromolecules*, 41(13):4856–4865, 2008.

33. J. Penfold and R. K. Thomas. The application of the specular reflection of neutrons to the study of surfaces and interfaces. *J. Phys. Condensed Matter*, 2(6):1369, 1990.

34. H. Motschmann, M. Stamm, and C. Toprakcioglu. Adsorption kinetics of block copolymers from a good solvent: a two-stage process. *Macromolecules*, 24(12):3681–3688, 1991.

35. H.-S. Lee and L. S. Penn. Polymer brushes make nanopore filter membranes size selective to dissolved polymers. *Macromolecules*, 43(1):565–567, 2010.

36. A. Friebe and M. Ulbricht. Cylindrical pores responding to two different stimuli via surface-initiated atom transfer radical polymerization for synthesis of grafted diblock copolymers. *Macromolecules*, 42(6):1838–1848, 2009.

37. J. Lyklema. *Fundamentals of interface and colloid science: solid-liquid interfaces*. Academic Press, New York, 1995.

38. F. Tessier and G. W. Slater. Modulation of electroosmotic flow strength with end-grafted polymer chains. *Macromolecules*, 39(3):1250–1260, 2006.

39. D. Marchal, C. Bourdillon, and B. Demé. Small-angle neutron scattering by highly oriented hybrid bilayer membranes confined in anisotropic porous alumina. *Langmuir*, 17(26):8313–8320, 2001.

40. A. G. Koutsioubas, N. Spiliopoulos, D. L. Anastassopoulos, A. A. Vradis, and G. D. Priftis. Nanoporous alumina enhanced surface plasmon resonance sensors. *J. Appl. Phys.*, 103:094521, 2008.

41. K. H. A. Lau, L.-S. Tan, K. Tamada, M. S. Sander, and W. Knoll. Highly sensitive detection of processes occurring inside nanoporous anodic alumina templates: a waveguide optical study. *J. Phys. Chem. B*, 108(30):10812–10818, 2004.

42. E. M. Sevick and D. R. M. Williams. Polymer brushes as pressure-sensitive automated microvalves. *Macromolecules*, 27(19):5285–5290, 1994.

43. S. P. Adiga and D. W. Brenner. Flow control through polymer-grafted smart nanofluidic channels: molecular dynamics simulations. *Nano Lett.*, 5(12):2509–2514, 2005.

44. D. J. Beebe, J. S. Moore, J. M. Bauer, Q. Yu, R. H. Liu, C. Devadoss, and B.-H. Jo. Functional hydrogel structures for autonomous flow control inside microfluidic channels. *Nature*, 404(6778):588–590, 2000.

45. L. Cheng and D. Cao. Designing a thermo-switchable channel for nanofluidic controllable transportation. *ACS Nano*, 5(2):1102–1108, 2011.

46. T. Kawai, K. Sugita, K. Saito, and T. Sugo. Extension and shrinkage of polymer brush grafted onto porous membrane induced by protein binding. *Macromolecules*, 33(4):1306–1309, 2000.

47. M. Tagliazucchi, O. Azzaroni, and I. Szleifer. Responsive polymers end-tethered in solid-state nanochannels: when nanoconfinement really matters. *J. Am. Chem. Soc.*, 132(35):12404–12411, 2010.

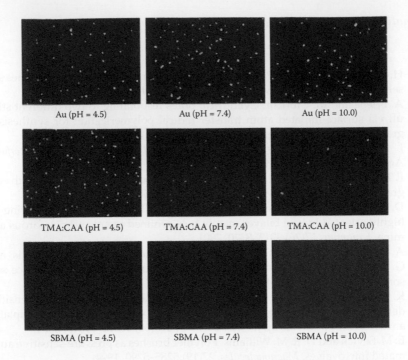

FIGURE 1.6
Fluorescence microscopy images showing *S. epidermidis* attachment to uncoated Au, TMA:CAA copolymer, and pSBMA (N-(3-sulfopropyl)-N-(methacryloxyethyl)-N,Ndimethylammonium betaine (SBMA)) surfaces at pH values of 4.5, 7.4, and 10.0 following a 3 h flow chamber adhesion assay. (Reproduced Mi, L., et al., *Biomaterials*, 31, 2919–25, 2010. With permission from Elsevier.)

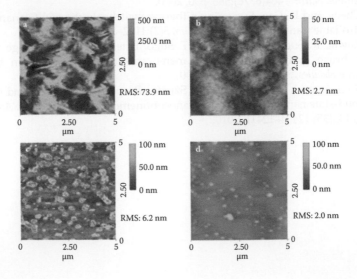

FIGURE 1.9
AFM images of (a) E9D and (c) E12D ZnO films and PNIPAm-modified (b) E9DPN and (d) E12DPN films. (Reproduced from Chang, C.-J., and Kuo, E.-H., *Thin Solid Films*, 519, 1755–60, 2010. With permission from Elsevier.)

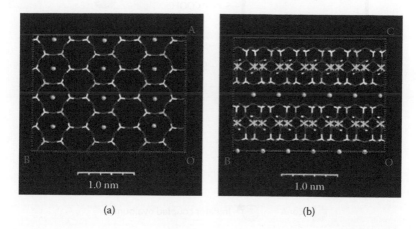

(a)   (b)

FIGURE 2.2

A simulated representation of mica unit cells. Al atoms are depicted in blue, Si in yellow, O in red, H in white, and K (or Li) ions in violet. (a) Top view showing only the upper tetrahedral layer and (b) side view showing that the cavities in the lamellae are fully occupied with alkali ions and stacked on each other. (Reprinted with permission from Heinz, H., et al., *Journal of American Chemical Society*, 125, 9500–10, 2003. Copyright (2003) American Chemical Society.)

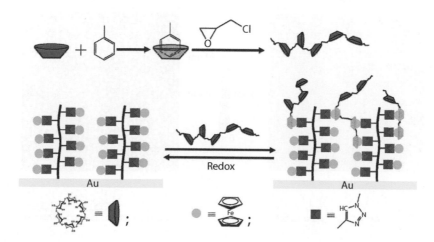

FIGURE 4.15

Redox-controlled reversible loading of the β-CD polymer on the gold substrates with surface-grafted and ferrocene-functionalized polymer brushes. (Reprinted with permission from Xu et al. 2010. One-Pot Preparation of "Ferrocene-Functionalized Polymer Brushes on Gold Substrates by Combined Surface-Initiated Atom Transfer Radical Polymerization and "Click Chemistry." *Langmuir* 26 (19):15376–15382, copyright (2010) American Chemical Society.)

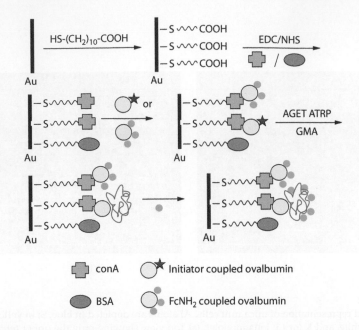

FIGURE 4.19
Fabrication of ferrocene-functionalized polymer brushes through protein-protein interaction and AGET SI-ATRP for the electrochemical detection of protein. (Reprinted with permission from Wu, Y. F., S. Q. Liu, and L. He. 2009. Electrochemical Biosensing Using Amplification-by-Polymerization. *Analytical Chemistry* 81 (16):7015–7021, copyright (2009) American Chemical Society.)

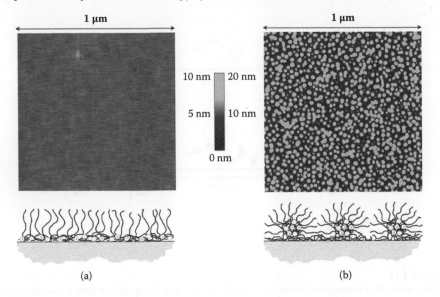

FIGURE 7.5
Tapping mode AFM pictures showing the adsorption of C3Ms to a silicon interface. (a) Upon adsorption of P2MVP, PAA-PAM micelles (pH 6, 10 mM NaNO₃), the cores are spread out on the interface to form a flat layer. (b) Adsorbed micelles of PAETB-PEO, PCETB-PEO (pH 7.7, 10 mM phosphate buffer) stay intact. (AFM Figure 7.5b from Hofs, B., et al., *J. Coll. Int. Sci.*, 325, 309–315, 2008. With permission from Elsevier. Schematic figures from Brzozowska, A. M., PhD thesis, Wageningen, 2010. With permission from author.)

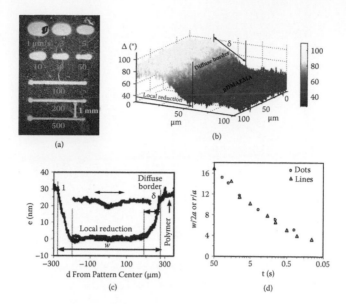

FIGURE 8.16

Patterns of PDMAEMA brushes formed by SECM from local erasing of initiator and ATRP from Si/SiO$_2$. (a) Condensation figure showing the patterns formed at different SECM tip writing speed then revealed by ATRP, as in Figure 8.14. (b,c) Local ellipsometry of polymer brush patterns: (b) image of the phase ellipsometric parameter Δ; (c) once reconstructed, Δ gives the evolution of the brush thickness along a pattern made by the tip electrogenerating (1) the 2,2′-dipyridyl (E^0 = −2.1 V vs. SCE) or (2) the nitrobenzene (E^0 = −1.08 V vs. SCE) anion radicals. (d) Evolution of the dimension of the patterns (width of lines or dots) with the etching time for the kinetic analysis of the ATRP initiator debromination (Equation 8.9). (Adapted from Slim, C., et al., *Chem. Mater.*, 20, 6677–6685, 2008.)

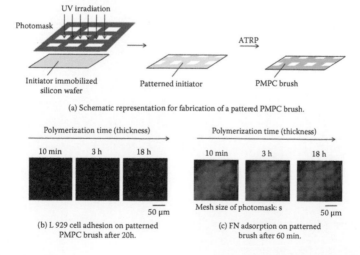

FIGURE 11.7

Protein and cell adhesions on PMPC brushes prepared by SI-ATRP. (a) Schematic illustration for fabrication of a patterned PMPC brush. (b) L929 fibroblast cell adhesion on patterned PMPC brush after 20 h. (c) FN adsorption on PMPC brush after 60 min. (Images (b) and (c). (Reprinted with permission from Iwata, R., et al., *Biomacromolecules*, 5, 2308, 2004. Copyright (2004) American Chemical Society.)

7

Polymer Brushes through Adsorption: From Early Attempts to the Ultra-Dense and Reversible "Zipper Brush"

Wiebe M. de Vos

Polymers at Interfaces Group, University of Bristol, Bristol, United Kingdom

J. Mieke Kleijn and Martien A. Cohen Stuart

Laboratory of Physical Chemistry and Colloid Science,
Wageningen University, Wageningen, The Netherlands

CONTENTS

7.1 Introduction

Polymer brushes have been shown in many investigations to be one of the most powerful systems to control interfacial properties [1–11]. Brushes can be used to coat colloidal particles, thereby strongly enhancing the colloidal stability as the brush prevents particles coming close enough to aggregate [2,3]. Polymer brushes, especially from poly(ethylene oxide) have been shown to protect interfaces from biofouling [4,5] and can be used on medical implants to reduce chances of inflammation [6]. Polymer brushes have been shown to have unique wetting properties [7,8] and can also greatly reduce the friction between two interfaces [9,10]. Especially, polyelectrolyte brushes have the potential to accommodate enzymes without affecting their structure and function [11].

With the huge potential that the polymer brush clearly has for a great number of applications, it is essential that good methods are available to produce them. The ideal method to produce a polymer brush would be one that is simple to do, cheap, but that still gives one good control over the brush properties, such as the grafting density (σ) and the chain length (N). Especially, the grafting density is seen as a key parameter for the many brush applications, and commonly a higher grafting density leads to better performance. Additionally, the technique should work on different substrates and be able to create brushes from chemically different polymers.

While over the years several methods have been developed to produce polymer brushes [4,12–18], it is also clear that this ideal method does not (yet) exist. All developed methods have their own advantages, but also their disadvantages. Methods to prepare brushes come in just three broad categories: (1) grafting existing polymers by means of noncovalent bonds (adsorption; Figure 7.1a), (2) covalently grafting existing polymers to the substrate (grafting to; Figure 7.1b), and (3) polymerizing chains from initiation sites on the substrate (grafting from; Figure 7.1c). Below, we will first briefly discuss the advantages and disadvantages of these methods, focusing on how well these perform in terms of control over N and σ. More extensive reviews have appeared [4,12–18] that discuss in more detail the workings of these methods.

The simplest and earliest method to make a brush is the adsorption of a soluble diblock copolymer with one of the blocks being a suitable anchoring block, while the other block will extend into the solution to form the brush layer. The right combination of diblock copolymer, solvent, and interface has indeed been shown to result in the spontaneous formation of polymer brushes [19,20]. This property makes this technique very attractive for use in applications, as large surface areas can be coated in a quick and simple way. However, there is also a large disadvantage connected to this technique: only low and uncontrolled density polymer brushes can be formed. In a nonselective solvent (in which both blocks are soluble) the problem lies with the anchor block. Short anchor blocks give too little adsorption energy to form a dense brush, but if the anchor blocks are made longer, then they crowd the interface, thereby preventing a dense brush from forming [3]. In selective solvents the problem is that the diblock copolymers will form micelles. To form a brush the micelles must adsorb and unfold on the surface. During the

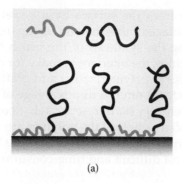

(a)

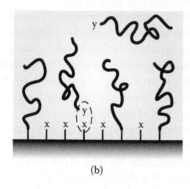

(b)

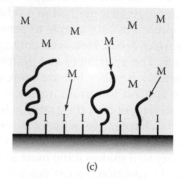

(c)

FIGURE 7.1

Schematic depiction of the three most common methods to produce polymer brushes: (a) adsorption, (b) grafting to, and (c) grafting from.

adsorption process, a sparse brush is formed that then effectively prevents any further micelles from reaching the interface, thus preventing the brush from becoming any denser.

In the grafting to method, an end-functionalized polymer is covalently connected to the interface by a suitable reaction between the end group and the interface. For example, hydroxyl end groups can react with silica [21], thiol groups with gold [22], and alkenes and alkynes with silicon [23]. This does, however, restrict this technique to reactive interfaces. Just as with the adsorption technique, while the brush layer is being prepared the polymers will have to diffuse through the forming layer, preventing the formation of very dense brushes. However, by performing the grafting from a polymer melt instead of from solution, this problem can be largely avoided and very dense polymer brushes can be prepared [23]. Grafting density and chain length are difficult to control independently, but a short polymer will tend to give dense brushes while a long polymer will give lower grafting densities.

The first step in the grafting from approach is to attach an appropriate initiator to the substrate [12,13]. Just as with the grafting to approach, a reactive interface is usually needed for this. Once the initiation sites are attached, polymers can be grown from the surface by a variety of polymerization schemes. The large advantage of this method, compared to adsorption and grafting to, is that during the synthesis of the brush the small monomers have no difficulty in reaching the interface to make for denser brushes. With this approach high degrees of polymerization and grafting densities [12] can be achieved, typically up to 0.8 nm^{-2}. A disadvantage of the technique is that after the reaction it is often very hard to separately determine N and σ. One can do this by cleaving off a sample of the polymer after the reaction to analyze it [24]. The grafting from method is generally considered to be the most powerful, but also the most difficult and time-consuming of the methods to produce brushes.

A conclusion that we can draw from the above is that at the moment there is no simple generic method to produce dense polymer brushes. The adsorption approach is very simple to perform, and could easily be performed on large surface areas. However, only sparse brushes can be formed using that approach. The grafting to and grafting from approaches are able to produce high-density brushes but are more difficult and time-consuming and can only be applied to certain interfaces. None of the three techniques provide complete control over the grafting density and chain length.

In this chapter we will focus on polymer brushes that are made by adsorption. Even though the technique is limited in grafting density, its simplicity still makes it the most likely candidate for any large-scale applications. In addition, recent years have seen the development of some very promising variations on the adsorption technique. For example, we will discuss a technique based on Langmuir–Blodgett (LB) transfer [25–27]. For that technique a sparse brush is first prepared at the air-water interface by simply applying an amphiphilic diblock copolymer. The brush can then be made to the desired grafting density by decreasing the surface area, and can subsequently be transferred to a solid substrate by LB transfer. The full control of grafting density and chain length makes this technique very powerful.

Another interesting system is based on the adsorption of so-called complex coacervate core micelles (also called polyion complex micelles) [28,29]. These micelles are formed in aqueous solution when two oppositely charged polyelectrolytes are mixed, with at least one of these polyelectrolytes being connected to an uncharged and water-soluble polymer. The complexed polyelectrolytes then form the complex coacervate core of the micelles, while the neutral chain forms the corona. These micelles have been shown to adsorb to surfaces with very different properties, such as silica and polystyrene. Although formed brushes are of low density, good antifouling properties have been observed [28,30].

Very recently, another technique was developed that is also based on electrostatic interactions [31,32]. First, a surface is covered with polyelectrolyte, for example, a poly(acrylic acid) brush layer, as this can accommodate many charges. In the second step a diblock copolymer with an oppositely charged block and a neutral block is adsorbed to this surface covered with polyelectrolyte, leading to the formation of a neutral brush. Due to the strong attraction between the oppositely charged polyelectrolytes and the large number of charges in the polyelectrolyte brush, a large amount of diblock copolymer can adsorb, and thus a very high grafting density can be reached. Another effect of the strong attraction between oppositely charged polyelectrolytes is that the adsorbed number of diblock copolymers is determined by full charge compensation between the charges in the brush and the charge of the adsorbed diblock copolymer. As a consequence, it is possible to tune the grafting density of the formed neutral brush by choosing the density and degree of polymerization of the polyelectrolyte brush and the degree of polymerization of the charged block of the diblock copolymer. Another advantage of this method is that complexes of oppositely charged polyelectrolytes can usually be broken by adding salt or changing the pH. Thus, in such a system a neutral brush cannot only be formed by adsorption, but can also be desorbed again. Therefore, we have named this new procedure to form neutral polymer brushes the "zipper brush" approach [31]. In this chapter we will pay special attention to this zipper brush approach.

7.2 Early Attempts

7.2.1 Block Copolymers on Surfaces: Theory

Ever since the days of Michael Faraday it has been known that hydrocolloids (soluble polymers) like gelatine can be powerful stabilizers of colloidal dispersions. The stabilizing effect was believed to be due to steric repulsion: polymer chains immersed in a good solvent tend to swell and resist compression. However, it was also known that long soluble homopolymers like, e.g., poly(acrylamide), could be very successfully used to flocculate colloidal particles. The underlying picture is that polymers can straddle the gap between two particles, forming "bridges" that keep the particles together. This apparent paradox was resolved by Fleer and Lyklema [33,34], who showed that the effect of an added homopolymer on colloidal stability depends on whether or not the particle surface is saturated with adsorbing polymer. As a consequence, achieving stability for a given dispersion by adding such a polymer becomes a balancing act between adsorption rates and particle collision rates, and this would not be a very robust approach.

Indeed, the homopolymer always induces both steric repulsion and bridging; the strength of either effect depends differently on chain length and adsorbed amount. The idea therefore came up that it would be better to have separate control over adsorption and steric repulsion, and that polymers consisting of two different blocks, one for adsorption and one for steric repulsion, would do a better job. Gradually, more kinds of diblock copolymers became available as new kinds of "living polymerization" saw the light, so this seemed a fertile ground for investigations.

In the meantime, Scheutjens and Fleer [35] had formulated their now classical numerical self-consistent field (lattice) (SF) theory for equilibrium adsorption of polymers, which provides a very detailed partition function from which structural and thermodynamic information can be derived. The key idea of a self-consistent field approach is that, on the one hand, the interactions between molecules are treated as if they constitute an external field on an individual molecule (thereby defining potential energies), and that each molecule finds its optimum distribution of conformations and locations as a Boltzmann distribution in that field. On the other hand, the distribution found has to be such that it will indeed produce the field, hence the term *self-consistent*. An iterative scheme is employed to reach consistency, and as soon as the solution is found, energy and entropy of the system are known as well.

Applying the SF theory to homopolymers between two surfaces, Scheutjens and Fleer noticed that they always predicted bridging attraction, unless both surfaces were fully covered with "irreversibly" adsorbed polymer [36]. Hence, for systems in equilibrium there seemed to be no escape from bridging with these molecules. The next logical step was therefore to consider diblock copolymers. The beauty of the SF scheme is that it can easily be generalized to cover different monomer units (segments) in any arbitrary sequence, so that all kinds of copolymers can be easily treated. The first project was to consider A-B diblocks with a fixed total number of segments N, but varying A/B ratio. Both A and B blocks were taken as soluble in the common solvent (a nonselective solvent), and only the A block was assigned an attractive interaction with the substrate. Representative results for the resulting grafting density and brush height, plotted as a function of the fraction n_A of "sticky" segments in the chain, are shown in Figure 7.2. Data is from Evers [37] for chains of 500 segments and an adsorption energy of 2 kT for the A segment. For comparison we present results of an analytical scaling theory proposed by Marques and Joanny [38] as a dotted line.

As can be seen in both diagrams, the grafting density σ has a pronounced maximum at around $n_A = 0.1$. Qualitatively, this can be understood by the following arguments. As should be expected, pure B chains have no segments that can bind to the surface, so there is no adsorption at $n_A = 0$. Upon introducing a few anchoring (A) segments, the adsorbed amount rapidly increases; due to the increased surface attraction, equilibrium shifts to a higher surface density of chains. As long as A blocks are short, the first thing that happens is that the swollen nonadsorbing B blocks begin to laterally interact. The response of anchored chains to strong lateral interaction is to stretch in the normal direction; a layer of such stretched chains is

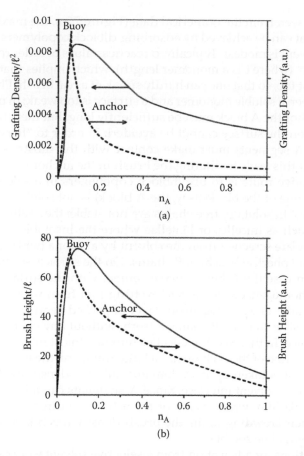

FIGURE 7.2
Comparison of SCF predictions [37] (solid lines) and scaling predictions [38] (dotted lines) for the adsorption of an A and B block diblock copolymer as a function of the fraction of A monomers (n_A, see text; $N = 500$, adsorption energy = 2 kT). (a) Resulting grafting density (per monomer length (ℓ) squared). (b) Brush height in monomer chain lengths.

commonly called a brush. Chain stretching lowers the average monomer density and concomitant osmotic pressure in the brush, but this comes at the price of a lower chain entropy. In equilibrium, the adsorption energy of the A segments must balance the stretching free energy in the brush of B segments. This regime, where the adsorption is essentially controlled by this balance, is called the buoy regime. As n_A becomes larger, interaction between A blocks kicks in: A segments start to crowd and the number of A blocks that can be accommodated on the surface begins to decrease. As a result, the overall adsorption decreases too. Now it is the lateral interaction between A blocks that controls the adsorbed amount; accordingly, this is called the anchor regime. The maximum grafting density occurs where both regimes meet.

As can be seen from the numerical data (Figure 7.2a), the maximum chain density σ that can be achieved by adsorbing diblock copolymers from a non-selective solvent is modest. Typically, σ reaches values of order roughly a few times $(\ell^2/N)^{-1}$ (where ℓ is a monomer length), which implies that the stretching is modest too, so that one can hardly speak of a real brush. Physical interaction between a soluble monomer and a surface is too weak; in order to have a strong anchor the A block must be sufficiently long. But then, the crowding of A blocks on the surface cannot be avoided: in order to "harvest" surface interaction, A segments must make contact with the substrate as much as possible, and this leads to diluting the brush in the anchor regime.

An alternative route is to use diblock copolymers in a selective solvent. In this case, one of the blocks (say, the A block) is not soluble. This has two consequences. In solution, free chains are not stable: they self-assemble into structures such as micelles or lamellae where the insoluble A chains form the core; they are screened from the solvent by a corona (which is, of course, some kind of spherical brush) of B chains. On the surface, several situations may arise. In case the A blocks make contact with the surface, either they can form a homogeneous (complete) wetting film from which a brush of B chains protrudes or a heterogeneous layer of packed hemimicelles can arise. It is also possible that the B chains adsorb, without any contact between A and the surface. In that case the micelles remain almost intact. The polymer-solvent interaction of the A chain has a large influence: the worse the solvent, the less solvent is taken up in A domains, and the denser the B chains are packed at the A-solvent interface. Since A segments can adsorb in the form of an insoluble film, where many of them have no direct contact with the substrate, their crowding limits the brush density much less severely than when they would be soluble.

SF calculations for adsorption from a selective solvent were carried out by van Lent and Scheutjens [39]. Of course, the calculations had to deal with the self-assembly process as well. For the case of a rather poorly soluble block connected to a soluble block and not too extreme block lengths the critical micellization concentration (CMC) is extremely low. For any practical situation it then can be safely assumed that adsorption takes place from a solution of almost constant chemical potential, so in equilibrium there is no significant effect of concentration. The longer the A blocks (and hence the shorter the B blocks), the thicker the A film and the higher the density of the brush of B blocks. These calculations make clear that it is favorable to use a strongly insoluble block as anchor, and that the wettability of the solvent-substrate pair by A is important. The case of a heterogeneous layer of hemimicelles requires a more elaborate two-dimensional SF scheme, which was not considered by Van Lent.

A similar picture arises from a scaling analysis by Marques et al. [40]. The situation where there is a wetting layer of anchor blocks from which the brush blocks protrude is called the van der Waals brush regime. The central result states that the brush density is rather weakly dependent on chain composition except for extreme cases, and the brush becomes more dilute with increasing length (as $N^{-3/4}$), but less so than for a brush made from a nonselective solvent where it scales as (N^{-1}).

The lesson to be taken from these studies is that if one attempts to make high brush densities by physical adsorption it is important to have strong anchoring while avoiding crowding of the anchor blocks as much as possible.

7.2.2 Block Copolymers on Surfaces: Experiments

There is a limited number of studies of adsorption of diblock copolymers from nonselective solvents. One practical problem is that it is not enough to find a nonselective solvent, but that a brush-like configuration also requires selective adsorption of the A block. Because polymers are notorious for adsorbing onto many substrates [3,41], the best way is to use a mixed solvent with a composition such that A adsorbs and not B. Some early data were obtained by Wu et al. [19] for diblocks of dimethyl-aminoethyl-methacrylate (A) and n-butylmethacrylate (B) from propanol onto silica. For chains of 200 and 700 monomers, Wu et al. found a clear maximum at around 10% of A monomers, where the adsorbed amount reaches a value of 13 mg/m^2. This behavior was thus very much in line with the theoretical predictions, as shown in Figure 7.2. The adsorbed amount of 13 mg/m^2 corresponds with a grafting density of about 1 chain per 10 nm^2, which is reasonable but not very dense. Later experiments on diblocks with a cationic block (dimethyl-aminoethyl-methacrylate (AMA)) and a neutral block (dihydroxy-propyl-methacrylate (HMA)) on silica and titania were carried out by Hoogeveen et al. [42] (Figure 7.3).

These also showed a maximum at about 10% of the chain in the cationic block, again very much in line with the predictions shown in Figure 7.2. In this case, the maximum adsorbed amount (on silica) was 3 mg/m^2, corresponding again to about one chain (but a shorter one) per 10 nm^2. On titania the maximum was half of that. Data by Guzonas [43] for poly(ethylene oxide-*b*-poly-styrene) adsorbing from the nonselective solvent toluene onto mica produced even lower maximum densities (up to one chain per 50 nm^2). Bijsterbosch [44] reported similar results for various diblocks involving poly(methyl vinyl ether) (PVME), poly(2-ethyl-2-oxazoline) (PEtOx), poly(2-methyl-2-oxazoline) (PMEOx), and poly(ethylene oxide) (PEO) from water onto silica. No combination of these blocks turned out to be selectively adsorbing, so that a simple anchor-buoy picture does not apply here. For diblocks with 20% PEO and 80% PMEOx a maximum in the adsorbed amount was found, suggesting that PEO acts as the anchoring block here, but the adsorbed amount was only 0.6 mg/m^2, which implies a rather low chain density. The same is the case for the combination PVME-PEtOx, where the latter block seems to be anchoring; here the maximum adsorbed amount was somewhat higher (1.3 mg/m^2). Incompatibility (repulsion) between the blocks leads to a somewhat more pronounced maximum but a dense brush is not obtained.

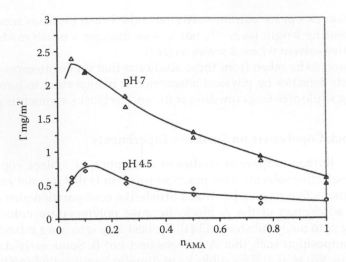

FIGURE 7.3
Adsorption of a diblock copolymer (AMA-HMA, $N = 120$), from aqueous solution to a silica surface as a function of the fraction of AMA monomers in the chain, for two pH values as indicated (in 0.5 mM barbital-acetate buffer, 10 mg/L polymer). (Data from Hoogeveen, N. G., et al., *Faraday Disc.*, 98, 161–172, 1994; guide for the eye lines not identical to original.)

Clearly, the success of making homogeneous and dense brushes by means of physical adsorption of diblocks from solution depends critically on the strength of the physical bonds one can establish between anchor block and the surface. Water is a particularly difficult solvent, as most inorganic substrates are typically hydrophilic so that the solvent itself is a strong competitor in the context of the polar interactions that one can use. As explained, increasing the block length is not a remedy since one soon runs into the crowding problem that limits chain density in the anchor-dominated regime.

Low dielectric solvents would seem to make a better case. One example was reported by Taunton et al. [45], who used polystyrene with a few strongly polar zwitterionic end groups and adsorbed that from toluene and xylene on mica. Polystyrene itself does not adsorb from these solvents. Even here, the adsorbed amount for chains of at least 1,000 monomer units reached at most 2 mg/m^2, implying a grafting density of order 0.02 chain/nm^2. Some stretching was observed, but this is far from a dense brush.

One may wonder whether adsorption from selective solvents leads to more dense brushes. As was pointed out above, this implies that the anchoring blocks attract each other so that they are prone to form self-assembled structures. One therefore has to bear in mind that not only does adsorption take place from a micellar solution, but also that there is a good chance that these micelles, once they are on the surface, remain micelles so that inhomogeneous adsorption layers are produced. Examples of adsorption studies of diblock copolymers from selective solvents are those by Marra and Hair [46] (polystyrene-poly(ethylene oxide) from heptane/toluene on mica),

Watanabe and Tirrell [47] (poly(2-vinylpyridine)-polystyrene and poly(2-vinylpyridine)-polyisoprene on mica), and Ansarifar and Luckham [48] (poly(2-vinylpyridine)-poly(tert-butylstyrene) from toluene on mica). No information can be found on the lateral homogeneity of these adsorption layers. However, the adsorbed amounts are strikingly low, being all in the range 0–3 mg/m². Apparently, the insoluble block does not help very much to enhance the (average) grafting density.

This was a quite surprising outcome, as the theoretical predictions on diblock copolymer adsorption from a selective solvent predicted much higher adsorbed amounts than for adsorption from nonselective solvents [39]. In a selective solvent there is no problem with crowding of the anchor blocks, as they can assemble in a three-dimensional phase separated layer on the surface, not having to compete for direct contact with the surface. Most likely a steric barrier, not taken into account in the theoretical predictions, prevents the polymer brush from becoming dense. For diblock copolymers in a selective solvent the concentration of free polymers will be practically zero. This means that the brush is most likely formed by the adsorption of micelles and, if the core is fluid enough, subsequent unfolding of the micelle. However, when during the brush-forming process a sparse brush is formed, it will act as a barrier, keeping further micelles from coming close enough to the interface to adsorb and unfold, thus preventing the brush from becoming any denser.

At this point we apparently have to conclude that preparing brushes by adsorbing diblock copolymers from solution, despite its advantage of simplicity, is not a very good approach because it does not allow varying and controlling the grafting density over a significant range, let alone that it allows making more special (e.g., patterned) brushes. Clearly, there is also some sort of crowding that stands in the way, either between the buoy blocks or between the anchor blocks, or some kind of steric barrier that prevents the brush from becoming denser. As we shall see later, however, the zipper method introduces a dramatic improvement.

7.2.3 Brush Formation through Adsorption of Complex Coacervate Core Micelles

One general problem with producing polymer brushes by adsorption, but also for the grafting to and grafting from methods, is that a certain method usually works only on a specific type of surface. For example, when adsorbing diblock copolymers, a positively charged anchor block will only work for a negatively charged surface, and an apolar anchor block will only stick to an apolar surface. There is, however, one system that has shown great potential on using the same polymers to produce brushes on very different surfaces. The method is based on the formation and subsequent adsorption of so-called complex coacervate core micelles (C3Ms; also called polyion complex micelles) [28–30,49,50].

Complex coacervate core micelles are spontaneously formed when one mixes aqueous solutions of two oppositely charged polyelectrolytes, if at least one of the polyelectrolytes is connected to a neutral water-soluble polymer [51,52]. The formation of such micelles, and one possibility for their subsequent adsorption, is schematically shown in Figure 7.4: the polyelectrolytes form the complex coacervate core of the micelle while the neutral block forms the corona.

Subsequent adsorption of the micelles to an interface can lead to two different modes of adsorption, as shown in Figure 7.5. If the core of the micelle is fluid enough, the micelle can unfold on the interface to form a continuously flat layer as confirmed by atomic force microscopy (AFM) data [30], but also by other techniques, such as neutron reflection [29]. In the other mode of adsorption the micelle does not unfold, and thus the surface is covered by complete micelles, as shown from the AFM data in the right-hand part of Figure 7.5. The planar, spread-out, adsorption is most desired, as this gives a uniform brush across the whole surface. Still, for both the planar adsorption [30] and the adsorption of complete micelles [49,53], significant reductions were observed in the subsequent adsorption of model proteins to the coated surface.

The main advantage of producing brushes with these micelles is that they have been shown to adsorb to interfaces with very different properties. For example, micelles prepared from mixing poly(acrylic acid)-poly(acrylamide) (PAA-PAM) and poly(N-methyl-2-vinyl pyridinium) (P2MVP) were found to adsorb on silica, polybutadiene, and polystyrene surfaces, but also on top of formed polyelectrolyte multilayers [28,30]. The reason for this uncommon adaptive behavior is believed to stem from the properties of the complex coacervate formed by the polyelectrolytes. When exposed to a charged surface, it will be the oppositely charged polyelectrolytes in the mixture that will provide the close contact with the interface, thus allowing adsorption to both positively and negatively charged surfaces. In case of an apolar surface,

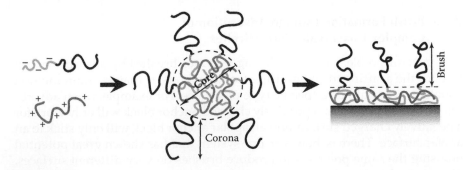

FIGURE 7.4
Schematic representation of the formation of a C3M micelle (see text), and how it can subsequently adsorb to form a brush layer.

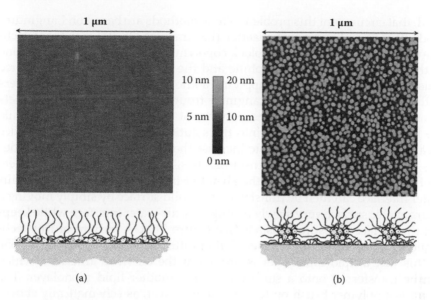

FIGURE 7.5 (See color insert.)
Tapping mode AFM pictures showing the adsorption of C3Ms to a silicon interface. (a) Upon adsorption of P2MVP, PAA-PAM micelles (pH 6, 10 mM NaNO$_3$), the cores are spread out on the interface to form a flat layer. (b) Adsorbed micelles of PAETB-PEO, PCETB-PEO (pH 7.7, 10 mM phosphate buffer) stay intact. (AFM Figure 7.5b from Hofs, B., et al., *J. Coll. Int. Sci.*, 325, 309–315, 2008. With permission from Elsevier. Schematic figures from Brzozowska, A. M., PhD thesis, Wageningen, 2010. With permission from author.)

the more hydrophobic backbone of the polyelectrolytes is thought to provide the closest contact [28].

The large disadvantage of this technique is, however, the same as that of brushes made from diblock copolymers in a selective solvent. To form a brush the micelles must adsorb and unfold on the surface. During the adsorption process, a sparse brush is formed that effectively prevents any further micelles from reaching the interface, thus preventing the brush from becoming any denser. Still, even for these low grafting densities, the brushes formed by adsorbed C3M micelles have shown to result in significant reductions of the fouling of interfaces with model proteins [28,30,49]. Some typical antifouling data can be found in Table 7.3, in the zipper brush section of this chapter, where it is compared to the antifouling capabilities of the zipper brush.

7.2.4 Langmuir–Blodgett Transfer

In Sections 7.2.1 to 7.2.3, we discussed the problems associated with making a polymer brush by simple adsorption of diblock copolymers from a selective solvent. When such a brush is being formed, it becomes very difficult for other copolymers, aggregated into micelles, to reach the interface and to adsorb there, strongly limiting the grafting density of the produced brush. However, methods are available, in which similar diblock copolymers are used, that circumvent this problem. These methods are based on Langmuir–Blodgett (LB) or Langmuir-Schaeffer (LS) transfer [25–27]. For these techniques it is essential that the diblock copolymers are amphiphilic; thus, one of the blocks should be hydrophilic and the other hydrophobic. A sparse brush can be made by simply applying a known amount of such polymers to the air-water interface in a Langmuir trough. The hydropobic block acts as an anchor block keeping the polymer at the air-water interface while the hydrophilic block extends into the solution. Subsequently, the surface area of the trough is decreased to increase the brush density. Not only does this allow for making much denser brushes, but it also provides complete control of the brush density at the air-water interface. The brush at the air-water interface can then be transferred to a solid surface by simply moving a hydrophobic surface very slowly through the air-water interface while keeping the surface pressure constant (see Figure 7.6a). A variant of this technique is to use a soluble polymer chain grafted to a lipid molecule [54,55]. In this way a lipid monolayer is formed at the air-water interface, which can be transferred onto a surface carrying another lipid monolayer. The result is a polymer brush on top of a lipid bilayer, as schematically shown in Figure 7.6b. Together, these transfer techniques have played an important role in investigations into the effect of brush grafting density on the prevention of protein adsorption [4].

When applying these techniques it is essential that the transfer is checked. Indeed, for some polymers transfer was found to be incomplete, leading to a lower grafting density than anticipated [27]. In Figure 7.7 we show for a number of different polymers how the brush grafting density at the air-water interface (σ_{AW}) relates to the grafting density on the solid surface after transfer (σ_{SA}). Although in all cases there is a linear dependence between σ_{AW} and σ_{SA}, there are some cases where σ_{SA} is smaller than σ_{AW}. For $PS_{40}PEO_{148}$

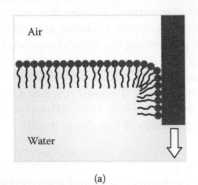

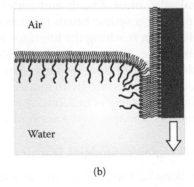

(a)   (b)

FIGURE 7.6
Schematic depiction of two methods to prepare polymer brushes via Langmuir–Blodgett transfer. (a) LB transfer of amphiphilic diblock copolymers to a hydrophobic interface. (b) LB transfer of a lipid monolayer, of which some lipids are connected to a hydrophilic polymer chain, to a surface coated with an opposite lipid monolayer.

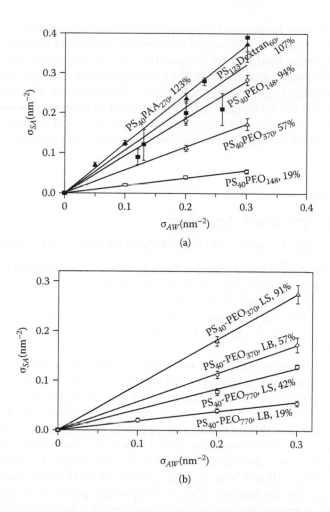

FIGURE 7.7
The brush grafting density obtained after transfer to a polystyrene substrate (σ_{SA}) as a function of the brush density at the air water interface (σ_{AW}). (a) For different diblock copolymers as indicated (PS = polystyrene, PAA = poly(acrylic acid), PEO = poly(ethylene oxide)) as prepared by Langmuir–Blodgett dipping. (b) For different diblocks as indicated and for different transfer methods (LB = Langmuir–Blodgett, LS = Langmuir-Schaeffer). (Data on PS-PEO from de Vos, W. M., et al., *Langmuir*, 25, 4490–4497, 2009; on PS-PAA from de Vos, W. M., et al., *Langmuir*, 24, 6575–6584, 2008; and on PS-Dextran from Bosker, W. T. E., et al., *Soft Matter*, 3, 754–762, 2007.)

(numbers represent numbers of monomers) and $PS_{123}Dextran_{37}$ the transfer ratio is close to 1 and thus nearly ideal [57]. However, for longer PEO chains ($PS_{40}\text{-}PEO_{370}$ and $PS_{40}\text{-}PEO_{770}$), the transfer ratio is considerably lower than 1. The reason for this is not well understood but speculated to be connected to long PEO chains competing with the PS blocks for adsorption at the polystyrene interface [27]. For our last example, $PS_{40}PAA_{270}$, the transfer ratio seems to exceed unity. In Figure 7.7 we show how the method of dipping affects the transfer ratio. Horizontal dipping (LS) instead of vertical dipping (LB) significantly increases the transfer ratio for the longer PEO chains.

The Langmuir–Blodgett technique has the tremendous advantage of allowing complete control of both the grafting density and the chain length, while reasonably high grafting densities can be reached. The main disadvantage is that the technique can only be used with very specific surfaces (smooth, hydrophobic, relatively small). For this reason, it is really only useful for laboratory investigations and not for practical applications.

7.3 The Zipper Brush Approach

7.3.1 Reversible Assembly of Dense Polymer Brushes: Basic Idea

From the above it is clear that to get high grafting densities from the adsorption of diblock copolymers one needs to meet three requirements: (1) a strong attraction between anchor block and surface, (2) avoidance of crowding of the anchor blocks at the surface as a limiting factor (keep out of the anchor regime), and (3) no high kinetic barrier built up for the polymers to reach the surface.

As already discussed, in a nonselective solvent a strong attraction between only one of the blocks of the copolymer and the substrate surface is generally hard to achieve. In aqueous solution, however, a good option is to use electrostatic interactions, letting a diblock copolymer with a charged block and a neutral block adsorb to an oppositely charged surface. The electrostatic interaction between the anchor block and the surface can be very large. However, as a flat surface can only accommodate a limited number of charges, this will not lead to high grafting densities: the anchor blocks will compete for the charged sites, a comparable problem to crowding. A solution to accommodate many more charges on a substrate is to go into the third dimension: cover it first with a charged layer of polyelectrolytes, for example, a charged brush layer. Subsequent adsorption of diblock copolymers with a neutral block and an oppositely charged block will result in the formation of a neutral brush. As already mentioned in the introduction, a large amount of the diblock copolymer is expected to adsorb as a result of the strong attraction between the oppositely charged polyelectrolytes and the large number of charges in the primary polyelectrolyte layer. Therefore, the formed neutral brush can have a very high grafting density. By adding salt or, in the case of weak polyelectrolytes, changing the pH the neutral brush can be desorbed again. The principle of this zipper brush approach is depicted in Figure 7.8. We have shown that with this new method, indeed very dense brushes by adsorption can be obtained [31,32]. Obviously, it meets two of the three requirements mentioned above. It is less obvious how the third one, i.e., no building up of a significant kinetic

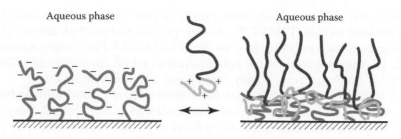

FIGURE 7.8
Schematic depiction of the formation of the zipper brush, the electrostatically driven assembly of a dense polymer brush. In this picture the polyelectrolyte block of the diblock copolymer has half the number of charged groups of the polyelectrolytes in the primary brush. Thus, assuming total charge compensation, the grafting density of the formed neutral brush is twice that of the original polyelectrolyte. Desorption of the diblock copolymer (e.g., by adding salt) restores the original brush layer. (Reproduced from from de Vos W. M., et al., *Angew. Chem. Int. Ed.*, 48, 5369–5371, 2009. With permission from Wiley.)

barrier during the adsorption process, is met. We will come back to that the end of this section.

An expected effect of the strong attraction between oppositely charged polyelectrolytes would be that the adsorbed number of diblock copolymers is determined by full charge compensation of the charges in the primary brush by the charges of the diblock copolymer. Therefore, it should be possible to tune the grafting density of the formed neutral brush, σ_{brush}, by choosing the density and degree of polymerization of the polyelectrolyte brush, σ_0 and N_0, and the degree of polymerization of the charged anchor block of the diblock copolymer, N_{anchor}:

$$\sigma_{brush} = \frac{\sigma_0 N_0}{N_{anchor}} \tag{7.1}$$

It is noted here that for the brush system obtained by adsorption of complex coacervate core micelles, as described in the previous section, the structure resembles very strongly the structure proposed for the zipper brush in Figure 7.8 (provided that the C3Ms spread their cores at the substrate surface). Both consist of a polyelectrolyte complex layer near the surface with, on top of that, a neutral polymer brush. However, the driving force for the formation of the neutral brush is very different for the two systems. In the case of the zipper brush the electrostatic attraction between oppositely charged polyelectrolytes is the driving force; in the micelle system it is the phase separation of the polyelectrolyte complex, which is much weaker.

7.3.2 Experimental Verification

We investigated the formation of a zipper brush and its expected special properties using poly(acrylic acid) (PAA) brushes as the primary brush to which the diblock copolymers poly(2-methyl vinyl pyridinium)-*block*-poly(ethylene oxide) (P2MVP-PEO) and poly(N,N-dimethyl amino ethyl methacrylate)-*block*-poly(ethylene oxide) (PDMAEMA-PEO) were adsorbed [31,32]. PAA is an anionic weak polyelectrolyte, while (quaternized) P2MVP and PDMEAMA are, respectively, strong and weak polycations.

PAA brushes were prepared on flat silicon wafers coated with polystyrene using the LB technique as described earlier in this chapter. Adsorption and desorption were monitored using fixed-angle optical reflectometry [58]. The kinetics and reversibility of the formation of a zipper brush are demonstrated in Figure 7.9, representing a typical reflectometry experiment of the adsorption and desorption of P2MVP-PEO to a PAA brush.

Supply of the diblock copolymer (P) to the PAA brush at pH 6 led to a strong adsorption of about 16 mg/m², much higher than the reported values for adsorption of P2MVP-PEO to a negatively charged flat surface [28]. Assuming that the diblock copolymer indeed adsorbs as depicted in Figure 7.8, this implies a high grafting density of 0.7 PEO chains per nm⁻².

Flushing the surface with a solution of pH 6 (R) did not result in desorption, showing that the diblock copolymer was strongly bound to the brush. However, flushing with a solution of pH 2, which results in almost complete decharging of the PAA chains in the primary brush, led to desorption of about 85% of the diblock copolymer. The small increase in the signal after again applying the pH 6 solution is not due to adsorption but most probably due to a change of the refractive index of the PAA brush layer with pH.

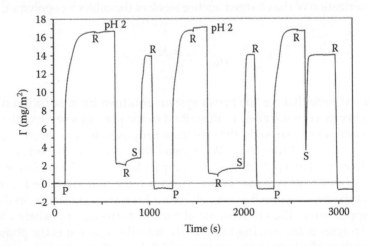

FIGURE 7.9

Adsorption of P2MVP$_{41}$-PEO$_{205}$ (0.1 g/l) to a PAA$_{270}$ brush ($\sigma = 0.1$ nm⁻²) and subsequent desorption. P represents addition of the diblock copolymer (at pH 6), R rinsing with a pH 6 solution, pH 2 rinsing with a 0.01M HNO₃ solution, and S rinsing with of a nonionic surfactant solution (C₁₂EO₅, 3′ CMC) of pH 2 (0.01 M HNO₃). Background electrolyte in all solutions is 10 mM KNO₃.

Not all P2MVP-PEO is desorbed by rinsing with the pH 2 solution. This is attributed to nonelectrostatic interactions: it has been shown that PEO and PAA interact due to hydrogen bond formation at low pH [59]. This is also an explanation for the strong adsorption of the nonionic surfactant to the PAA brush (at pH 2) that was observed.

It was found that the initial adsorption rate is proportional to the copolymer concentration in solution, indicating that the rate of adsorption is limited by the rate of diffusion of the polymer to the surface (in the stagnant layer flow cell of the reflectometer).

In Figure 7.10 the adsorption of P2MVP-PEO to PAA brushes of different grafting densities and chain lengths is shown as a function of pH. With increasing grafting density or chain length the adsorbed amount strongly increases. For all grafting densities and chain lengths the adsorption as a function of pH shows the same behavior. At pH 2, when the PAA brush is almost uncharged, only a small adsorbed amount was measured, probably mainly due to hydrogen bonding between PEO and the uncharged PAA. From pH 2 to pH 6 the adsorption increased strongly, while between pH 6 and pH 10 there was only a slight increase in adsorption. The increased adsorption with increasing pH is obviously the result of the charging of the PAA brush so that more P2MVP can bind.

From these results it appears that due to complex formation with P2MVP, the PAA brush at pH 6 already has a high degree of dissociation (we estimate about 90%, based on the difference in adsorbed amounts between pH 6 and pH 10). For a bare PAA brush the degree of dissociation at pH 6 would be much lower, considering that PAA dissolved in a 10 mM NaCl aqueous solution at pH 6 has a degree of dissociation of about 50% [26], and in a brush,

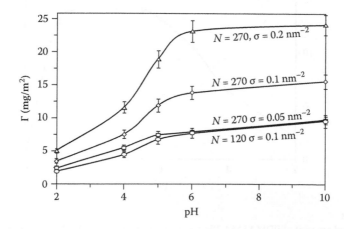

FIGURE 7.10
The adsorbed amount of $P2MVP_{42}$-PEO_{205} to PAA_{270} and PAA_{120} brushes of different grafting densities as indicated as a function of pH (background electrolyte 10 mM KNO_3).

due to the high negative local potential, the degree of dissociation is even lower than in bulk solution.

When we compare the adsorption of PDMAEMA-PEO to a PAA brush as a function of pH (Figure 7.11) with that of P2MVP-PEO (Figure 7.10) strong similarities are observed. At low pH, when the PAA is uncharged, a small amount of adsorption was found. With increasing pH, the PAA charges up and more diblock copolymer is adsorbed until at pH 6.5 a maximum is reached. The difference between the adsorption of PDMAEMA-PEO and P2MVP-PEO is that for PDMAEMA-PEO adsorption abruptly drops to zero between pH 9 and 10. At this pH, PDMAEMA becomes uncharged, and as there is no electrostatic interaction anymore, no adsorption occurs. As for P2MVP-PEO, the combination of low pH and nonionic surfactants could be used to rinse off all adsorbed polymer. For PDMAEMA-PEO this was also the case. In addition, a solution of pH 10 or higher also removed all adsorbed polymer.

In Figure 7.12 the effect of the salt concentration on the adsorption of P2MVP-PEO is shown for three different pH values. For pH 6 and 10 the high adsorption remains fairly constant up to a salt concentration of 250 mM. Above this, the adsorption decreases to only a few mg/m² at 500 mM and almost zero at 1 M. This shows how strong the attraction is between the oppositely charged polyelectrolytes; high salt concentrations are needed to significantly reduce this interaction.

At pH 4 the adsorption decreased gradually from about 8 mg/m² at low salt concentration to 4 mg/m² at 1 M of salt. As mentioned above, at low pH probably hydrogen bonds are formed between the PEO and the uncharged PAA. This is why even at very high salt concentrations, when all electrostatic

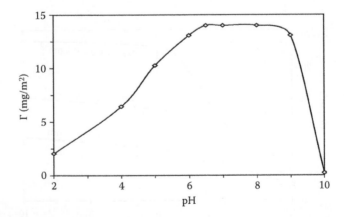

FIGURE 7.11
The adsorbed amount of PDMAEMA-PEO as a function of pH to a PAA$_{270}$ brush ($\sigma = 0.2$ nm^{-1}). Background electrolyte 10 mM KNO$_3$. (Reproduced from de Vos W. M., et al., *Angew. Chem. Int. Ed.*, 48, 5369–5371, 2009. With permission from Wiley.)

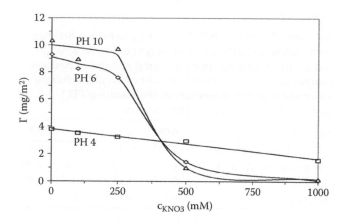

FIGURE 7.12
Adsorption of P2MVP$_{41}$-PEO$_{205}$ to a PAA$_{270}$ brush ($\sigma = 0.2$ nm^{-2}) as a function of KNO$_3$ concentration at different pH values, as indicated. (Reproduced from de Vos, W. M., et al., *Soft Matter*, 6, 2499–2507, 2010. With permission of the Royal Society of Chemistry.)

interactions are screened, still a finite adsorption of the diblock copolymer was found.

7.3.3 Charge Compensation

Analysis shows that adsorption stops when (almost) all charges of the PAA brush are neutralized by the P2MVP. Table 7.1 gives the charge compensation by the adsorption of P2MVP-PEO at pH 6 for different grafting densities of the PAA brush. For simplicity, complete dissociation of PAA was assumed in the calculations. However, at pH 6 after complex formation with P2MVP, still a small fraction of the acid groups is uncharged; this is probably the reason that most of the calculations give a charge compensation somewhat smaller than 1.

According to Equation 7.1 the grafting density of the neutral brush should be a multiplication of that of the PAA brush by a factor equal to the ratio between the length of the PAA chain and the length of the P2MVP chain. For the system in Table 7.1 this ratio is 270/41 = 6.6. In this table the obtained multiplication factors are indicated; most of them are somewhat smaller than 6.6, again because at pH 6 full dissociation of the PAA brush is not yet entirely achieved. The maximum PEO grafting density that was reached is 1.59 nm^{-2}. To our knowledge such a high grafting density has never been reported for any other brush systems consisting of long neutral polymer chains.

In Table 7.2 results of adsorption experiments are given for diblock copolymers of different anchor and buoy block lengths. We first compare the adsorption within two sets of polymers with the same PEO length and a different cationic block length, i.e., P2MVP$_{42}$-PEO$_{450}$ and P2MVP$_{72}$-PEO$_{450}$, and PDMAEMA$_{35}$-PEO$_{120}$ and PDMAEMA$_{77}$-PEO$_{120}$. In both cases, a longer

TABLE 7.1

Adsorbed Amount of $P2MVP_{41}$-PEO_{205} to PAA_{270} Brushes with Different Grafting Densities (pH 6, 10 mM KNO_3) and Charge Compensations (the total amount of positive charges divided by the total amount of negative charges assuming complete dissociation; the resulting PEO grafting densities are indicated as well)

σ_{PAA} (nm^{-2})	0.05	0.1	0.2	0.3
Adsorbed amount (mg m^{-2})	7.73	13.7	23.1	36.1
Charge compensation (–)	1.02	0.90	0.76	0.79
σ_{PEO} (nm^{-2})	0.34	0.60	1.01	1.59
$\sigma_{PEO}/\sigma_{PAA}$	6.8	6.0	5.1	5.3

TABLE 7.2

Adsorbed Amount of Diblock Copolymers with Different Block Lengths to PAA_{270} Brushes ($\sigma = 0.2$ nm^{-1}): The Resulting Charge Compensation and PEO Grafting Density (pH 6, 10 mM KNO_3)

	Adsorbed Amount (mg m^{-2})	Charge Compensation(–)	σ_{PEO} (nm^{-2})	N_{PAA}/N_{anch} (–)	$\sigma_{PEO}/\sigma_{PAA}$ (–)
$P2MVP_{42}$	18.4	1.50	—	—	—
$P2MVP_{42}$-PEO_{205}	25.9	0.90	1.16	6.4	5.8
$P2MVP_{42}$-PEO_{450}	33.0	0.63	0.81	6.4	4.1
$P2MVP_{72}$-PEO_{450}	26.2	0.74	0.56	3.8	2.8
$PDMAEMA_{35}$-PEO_{120}	26.4	0.98	1.51	7.7	7.6
$PDMAEMA_{77}$-PEO_{120}	21.6	1.09	0.76	3.5	3.8

cationic block led to a reduction in the adsorbed amount while the degree of charge compensation remained almost the same. As predicted by Equation 7.1, the grafting density of the neutral block is inversely proportional to the length of the cationic block. Within experimental error, these results strongly support this.

It is also interesting to compare $P2MVP_{42}$, $P2MVP_{42}$-PEO_{205}, and $P2MVP_{42}$-PEO_{450} as in this set of (co)polymers; the size of the buoy block is varied while that of the anchor block is kept the same. In the case of the cationic polymer without neutral block a degree of charge compensation much larger than unity is found. From experiments on polyelectrolyte multilayers (see [60], specifically for the PAA and P2MVP system) it is known that the adsorption of a polyelectrolyte to a layer of oppositely charged polyelectrolyte usually leads to overcompensation. When a neutral block is attached to the cationic block ($P2MVP_{42}$-PEO_{205}) the degree of charge compensation is reduced to close to unity. However, when the PEO length becomes longer ($P2MVP_{42}$-PEO_{450}) the degree of charge compensation becomes significantly lower.

We believe that for such a long chain, the very high pressure in the thick and dense neutral polymer brush is the cause of adsorption of diblock copolymer stopping before all charges in the brush have been compensated. This shows that there is a limit to the leverage that one can achieve using the zipper method, and that Equation 7.1 is no longer accurate if the neutral brush density and degree of polymerization exceed a critical value.

7.3.4 Antifouling Properties of the Zipper Brush

The antifouling properties of the zipper brush (PAA_{270}, $\sigma = 0.2$ nm^{-1}; $P2MVP_{41}$-PEO_{205}) were tested by exposure to a number of different proteins. The resulting adsorbed amounts are listed in Table 7.3 and compared to the adsorption onto various other surfaces: a polystyrene surface, a polystyrene surface covered with C3Ms (complexes of $P2MVP_{41}$-PEO_{205} and PAA), and a polystyrene surface covered with a PAA_{270} brush ($\sigma = 0.2$ nm^{-1}). The grafting density of PEO chains for the C3M system (see Section 7.2.3) on polystyrene was approximately 0.05 nm^{-2} (as calculated from the measured adsorbed amount), thus much lower than the grafting densities reached with the zipper brush approach.

Adsorption of protein to a negatively charged PAA brush can lead to very high-adsorbed amounts (for a number of proteins too high to measure accurately). The only protein that did not adsorb strongly to the PAA brush is β-lactoglobulin, which is negatively charged at pH 6. At this pH BSA is also negatively charged, but is known to adsorb to negatively charged brushes at pH values just above its isoelectric point [11,56]. This adsorption has been explained by charge regulation and patchiness.

When P2MVP-PEO is adsorbed to such a PAA brush, forming a high-density PEO brush, the adsorption is reduced to practically zero for almost all investigated proteins. This shows that the zipper brush has very good antifouling properties. Of the investigated proteins only cytochrome c adsorbs weakly to the zipper brush. By choosing another buoy block for the diblock, antifouling properties may be tuned to specific requirements.

TABLE 7.3

Adsorbed Amount of Various Proteins on a Number of Different Surfaces as Indicated, at pH 6 and 10 mM KNO_3 (for the PAA brush, $N = 270$, $\sigma = 0.2$ nm^{-1})

	PS [30] (mg m^{-2})	C3M [30] (mg m^{-2})	PAA Brush (mg m^{-2})	Zipper Brush (mg m^{-2})
β-Lactoglobulin	0.91	0.45	0	0
BSA	1.21	0.37	25	0.03
Fibrinogen	8.37	3.11	>30	0
Lysozyme	0.47	0	>30	0
Cytochrome c	—	—	>30	0.8

7.3.5 Structure of the Zipper Brush and the Kinetic Barrier in Its Formation

At the end of this section it is appropriate to revisit some specific aspects of the zipper brush. First, one may question if the structure of the zipper brush as schematically depicted in Figure 7.8 is appropriate. Although there is no direct evidence, there are various strong indications that it indeed correctly represents the formed structure. The strongest one is probably the good protein repelling properties observed for this system. This can only be explained by the existence of a thick neutral polymer brush. Furthermore, almost complete charge compensation was found for the formation of the zipper brush from different types of diblocks. Complexes of oppositely charged polyelectrolytes close to charge compensation are known to form separate polyelectrolyte-rich phases or even a precipitate [61]. Thus, the PAA brush is definitely expected to collapse after complex formation with the diblock copolymer, thus forming a dense phase of oppositely charged polyelectrolytes. Atomic force microscopy (AFM) experiments on a dry layer of $P2MVP_{41}$-PEO_{205} to a PAA brush showed a flat layer with a roughness of approximately 3 nm (on a total thickness of approximately 30 nm), proving that the polymer is evenly spread over the surface and indeed adsorbs as a flat layer.

Finally, what do our findings imply for the kinetic barrier argument that ruled out high densities for brushes prepared by adsorption? It seems that such a barrier hardly exists for the formation of a zipper brush. A plausible explanation seems that adsorbing chains do not have to go all the way through the brush before they can attach, because the bare PAA chains themselves stand out from the surface so that they are able to capture incoming chains at the periphery of the brush. Once attached the anchor blocks can "crawl" along the PAA chains to increase their favorable electrostatic interactions on a larger timescale.

7.4 Conclusions and Perspectives

With the zipper brush approach, we have found a method to produce polymer brushes by adsorption that circumvents all the normal problems associated with diblock copolymer adsorption. The driving force for the adsorption is very large, while there is no crowding of anchor blocks or a large steric barrier that prevents the brush from becoming dense. This means that, as we have shown above, ultra-dense polymer brushes can be prepared that have excellent antifouling properties. The density of the formed zipper brush is determined by charge compensation of the polyelectrolyte attached to the interface. As such, control over the length of the charged block of the adsorbing diblock copolymer gives one complete control over the grafting density of the formed brush. Control over the length and chemical makeup of the neutral block of the adsorbing diblock copolymer, on the other hand, makes it possible to tune the desired properties of the formed brush. For these reasons one could argue that of the many ways to produce polymer brushes, the zipper brush approach is one of the few that comes close to that of the ideal method described in the introduction of this chapter. Still, there are clearly aspects that can be improved in the current method to prepare the zipper brush, and there are also very promising variations on the technique that have not been investigated yet.

While the formation of the zipper brush by adsorption of the diblock copolymer is a very simple procedure, one needs to first prepare a surface coated with attached polyelectrolyte. In the current procedure, a polyelectrolyte brush is prepared by the LB method, and this is very time-consuming and not suitable for large-scale applications. We stress, however, that this polyelectrolyte brush could also be prepared by any other convenient method to produce polymer brushes such as grafting from and grafting to. Moreover, the polyelectrolyte does not need to be present in the form of a polyelectrolyte brush; a thin polyelectrolyte gel layer would, for example, also suffice. We believe that perhaps the simplest procedure to attach a polyelectrolyte would be to produce a sparse but long polyelectrolyte brush in a simple adsorption step, as, for example, done by Balastre et al. [62]. By subsequently adsorbing a diblock copolymer with a relatively short charged block, one could still achieve a very dense zipper brush. Admittedly, the very precise control over the grafting density inherent to the LB method is partly lost by using these simpler alternatives. However, it is still possible to prepare a very dense brush in just two simple adsorption steps, a huge advantage for many applications.

A new step in the development of the zipper brush method would be to create mixed polymer brushes, i.e., polymer brushes that consist of two (or more) chemically different polymers. For example, exposing a layer of attached negative polyelectrolyte to an aqueous solution of P2MVP-PEO and P2MVP-PVA might result in the formation of a mixed PEO and PVA brush. Using a mask or some printing technique, one might also selectively expose parts of the surface to one type of diblock copolymer, and subsequently the other parts to another to create patterned surfaces.

The reversibility is another property of the brush that warrants more investigations. As it is possible to remove the brush layer by a pH change or by applying high salt concentrations, one can easily replace a brush when it is damaged, or change the brush to another grafting density, chain length, or chemical makeup when desired. This also means that the zipper brush could be employed as a sacrificial layer [63], which can be easily removed together with any dirt that it has collected over time. As such, the brush would not only help to keep surfaces clean, but also help to clean surfaces if they eventually do get dirty.

Clearly our investigations have shown that the zipper brush is a powerful system for surface modifications. Future investigations will hopefully reveal even more interesting properties and possibilities for application.

References

1. Milner, S. T. 1991. Polymer brushes. *Science* 251:905–914.
2. Napper, D. H. 1983. *Polymeric stabilization of colloidal dispersions.* Academic Press, London.
3. Scheutjens, J. M. H. M., Fleer, G. J., Cohen Stuart, M. A., Cosgrove, T., and Vincent, B. 1993. *Polymers at interfaces.* Chapman & Hall, London.
4. Currie, E. P. K., Norde, W., and Cohen Stuart, M. A. 2003. Tethered polymer chains: surface chemistry and their impact on colloidal and surface properties. *Adv. Colloid Interface Sci.* 100–102:205–265.
5. Halperin, A., and Leckband, D. E. 2000. From ship hulls to contact lenses: repression of protein adsorption and the puzzle of PEO. *C. R. Acad. Sci. Paris* IV:1171–1178.
6. Raynor, J. E., Capadona, J. R., Collard, D. M., Petrie, T. A., and García A. J. 2009. Polymer brushes and self-assembled monolayers: versatile platforms to control cell adhesion to biomaterials. *Biointerphases* 4:FA3–FA16.
7. Cohen Stuart, M. A., de Vos, W. M., and Leermakers F. A. M. 2006. Why surfaces modified by flexible polymers often have a finite contact angle for good solvents. *Langmuir* 22:1722–1728.
8. Cohen Stuart, M. A., Huck, W. T. S., Genzer, J., Müller, M., Ober, C., Stamm, M., Sukhorukov, G. B., Szleifer, I., Tsukruk, V. V., Urban, M., Winnik, F., Zauscher, S., Luzinov, I., and Minko, S. 2010. Emerging applications of stimuli-responsive polymer materials. *Nature Mat.* 9:101–113.
9. Klein, J., Kumacheva, E., Mahalu, D., Perahla, D., and Fetters, L. J. 1994. Reduction of frictional forces between solid surfaces bearing polymer brushes. *Nature* 370:634–636.
10. Klein, J. 1996. Shear, friction, and lubrication forces between polymer bearing surfaces. *Annu. Rev. Mater. Sci.* 26:581–612.
11. Wittemann, A., and Ballauff, M. 2006. Interaction of proteins with linear polyelectrolytes and spherical polyelectrolyte brushes in aqueous solution. *Phys. Chem. Chem. Phys.* 8:5269–5275.
12. Tsujii, Y., Ohno, K., Yamamoto, S., Goto, and A., Fukuda, T. 2006. Structure and property of high-density polymer brushes prepared by surface-initiated living radical polymerization. *Adv. Polym. Sci.* 197:1–45.
13. Edmondson, S., Osborne, V. L., and Huck, W. T. S. 2004. Polymer brushes via surface-initiated polymerizations. *Chem. Soc. Rev.* 33:14–22.
14. Halperin, A., Tirrell, M., and Lodge, T. P. 1992. Tethered chains in polymer microstructures. *Adv. Polym. Sci.* 100:31–71.
15. Zhao, B., and Brittain, W. J. 2000. Polymer brushes: surface-immobilized macromolecules. *Prog. Polym. Sci.* 25:677–710.
16. Rühe, J., Ballauff, M., Biesalski, M., Dziezok, P., Gröhn, F., Johannsmann, D., Houbenov, N., Hugenberg, N., Konradi, R., Minko, S., Motornov, M., Netz, R. R., Schmidt, M., Seidel, C., Stamm, M., Stephan, T., Usov, D., and Zhang, H. 2004. Polyelectrolyte brushes. *Adv. Polym. Sci.* 165:79–150.
17. Advincula, R. C., Brittain, W. J., Caster, K. C., Rühe, J. 2004. *Polymer brushes.* Wiley-VHC, Weinheim.

18. de Vos, W. M., Kleijn, J. M., de Keizer, A., Cosgrove, T., and Cohen Stuart, M. A. 2010. Polymer brushes. *Kirk-Othmer Encyclopedia*, New York: John Wiley & Sons.
19. Wu, D. T., Yokoyama, A., and Setterquist, R. L. 1991. An experimental study on the effect of adsorbing and nonabsorbing block sizes on diblock copolymer adsorption. *Polym. J.* 23:709–714.
20. Hadziioannou, G., Patel, S., Granick, S., and Tirrell, M. 1986. Forces between surfaces of block copolymers adsorbed on mica. *J. Am. Chem. Soc.* 108:2869–2876.
21. Tran, Y., Auroy, P., and Lee, L.-T. 1999. Determination of the structure of polyelectrolyte brushes. Macromolecules 32:8952–8964.
22. Wuelfing, W. P., Gross, S. M., Miles, D. T., and Murray, R. W. 1998. Nanometer gold clusters protected by surface-bound monolayers of thiolated poly(ethyleneglycol) polymer electrolyte. *J. Am. Chem. Soc.* 12:12696–12697.
23. Maas, J. H., Cohen Stuart, M. A., Sieval, A. B., Zuilhof, H., and Sudhölter, E. J. R. 2003. Preparation of polystyrene brushes by reaction of terminal vinyl groups on silicon and silica surfaces. *Thin Solid Films* 426:135–139.
24. Jones, D. M., Brown, A. A., and Huck, W. T. S. 2002. Surface-initiated polymerizations in aqueous media: effect of initiator density. *Langmuir* 18:1265–1269.
25. Currie, E. P. K., van der Gucht, J., Borisov, O. V., and Cohen Stuart, M. A. 1999. Stuffed brushes: theory and experiment. *Pure Appl. Chem.* 71:1227–1241.
26. Currie, E. P. K., Sieval, A. B., Avena, M., Zuilhof, H., Sudhölter, E. J. R., and Cohen Stuart, M. A. 1999. Weak polyacid brushes: preparation by LB-deposition and optically detected titrations. *Langmuir* 15:7116–7118.
27. de Vos, W. M., de Keizer, A., Kleijn, J. M., Cohen Stuart, M. A. 2009. The production of PEO polymer brushes via Langmuir-Blodgett and Langmuir-Schaeffer methods: incomplete transfer and its consequences. *Langmuir* 25:4490–4497.
28. van der Burgh, S., Fokkink, R., de Keizer, A., and Cohen Stuart, M. A. 2004. Complex coacervation core micelles as anti-fouling agents on silica and polystyrene surfaces. *Coll. Surf. A* 242:167–174.
29. Voets, I. K., de Vos, W. M., Hofs, B., de Keizer, A., and Cohen Stuart, M. A. 2008. Internal structure of a thin film of mixed polymeric micelles on a solid/liquid interface. *J. Phys. Chem. B* 112:6937–6945.
30. Brzozowska, A. M., Hofs, B., de Keizer, A., Fokkink, R., Cohen Stuart, M. A., and Norde, W. 2009. Reduction of protein adsorption on silica and polystyrene surfaces due to coating with complex coacervate core micelles. *Coll. Surf. A* 347:146–155.
31. de Vos W. M., Kleijn, J. M., de Keizer, A., and Cohen Stuart, M. A. 2009. Ultra dense polymer brushes by adsorption. *Angew. Chem. Int. Ed.* 48:5369–5371.
32. de Vos, W. M., Meijer, G., de Keizer, A., Cohen Stuart, M. A., and Kleijn, J. M. 2010. Charge-driven and reversible assembly of ultra-dense polymer brushes: formation and anti-fouling properties of a zipper brush. *Soft Matter* 6:2499–2507.
33. Fleer, G. J., and Lyklema, J. 1974. Polymer adsorption and its effect on the stability of hydrophobic colloids. II. The flocculation process as studied with the silver iodide-polyvinyl alcohol system. *J. Coll. Int. Sci.* 46:1–12.
34. Fleer, G. J., and Lyklema, J. 1976. Polymer adsorption and its effect on the stability of hydrophobic colloids. III. Kinetics of the flocculation of silver iodide sols. *J. Coll. Int. Sci.* 55:228–238.

35. Scheutjens, J. M. H. M., and Fleer, G. J. 1979. Statistical theory of the adsorption of interacting chain molecules. 1. Partition function, segment density distribution, and adsorption isotherms. *J. Phys. Chem.* 83:1619–1635.
36. Scheutjens, J. M. H. M., and Fleer, G. J. 1985. Interaction between 2 adsorbed polymer layers. *Macromolecules* 18:1882–1900.
37. Evers, O. A., Scheutjens J. M. H. M., and Fleer, G. J. 1990. Statistical thermodynamics of block copolymer adsorption. Part 2. Effect of chain composition on the adsorbed amount and layer thickness. *J. Chem. Soc. Faraday Trans.* 86:1333–1340.
38. Marques C. M., and Joanny, J. F. 1989. Block copolymer adsorption in a nonselective solvent. *Macromolecules* 22:1454–1458.
39. van Lent, B., Scheutjens, J. M. H. M. 1989. Influence of association on adsorption properties of block copolymers. *Macromolecules* 22:1931–1937.
40. Marques, C. M., Joanny, J. F., and Leibler, L. 1988. Adsorption of block copolymers in selective solvents. *Macromolecules* 21:1051.
41. Cohen Stuart, M. A., Fleer, G. J., Scheutjens, J. M. H. M. 1984. Displacement of polymers. II. Experiment. Determination of segmental adsorption energy of poly(vinylpyrrolidone) on silica. *J. Coll. Int. Sci.* 97:526–535.
42. Hoogeveen, N. G., Cohen Stuart M. A., and Fleer, G. J. 1994. Adsorption of charged block copolymers with two adsorbing blocks. *Faraday Disc.* 98:161–172.
43. Guzonas, D. A., Boils, D., Tripp, C. P., and Hair, M. L. 1992. Role of block size asymmetry on the adsorbed amount of polystyrene-b-poly(ethylene oxide) on mica surfaces from toluene. *Macromolecules* 25:2434–2441.
44. Bijsterbosch, H. D., Cohen Stuart, M. A., and Fleer, G. J. 1998. Nonselective adsorption of block copolymers and the effect of block incompatibility. *Macromolecules* 31:7436–7444.
45. Taunton, H. J., Toprakcioglu, C., Fetters, L. J., and Klein, J. 1990. Interactions between surfaces bearing end-adsorbed chains in a good solvent. *Macromolecules* 23:571–580.
46. Marra, J., and Hair, M. L. 1988. Interactions between two adsorbed layers of poly(ethylene oxide)/polystyrene diblock copolymers in heptane-toluene mixtures. *Coll. Surf.* 34:215–226.
47. Watanabe, H., and Tirrell, M. 1993. Measurement of forces in symmetric and asymmetric interactions between diblock copolymer layers adsorbed on mica. *Macromolecules* 26:6455–6466.
48. Ansarifar, M. A., and Luckham, P. F. 1988. Measurement of the interaction force profiles between block copolymers of poly(2-vinylpyridine)/poly(t-butylstyrene) in a good solvent. *Polymer* 29:329–335.
49. Hofs, B., Brzozowska, A. M., de Keizer, A., Norde, W., and Cohen Stuart, M. A. 2008. Reduction of protein adsorption to a solid surface by a coating composed of polymeric micelles with a glass-like core. *J. Coll. Int. Sci.* 325:309–315.
50. Brzozowska, A. M., Zhang, Q., de Keizer, A., Norde, W., and Cohen Stuart, M. A. 2010. Protein adsorption on silica and polysulfone surfaces coated with complex coacervate core micelles with poly(vinyl alcohol) as a neutral brush forming block. *Coll. Surf. A* 368:96–104.
51. Cohen Stuart, M. A., Besseling, N. A. M., and Fokkink, R. G. 1998. Formation of micelles with complex coacervate cores. *Langmuir* 14:6846–6849.
52. Voets, I. K., de Keizer, A., and Cohen Stuart, M. A. 2009. Complex coacervate core micelles. *Adv. Coll. Int. Sci.* 147:300–318.
53. Brzozowska, A. M. 2010. PhD thesis. Wageningen.

54. Kuhl, T. L., Leckband, D. E., Lasic, D. D., and Israelachvili, J. N. 1994. Modulation of interaction forces between lipid bilayers exposing short chained ethylene oxide. *Biophysical J.* 66:1479–1488.

55. Efremova, N. V., Sheth, S. R., and Leckband, D. E. 2001. Protein-induced conformational changes in poly(ethylene glycol) brushes: molecular weight and temperature dependence. *Langmuir* 17:7628–7636.

56. de Vos, W. M., Biesheuvel, P. M., de Keizer, A., Kleijn J. M., and Cohen Stuart, M. A. 2008. Adsorption of the protein bovine serum albumin in a planar poly(acrylic acid) brush layer as measured by optical reflectometry. *Langmuir* 24:6575–6584.

57. Bosker, W. T. E., Patzsch, K., Cohen Stuart M. A., and Norde W. 2007. Sweet brushes and dirty proteins. *Soft Matter* 3:754–762.

58. Dijt, J. C., Cohen Stuart, M. A., Hofman J. E., and Fleer G. J. 1990. Kinetics of polymer adsorption in stagnation point flow. *Coll. Surf.* 51:141–158.

59. Lutkenhaus, J. L., Mcennis, K., and Hammond, P. T. 2007. Tuning the glass transition of and ion transport within hydrogen-bonded layer-by-layer assemblies. *Macromolecules* 40:8367–8373.

60. Kovacevic, D., van der Burgh, S., de Keizer, A., and Cohen Stuart, M. A. 2003. Specific ion effects on weak polyelectrolyte multilayer formation. *J. Phys. Chem. B* 107:7998–8002.

61. Bungenberg de Jong, H. G. 1949. *Colloid science.* Vol. II. Elsevier, Amsterdam.

62. Balastre, M., Li, F., Schorr, P., Yang, J., Mays, J. W., and Tirrell M. V. 2002. A study of polyelectrolyte brushes formed from adsorption of amphiphilic diblock copolymers using the surface forces apparatus. *Macromolecules* 35:9480–9486.

63. de Vos W. M., de Keizer, A., Cohen Stuart, M. A., and Kleijn, J. M. 2010. Thin polymer films as sacrificial layers for easier cleaning. *Coll. Surf. A* 358:6–12.

8

Scanning Electrochemical Microscopy: Principles and Applications for the Manipulation of Polymer Brushes

Frédéric Kanoufi

Physicochimie des Electrolytes Colloïdes et Sciences Analytiques,
CNRS UMR 7195, ESPCI ParisTech, Paris, France

CONTENTS

8.1 Introduction

Over the past two decades surface-anchored polymer chains, also known as polymer brushes, have been tremendously used to create smart or responsive interfaces or surfaces. Indeed, as one end of the macromolecular chain is tethered to the surface while the other is freely moving in solution, polymer brushes present very interesting properties based on their ability to reorganize upon environmental changes.[1] It is then possible to trigger the alteration of the chain's conformation upon a wide range of stimuli: pH, ionic strength, electrical, thermal, photo- or electrochemical actuations. It explains the increasing number of applications using polymer brushes, ranging from the design of thermo- or solvent-responsive materials to materials for the design of biosensors for diagnostics or for drug delivery.[2,3] The most efficient route to generate thick polymer brushes with high grafting densities is afforded by surface-initiated polymerizations where the polymerization is directly processed from initiators of polymerization anchored to surfaces. Among these grafting from living polymerizations, the method of choice is likely the highly controlled atom transfer radical polymerization (ATRP). It allows the generation of dense architectures of polymer brushes with a maximum control over brush density and polydispersity.[4–10]

Surfaces tethered with such controlled polymer brushes and their stimuli-responsive behaviors have been characterized using a large number of techniques that have been recently reviewed.[4] Among them, electrochemical techniques are particularly appealing as they could be used both for the characterization of the materials and for their actuation. For instance, polymer brushes are appealing materials for the design of electrochemical biosensors. In this context, polymer brushes are commonly characterized by traditional electrochemical techniques (like cyclic voltammetry), but so far, their characterization by electrochemical imaging techniques is sparse.

Within a few years, the invention of scanning tunneling microscopy opened a route to the development of very promising new tools named scanning probe microscopes (SPMs). The general principle of these tools is to use a sharp probing tip to interact in the close vicinity of a surface in order to get local information about its topographical but also physical or chemical characteristics. Among these SPMs, scanning electrochemical microscopy (SECM; the acronym also designates the tool) is able to probe quantitatively (electro-)chemical reactions in solution at various interfaces with a high spatial resolution. It is therefore a real chemical microscope that provides not only the topography of a substrate surface but also the interrogation of its local chemical reactivity.

The combination of electrochemistry and controlled polymerization reactions is common, first because electrochemistry provides pertinent techniques for the polymer brushes' characterization, and second because the intimate mechanisms beyond some polymerization reactions, and more particularly ATRP,[11] are based on electron transfer processes, and electrochemistry should provide an elegant route for the initiation and control of such polymerizations, as recently demonstrated.[12]

This chapter presents the advantages of combining the methodological developments of both local electrochemical techniques and more particularly SECM and of surface-anchored polymer brushes. First, the principles of operation modes of the SECM are introduced. Then, examples of its applications with polymer brush materials are described. It concerns mainly the use of the SECM for the description of the transport of substances in anchored brush layers and for the assessment of the chemical reactivity of these interfaces. As these fields have been investigated only recently by SECM, the method is compared with alternative electrochemical approaches. Readers who are interested in the broader applications of the SECM for the investigation of charge transfer at various interfaces are referred to recent reviews,[13-15] to the SECM monograph,[16] or its forthcoming second edition.[17]

8.2 SECM—Principle and Operation Modes

8.2.1 Principle

Scanning electrochemical microscopy (SECM) is a scanning probe technique that provides a local chemical interrogation of an interface. In a typical experimental setup (Figure 8.1), the probe, also named the tip, is an ultramicroelectrode (UME), a microelectrode of radius <25 μm, which is rastered with micropositioners in the vicinity of an interface in order to detect an electrochemically active species. Contrary to other scanning probe microscopies,

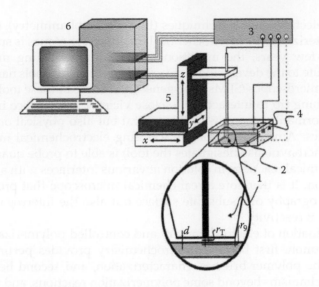

FIGURE 8.1

Experimental SECM setup: (1) disk-shaped amperometric UME, (2) substrate, (3) potentiostat, (4) reference and auxiliary electrodes, (5) micropositioning system, (6) computer for setup control and data acquisition (From Wittstock, G., et al., *Angew. Chem. Int. Ed.*, 46, 1584–1617, 2007.)

SECM is not only used for local topographic information of the inspected interface, but also for the interrogation of its local chemical or electrochemical reactivity.

Different modes of operation of the SECM, presented in Figure 8.2, have been proposed to interrogate the local electrochemical or chemical activity of an interface. The most used ones are based on the current measurement at an amperometric UME probe. The strategy relies on the ability of microelectrodes[18] to detect an electroactive species, Ox, present in a solution by its transformation (Equation 8.1) and through the rapid establishment of a steady-state mass-transfer-limited current.

$$Ox + n e \rightarrow Red \tag{8.1}$$

For the most popularly used disk UME of radius a, as illustrated in Figure 8.3a, the steady-state current associated to the steady mass transfer of Ox to the tip surface is given by

$$i_{T,\infty} = 4 \, n \, F \, D \, C_O \, a \tag{8.2}$$

where F is the Faraday, D the diffusion coefficient, and C_O the concentration of the electroactive species in the solution. Henceforth, because of the rapid establishment of such steady mass transfer to microelectrodes, a SECM tip

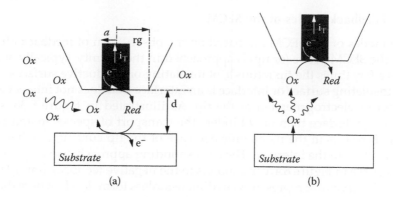

FIGURE 8.2

Most popular modes of operation of the SECM. (a) Feedback mode. (b) Generation-collection mode (herein the substrate is generating an electroactive species Ox, which is collected at the SECM probe (tip) by its reduction according to Equation 8.1. (Adapted from Kanoufi, F., *Acta Chim.*, 311, 36–48, 2007.)

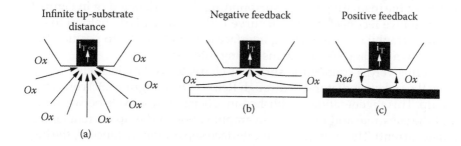

FIGURE 8.3

Principle of the feedback mode of the SECM. (a) Schematic representations of the mass transfer of Ox at the SECM tip generating Red from (1) and held far from a substrate, or in the vicinity of (b) an insulating substrate (negative feedback) or (c) a conducting substrate (positive feedback). (Adapted from Kanoufi, F., *Actual Chim.*, 311, 36–48, 2007.)

acts as a micrometric source of Red, whose transport is mainly developed in a hemisphere of radius about five times the tip radius.

When the tip, or the micrometric source of a reactant, is brought in the vicinity of an interface, the mass transfer of the redox active species (Ox and Red) is perturbed by the interface: its presence, but also its possible reactivity. The approach of the tip then results in a change of the tip current, i_T, with the tip-interface separation distance, d. The establishment of such $i_T - d$ curves, also named approach curves, allows the quantitative estimate of the chemical reactivity of the interface from its aptitude to alter the redox species mass transfer in the gap between the tip and the interface.

8.2.2 Feedback Modes of the SECM

These modes of the SECM are based on the observation of feedback effects when the SECM probe (or tip) is approached in the vicinity, typically at distances a few times the tip radius a, of insulating or conductive surfaces.

An insulating surface or interface is an interface that does not interact with the species electrogenerated at the tip. As illustrated in Figure 8.3b, such insulating interface then only hinders the transport of species to and from the tip. It results in the monotonic decrease of the tip current, i_T, when it is brought close to the interface. The corresponding approach curve, $i_T - d$, the lowest curve in Figure 8.4, corresponds to the negative feedback limit. These approach curves are represented in dimensionless form, $I_T - L$, where the tip current i_T is normalized by its value at infinite distance from the substrate, $i_{T,\infty}$ given by Equation 8.2, $I_T = i_T/i_{T,\infty}$, and the tip-substrate separation distance, d, is normalized by the tip radius, a, $L = d/a$. The negative feedback approach curve is then only governed by the geometrical configuration of the electrochemical cell formed by the tip confined in the vicinity of the substrate. It then depends only on the tip and substrate geometric characteristics (like the radius of the insulating sheath surrounding the tip, denoted rg in Figure 8.2). The negative feedback approach curve is then particularly helpful to image the topography of an insulating interface (or substrate) or to position the tip at a known distance from any insulating interface.

Alternatively, a conducting surface is characterized by its ability to regenerate the redox mediator (Ox through Red → Ox in Figure 8.3c) consumed at the tip. This regeneration establishes an efficient redox cycle between the tip and the substrate and results in an amplification of the tip current (and substrate current). The shorter the tip-interface separation distance, d, the higher

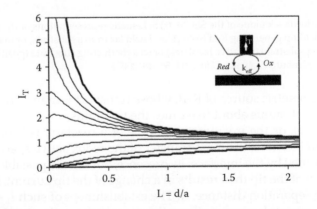

FIGURE 8.4

Approach curves for a disk SECM tip ($a = 12.5$ μm) to a substrate that presents finite heterogeneous charge transfer with effective first-order rate constant, from top to bottom, k_{eff} (in cm s^{-1}) $= \infty$ (>3, positive feedback), 0.3, 0.1, 0.05, 0.03, 0.02, 0.01, 5 10^{-3}, 2 10^{-3}, 5 10^{-4}, 0 (<3 10^{-4}, negative feedback). Insert: Schematic description of the finite kinetics at the substrate.

the redox cycle turnover efficiency and the higher the tip current amplification. At best, the mediator regeneration at the interface is controlled by the redox active species transport in the gap; this limiting situation corresponds to the observation of the positive feedback approach curve (upper limiting curve in Figure 8.4). For lower but detectable regeneration or feedback, the approach curve lies between the two extreme limiting cases of the negative and positive feedback (Figure 8.4). The higher the turnover of the redox cycle, the higher the current amplification and the approach curve.

In these intermediate situations of finite kinetics at the interface, it is possible to quantify the rate constant corresponding to the kinetics of the redox cycling reaction at the interface. The procedure consists in comparing the experimental approach curve to theoretical ones. The theoretical approach curves of a model situation may be obtained from the computation of the transport and reactivity processes within the given geometrical constraints, and this is accessible with easy-to-use commercial solvers using finite element methods, such as COMSOL®.[19] However, most of the encountered situations can be fitted by considering that the interfacial reaction follows a first-order rate law (with respect to Red) and depicted by an effective first-order heterogeneous rate constant k_{eff} (in cm s^{-1}) at the interface, as shown in Figure 8.4. Analytical expressions have been proposed to describe such situation and allow the easy estimate of the local interface reactivity, k_{eff}. To envision the complexity of the situation, these curves depend on three parameters, L, RG, and Λ. The first two geometric parameters correspond to the dimensionless tip-substrate separation distance, $L = d/a$, and the dimensionless size of the insulating part of the tip (RG = rg/a). They characterize the transport within the electrochemical cell. The latter $\Lambda = k_{eff}a/D$ rules the competition between the heterogeneous redox cycling and the diffusive transport of the species to the interface. Different expressions have been proposed;[20,21] the most accurate expression[21] uses the combination of Equations 8.3 to 8.5, given here for RG = 10:

$$I_{T,ins}(L) = \frac{0.912L + 1.57}{0.912L + 1.59 + \frac{2.3}{L} 0.0636 Ln(1 + \frac{15.7}{L})} \tag{8.3}$$

$$I_{T,cond}(L) = 0.652 + \frac{0.772}{ArcTan(L)} - 0.0911 ArcTan(L) \tag{8.4}$$

$$I_{T,fin}(L,\Lambda) = I_{T,cond}\left(L + \frac{1}{\Lambda}\right) + \frac{I_{T,ins}(L) - 1}{(1 + 5.6L\Lambda)(1 + L^{0.177}\Lambda^{0.64})} \tag{8.5}$$

where Equations 8.3 and 8.4 are the expressions for, respectively, the negative and positive feedback approach curves, and Equation 8.5, which uses both Equations 8.3 and 8.4, corresponds to the expression for the tip current derived for finite heterogeneous kinetics. The general expressions of Equations 8.3 to 8.5 for other RG values may be found in the original work.[21] Figure 8.4 presents the typical I_T – L approach curves in their dimensionless form generated from these analytical expressions.

8.2.3 Generation-Collection Modes

In these modes, the substance that has to be detected is not initially present in the solution. For instance, in the substrate generation–tip collection mode (SG/TC)[22] depicted in Figure 8.2b, the substrate (or any of its local active regions) generates a substance that diffuses in the solution and is detected at the SECM tip. For an amperometric UME tip, the substance generated at the substrate is electrochemically active, and it is detected at the tip held at a potential, E_{tip}, from its electrochemical consumption. For instance, in Figure 8.2b, the collection of Ox at the probe from its mass-transfer-limited reduction requires the application of $E_{tip} < E^0_{Ox/Red}$, the standard reduction potential of the redox couple defined in Equation 8.1. The measurement of the tip current associated to this transformation allows, in principle, for the determination of the local concentration of the substance at the tip positioning. For small disk tips positioned at distances, d, larger than the tip radius (d > 3a) the probe is considered passive, as it does not perturb the source of substance. The tip current is then given by Equation 8.2, where C_O refers to the concentration detected at the position of the probe. The concentration at the probe can be related to that at the source surface if the collection does not perturb the generation at the source. Different situations have been studied that model (sub-)micrometric[23–25] or planar[26] sources generating, respectively, steady-state hemispherical or transient one-dimensional (1D) linear diffusional fluxes.

The use of small noninterfering probes is pertinent for the inspection of transient experiments because of the fast (<0.1 s) quasi-steady-state response of UME to local concentration change. This is particularly interesting to follow, in real-time, time-dependent processes, in order to analyze fluctuations of the flux of the substances produced at a given source.[27,28]

At smaller values of the tip-substrate separation distance, d, feedback and hindering effects, as those observed in the feedback modes at amperometric tips, perturb the transport processes. The tip then starts to interfere with the source, which complicates the quantitative data analysis without numerical modeling.[29] Such perturbing effects are not observed when passive probes such as potentiometric[30–32] or biosensor[33,34] microelectrodes are used, but then the positioning of these substance-selective sensors is difficult. An alternative would be to consider the use of the scanning ion conductance microscopy (SICM). Indeed, recently, the SICM afforded the opportunity to image and quantify precisely local K^+ and Cl^- ionic fluxes.[35]

8.2.4 Lithographic Mode

This mode of operation corresponds to the use of SECM as a tool for patterning or fabricating microstructures on a surface. One of the advantages of the SECM resides mainly in its multifunctionality, as it is able, all at once, to transform locally a surface, to understand the mechanistic processes of the surface transformation and finally to characterize the patterned surface. Moreover, electrodes and their electrochemical activation allow potentially the production of a high diversity of chemical reagents (oxidants, reducers, but also acids, bases, radicals, etc.).[36] The use of small tips, *a*, and tip-substrate separation distances, d, is also favourable to the handling of unstable electrogenerated reagents for the local chemical transformations of a substrate surface.

The main approaches used for surface patterning by SECM are the direct mode and the feedback mode presented, respectively, in Figure 8.5a and b. In the direct mode, the substrate surface is used as an electrode and polarized to drive its electrochemical transformation, and the SECM tip serves as a local counterelectrode to confine the transformation on a small area of the substrate facing the tip. In the feedback mode surface patterning the tip electrogenerates a reagent that is transported in solution to the substrate where it induces its local chemical transformation. The feedback mode then applies to any kind of surfaces (insulating to conducting).

The potentiality of the lithographic properties of the SECM was illustrated in different configurations to fabricate at the micrometric or submicrometric scale various patterns on a substrate.[37,38] The SECM has been used to decorate surfaces with various organic functionalities. For example, it was successfully devoted to the local deposition[39-43] or etching[44-47] of polymers, to the local desorption,[48,49] etching,[50-57] or derivatization by click chemistry[58] of organic layers, and to induce the local decoration of conducting surfaces with electrogenerated radicals.[59,60]

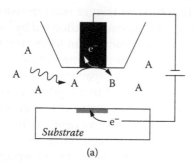
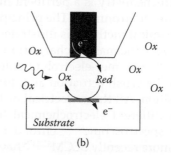

(a)　　　　　　　　　　　　　(b)

FIGURE 8.5

Lithographic mode of the SECM. The SECM tip is used to etch locally a substrate surface and transform it, by local reduction, into a grey microstructure. (a) Direct mode—the tip is used as the counterelectrode of the electrochemical cell. (b) Feedback mode—the tip is used as a source of an electrogenerated etchant. (Adapted from Kanoufi, F., *Actual Chim.*, 311, 36–48, 2007.)

Among those examples, the SECM was also used to get mechanistic insight and then to access the local chemical or electrochemical reactivity of the surfaces. The interrogation of the surface reactivity is obtained as discussed previously (1) from the standard procedure using $i_t - d$ approach curves, or (2) from the measurement of transient feedback changes available through variations of the tip current, i_T, at fixed d, or (3) when no change in feedback is detected, from the (ex situ) observation of the growth rate of the patterns.[57,61]

8.3 Transport Processes of Electroactive Species within Polymer Brushes

8.3.1 Presentation

Owing to their ability to reorganize upon environmental changes, polymer brushes are very interesting candidates for the design of functional surfaces. Such "responsive" behavior of polymer brushes is promising for the creation of "smart" interfaces where the change in chain conformation may be triggered by a wide variety of stimuli: pH, ionic strength, solvent, light, or temperature. This prospect is particularly appealing for the design of electrochemical (bio-)sensors[2] that use an electric signal to trigger or to detect such conformational change. For example, polyelectrolyte brushes may display significant reorganization processes upon incorporation of free counterions of controlled chemistry (hydrophobicity)[62] or upon protonation/dissociation of their acidic groups.[63] It may lead to the extreme situation where the brush conformation can change from a swollen to a totally collapsed state. A number of experimental methods have been proposed to study the swelling-collapse transition in polymer brushes.[4] Electrochemistry is a pertinent method to follow or trigger such transition in a wet environment. The principle of the characterization afforded by electrochemical methods is illustrated in Figure 8.6. It consists in following the change in the mass or charge transfer properties within the macromolecular film, the polymer-electrolytic solution resistance and capacitance, resulting from the brush actuation and, for example, associated to a conformational transition. All these electrical or transport properties are readily accessible from different electrochemical techniques,[64] and electrochemical impedance spectroscopy,[65–69] chronoamperometry,[70] cyclic voltammetry,[52,53,70–73] and more recently SECM[52,53,74] have been proposed to probe the transport of substances within polymer brushes.

In typical examples a polymer brush is grafted from an electrode surface and a species is allowed to be transported within the brush. The electrode is polarized and owing to the organization and constrained structure of the polymer brush, the response of the whole system to the electrode

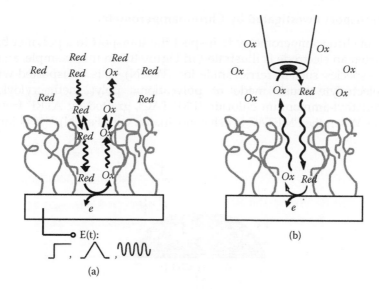

FIGURE 8.6
Using electrochemistry to probe the transport of Ox/Red within polymer brushes grafted onto substrates. (a) Direct electrochemical measurement at a grafted electrode by chronoamperometry (CA), cyclic voltammetry (CV), electrochemical impedance spectroscopy (EIS) represented by their respective waveform. (b) Indirect measurement by SECM. The overall process is decomposed in (a) into the subsequent steps (from top to bottom): solution diffusion, partition at the solution-polymer brush interface, diffusion in the brush, and electron transfer at the electrode. (Adapted from Matrab, T., et al., *ChemPhysChem*, 11, 670–682, 2010.)

polarization is ruled by the response of the polymer brush. The electrochemical time-dependent current response of the electrode depicts the overall steps involved in this process. For instance, the process presented in Figure 8.6a can be divided into four subsequent steps that correspond respectively to (1) the species diffusion in the solution with a diffusion coefficient D, (2) the partition of the species between both solution and polymer brush phase described for Red by the equilibrium constant P (Equation 8.6), (3) the diffusion of species within the polymer brush structure with a diffusion coefficient D_p, and (4) a possible heterogeneous electron transfer (electrochemical transformation) at the brush-electrode interface characterized by a heterogeneous electron transfer rate constant k_{et} (in cm s^{-1}).

$$Red_{sol} \rightleftharpoons Red_p \quad P = \frac{[Red]_p}{[Red]_{sol}} \tag{8.6}$$

The electrode current then describes all these possible phenomena of the polymer brush response and may account for possible structural changes associated with this electrochemical activation.

8.3.2 Transport Investigated by Chronoamperometry

The use of chronoamperometry to inspect the transport in a polymer brush grown from an electrode is illustrated in Figure 8.7. In this example, an electroactive species, such as ferricyanide ion ($Fe(CN)_6^{3-}$), is transported within a polyelectrolyte brush made of polycationic poly(2-methacryloyloxy)-ethyl-trimethyl-ammonium chloride (PMETAC), grafted by ATRP from an electrode. When the PMETAC brush coordinated with $Fe(CN)_6^{3-}$ is polarized

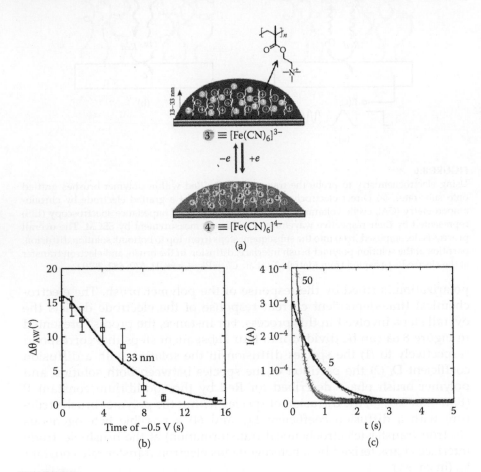

(a)

(b)

(c)

FIGURE 8.7
Reversible electrochemical switching of the surface energy of a PMETAC polycationic brush with $Fe(CN)_6^{4-/3-}$ ions. (a) Schematic principle of the surface energy switching. (b) Changes in the surface contact angle during the application of E = −0.5 V at the electrode corresponding to the transformation $Fe(CN)_6^{3-} \rightarrow Fe(CN)_6^{4-}$ in a 33 nm thick PMETAC brush in the presence of 5 mM KCl supporting electrolyte. (c) Chronoamperometry (changes in the oxidation current) at the electrode during the reverse transformation in a 21 nm thick brush: $Fe(CN)_6^{4-} \rightarrow Fe(CN)_6^{3-}$ by application of E = + 0.5 V in 50 or 5 mM KCl supporting electrolyte; symbols: experiments, lines: best fit to diffusion-controlled decays. (Adapted from Spruijt, E., et al., *Langmuir*, 24, 11253–11260, 2008.)

at a potential E_T more reducing (negative) than the reduction potential of the electroactive species E^0, the $[PMETAC^+, Fe(CN)_6^{3-}]$ brush is completely transformed into $[PMETAC^+, Fe(CN)_6^{4-}]$. Owing to the difference in hydrophobicity of both iron ions, one can easily and reversibly switch the material from a hydrophilic to a hydrophobic surface state (as illustrated in Figure 8.7a) upon simple change of the electrode potential (change of the oxidation state of the electroactive ion).[70] During the chronoamperometric measurement, while the electrode is polarized at the potential for the redox transformation, the continuous change in surface wetting properties (Figure 8.7b) is ascertained by water contact angle measurement. When the electrode is polarized at $E = -0.5$ V, it depicts the kinetics for the transformation of the electroactive species $Fe(CN)_6^{3-} \rightarrow Fe(CN)_6^{4-}$ coordinated in the PMETAC brush (Figure 8.7b). Moreover, the redox transformation results in a charge transfer at the electrode where a current associated with this transformation is measured. Generally, the phase transformation is constrained by transport limitations, such as the diffusion of (electroactive) species within the polymer brush. The time evolution of the electrode current presented in Figure 8.7c then reflects the kinetics of the transformation process, more particularly the reverse transformation $Fe(CN)_6^{4-} \rightarrow Fe(CN)_6^{3-}$ when $E = +0.5V$ is applied at the electrode. The current decay observed is paralleled to the contact angle variations, and both result from the limited diffusive transport of the electroactive ion within the film. The modeling of the time evolution of the electrode current (lines in Figure 8.7c) allows the estimate of the $Fe(CN)_6^{4-/3-}$ diffusion coefficient within the brush D_p. It depends strongly on the electrolyte solution concentration (KCl) and ranges from 10^{-12} to 10^{-10} cm^2 s^{-1} (compare to $D = 7 \ 10^{-6}$ cm^2 s^{-1} in solution) for 10^{-4} to $5 \ 10^{-2}$ M KCl solutions, respectively. The slower diffusion in brushes in contact with more dilute electrolyte shows the extent of screening by the electrolyte of the PMETAC cationic charges.

8.3.3 Transport Investigated by Cyclic Voltammetry

8.3.3.1 *Qualitative Approach*

Similar information about the transport of electroactive species in a polymer membrane is also accessible from cyclic voltammetric experiments. Cyclic voltammetry is the most figurative electrochemical technique, as it shows the evolution of the electrode current as the electrode potential is changed continuously; this is a considerable advantage at least for an easy qualitative interpretation of the phenomenon associated with the electrode potential change. For example,[72] the technique has been used to detect the collapse of polymer brushes due to the change of pH of an electrolytic solution containing the electroactive $Fe(CN)_6^{4-}$, as illustrated in Figure 8.8. When the polycationic brush of poly-4-vinylpyridine (P4VP) is held in acidic pH, the electroactive species is freely transported into the polycationic brush, and the cyclic voltammogram (Figure 8.8b) shows the diffusive transport of $Fe(CN)_6^{4-}$ in the brush toward the underlying indium tin oxide (ITO) surface, where it

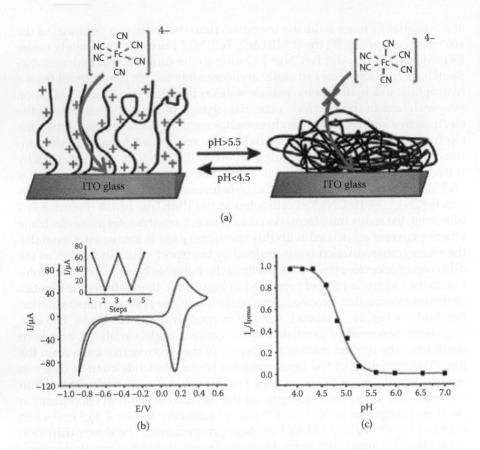

FIGURE 8.8
pH-controlled reversible switching of P4VP brush between a swelled and a collapse state activating or inhibiting the transport in the brush and oxidation at the electrode of Fe(CN)$_6^{4-}$. (a) Schematic principle (left: swelled or ON state; right: collapse or OFF state). (b) Cyclic voltammogram for the allowed transport of Fe(CN)$_6^{4-}$ in the P4VP brush (acidic pH); insert: peak current variation along repeated cycles of ON/OFF states. (c) Variations of the peak current with the solution pH showing the ON/OFF with the electrolytic solution pH. (Adapted from Tam, T.K., et al., *Langmuir*, 26, 4506–4513, 2010.)

is subsequently transformed. When the pH is increased (either by changing the electrolytic solution or by electrogenerating OH⁻ from electrochemical activation of the ITO electrode), the deprotonated brush collapses onto the electrode (Figure 8.8a, right). For pH > 5.5, the collapse is very efficient, as it prevents the transport of the electroactive species and a very low level of current is recorded at the electrode (Figure 8.8c).

8.3.3.2 Quantitative Analysis of Permeation Processes

The cyclic voltammogram can also be interpreted from a more quantitative point of view, and different reaction mechanisms have been deciphered by this powerful technique. In cyclic voltammetry, it is possible to modulate the frequency of observation of electrochemical processes by varying v, the rate at which the electrode potential is scanned. Modulating v allows for modulating the size of the region near the electrode that is explored by species diffusion. The diffusion length, δ, explored by cyclic voltammetry performed at the scan rate v is of the order of $\delta \approx (DRT/Fv)^{1/2}$, where D is the species diffusion coefficient, R the constant gas, T the temperature in K, and F the Faraday. Then, the faster the scan rate, v, the higher the frequency of observation, and therefore the thinner δ, the region in the vicinity of the electrode explored by species diffusion. Cyclic voltammetry has then been used for inspecting transport of electroactive species by permeation in thin layers immobilized on electrodes. Appropriate modeling[75–77] shows that the cyclic voltammetric response depends on the competition between four dimensionless parameters that characterize the permeability of the film, characterized by the product $(PD)_p$, the film thickness, e (compared to δ, the cyclic voltammetry characteristic diffusion length), the diffusion rate within the film, D_p, and finally the kinetics of the electron transfer to the electroactive species at the electrode, k_{et}. For fast electron transfer at the electrode, meaning when reversible outer-sphere electroactive species presenting fast electron transfer at the electrode are used (likely not $Fe(CN)_6^{4-/3-}$), the cyclic voltammetry response does not depend on the electron transfer step, k_{et}, but on the three other parameters.

Depending on the potential scan rate, v, three limiting situations may be observed:

1. At a slow scan rate, the diffusion length explored during the cyclic voltammetry is larger than the film thickness and the classical diffusive shape of the cyclic voltammogram is observed characteristic of solution diffusion. This situation is met, for example, for ferrocenemethanol (FcMeOH) transported in 12 nm thin brush of polyglycidylmethacrylate (PGMA), as shown in Figure 8.9a.

2. When the scan rate is increased and the characteristic diffusion length explored is comparable to the film thickness, the electrochemical response is governed by the permeation reaction (Equation 8.6) in the film and the cyclic voltammogram adopts a sigmoidal shape characteristic of steady mass-transfer-limited processes. This situation is also met at a fixed scan rate when exploring the transport of species in thicker layers, as illustrated by the sigmoidal shape of the cyclic voltammogram for FcMeOH transport in a 110 nm thick PGMA brush in Figure 8.9b.

3. When finally the scan rate is still increased such that the diffusion length is much smaller than the film thickness a diffusive shape is recovered, as the species diffusion within the film is now kinetically limiting, and the voltammogram analysis allows the estimate of the product $PD_p^{1/2}$.

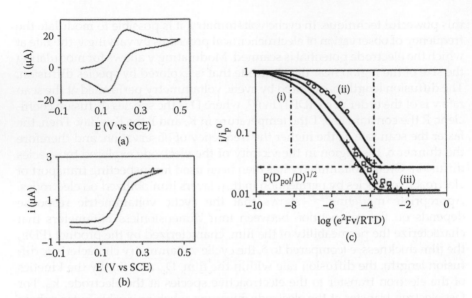

FIGURE 8.9
Cyclic voltammetric inspection of the permeation of electroactive species in thin layer immobilized on an electrode illustrated by the permeation of FcMeOH in PGMA brushes. Typical voltammograms of 3 mM FcMeOH solution at v = 20 mV s^{-1} on 0.03 cm^2 Au surfaces showing (a) permeation in a 12 nm thick brush controlled by diffusive transport (here diffusion in the solution phase) or (b) permeation in a 110 nm thick brush: steady-state transport controlled by the partition reaction at the brush-solution interface. (c) Interpretation of FcMeOH permeation into (from top to bottom) e = 12, 70, 110, and 170 nm thick PGMA brushes. Comparison with theoretical prediction of the different kinetic regimes observed for permeation in thin layers (transition from control by (i) solution diffusion to (ii) partition to (iii) diffusion in the layer). (Adapted from Matrab, T., et al., *ChemPhysChem*, 11, 670–682, 2010.)

The exploration of a wide range of diffusion lengths or scan rates, v, allows the complete characterization of species transport within polymer brushes by observing the variation of the peak or plateau current, i, for the electrode transformation of the electroactive species with the scan rate, v. This is illustrated in Figure 8.9c, which summarizes the cyclic voltammetric investigation of the transport of FcMeOH in PGMA brushes of different thicknesses. It shows that within 12 nm thin brushes the behavior observed goes from situation 1 to situation 2, allowing for the estimate of the brush permeability PD_p = 2 10^{-9} cm^2 s^{-1}. For thicker brushes (e > 70 nm) the behavior observed goes from situation 2 to situation 3, allowing for the estimate of the permeability PD_p = 6.6 10^{-10} cm^2 s^{-1} of FcMeOH and for the separation of the contribution of diffusive transport D_p = 2 10^{-10} cm^2 s^{-1} and of the partition equilibrium P = 3.3 for FcMeOH between the solution and the polymer brush phases. The values of the permeabilities or of the diffusion coefficient of FcMeOH within PGMA brushes are of the same order as those estimated by chronoamperometry for Fe(CN)$_6^{3-}$ in PMETAC polyelectrolyte brushes.

8.3.4 Transport Investigated by Electrochemical Impedance Spectroscopy

If insightful mechanistic information is obtained by modulating the frequency of actuation of a system (the potential of an electrode) and observing the response of the system (the current flowing through the electrode), a more precise and complex picture of the processes associated with electrode actuation may be obtained from electrochemical impedance spectroscopy. In this technique a sinusoidal modulation of amplitude, ΔE, and frequency, $w/2\pi$, of the electrode potential is applied around a given value, E_i ($E(t) = E_i + \Delta E \sin(wt)$). The electrode current, $i(t)$, and therefore the impedance of the system, defined as $Z = \Delta E/i$, are then measured. The impedance of the electrochemical process is a complex number whose real contribution, $Re(Z)$ (analogous to a resistance), and imaginary one, $Im(Z)$ (analogous to the inverse of a capacitance), can be easily obtained experimentally by detecting the electrode current in phase with $E(t)$ or with quarter-period phase retard relative to $E(t)$. The impedance spectroscopy consists in obtaining the evolution of the impedance $Z(w)$ over a wide range of frequencies (from fraction of Hz to MHz). The mechanistic characterization of any electrochemical process is based on its decomposition into elementary steps, which can be compared analogously to equivalent individual electronic components. Typically, a charge transfer process at an electrode is associated with a charge transfer resistance. Diffusive processes are associated with a so-called Warburg (Z_W) impedance component that corresponds to mixed frequency-dependent capacitance and resistance. More details are to be found in a general electrochemistry monograph.[64]

The equivalent electronic circuitry used to depict the physicochemical processes occurring in the electroactive species transport and electrochemical transformation at an electrode grafted with a polymer brush is given in Figure 8.10a. The theroretical response of such an equivalent circuit is known in terms of the frequency-dependent variations of $Re(Z)$ and $Im(Z)$. Typical examples of such variations are presented in Nyquist plots ($Im(Z)$ as a function of $Re(Z)$ for different frequencies) as in Figure 8.10b. These plots allow for the observation of $Fe(CN)_6^{4-}$ transport within poly2vinylpyridine (P2VP) brushes grafted onto ITO.[65] From these frequency-dependent variations of the imaginary part $Im(Z)$ vs. the real part $Re(Z)$ of the impedance one could extract the charge transfer resistance, R_{et}, identified as the diameter of the semicircle observed at high frequencies, or diffusive limitations detected by the diagonal line observed at the lowest frequencies. In the example presented in Figure 8.10, it is possible to identify, with such technique, the change in electron transfer rate at the electrode associated with the brush structural changes due to variations of the electrolytic solution pH (Figure 8.10c). It is then used to reveal the presence of nanostructured domains of segregated polymer brushes within mixed polymer brushes.[65] The technique has also been successfully devoted to the characterization of the transport properties affected by conformational changes in different polymer brushes.[66-67]

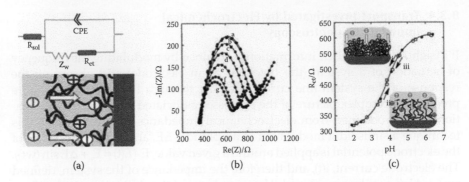

(a)   (b)   (c)

FIGURE 8.10
Faradaic impedance spectroscopy investigation of $Fe(CN)_6^{4-/3-}$ transport in P2VP brush-modified ITO. (a) Equivalent electronic circuit used to model physicochemical processes associated with the faradaic impedance spectra. (b) Typical impedance spectra (Nyquist plot), Im(Z) vs. Re(Z), for decreasing pH values (from a to g); line: best fit to the equivalent circuitry. (c) Titration curve showing changes of the electron transfer resistance derived from (b) and fitted by (a) upon variation of the pH going in the (i) acidic then (ii) basic and (iii) acidic directions. (Adapted from Motornov, M., et al., *ACS Nano*, 2, 41–52, 2008.)

8.3.5 Transport Investigated by Local Electrochemical Probes—SECM

As discussed in Section 8.2, the SECM is a pertinent tool for the characterization of charge or mass transfer processes at various interfaces and should allow the characterization of transport processes within polymer brushes. It has been already used to detect ion transport into various polymeric membranes deposited onto electrodes.[78–80]

8.3.5.1 Strategies

As sketched in Figure 8.8b, the SECM tip electrogenerates a redox species that is transported to the substrate through the polymeric film. The measurement of a redox cycling or feedback allows for the quantification of the permeation of the redox species within the membrane. Steady-state measurements such as those obtained from these approach curves, $i_T - d$, for example, presented in Figure 8.11, have been proposed. A complete theoretical analysis of the approach curve shape depending on the membrane characteristics (permeation and thickness) has been described.[81] For thin membrane layers of thickness e, the approach curves follow the first-order approximation, and when fitted by Equations 8.3 to 8.5, an effective heterogeneous rate constant $k_{eff} = PD_p/e$ is obtained from which the permeability of the electroactive species PD_p ensues.

This strategy has been used at artificial or natural membranes to highlight size-selective transport paths such as nanopores of a dimension comparable to that of the redox species.[82–85] It also allows for the electrochemical imaging of micrometric regions of inhomogeneous transport properties on a surface. These can be domains generated on purpose by lithographic techniques where the permeable membrane has been immobilized.[82] They can

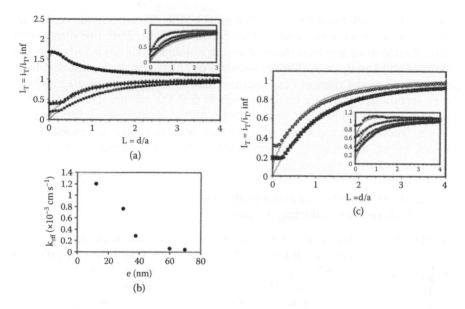

FIGURE 8.11

Permeation of electroactive species within PGMA and PS brushes grafted from Au surfaces as measured by SECM. SECM approach curves, i_T – d, obtained at a Pt disk UME tip (radius a = 12.5 μm) on (a) a 38 nm thick PGMA brush (■, ▲, ●) and (c) a 90 nm thick PS brush (□, Δ, O) grown by ATRP from Au substrates. Approach curves obtained in the presence of 5 mM of (□, ■) $Fe(CN)_6^{4-}$, (Δ, ▲) $Ru(NH_3)_6^{3+}$, or (x, +) FcMeOH in H_2O + 0.1 M KCl; or (O, ●) Fc in DMF + 0.1 M NBu_4BF_4. The solid lines are the theoretical curves for finite irreversible heterogeneous electron transfer. Inserts: SECM approach curves on brushes of increasing thicknesses from top to bottom for (a) aqueous FcMeOH on 12, 38, and 70 nm thick PGMA brushes; (b) Fc in DMF on 43, 60, 80, and 90 nm thick Au-1-PS films. (b) Variation of the effective first-order heterogeneous rate constant estimated from the approach curves with the brush thickness (aqueous FcMeOH on PGMA brushes). (Adapted from Matrab, T., et al., *ChemPhysChem*, 11, 670–682, 2010.)

also be regions of facilitated transport in a membrane, and this is exploited for the characterization of the heterogeneity of proton transport in fuel cell membranes.[86]

It is, however, difficult to decouple both partition and diffusion contributions from the single approach curves strategy, as with other electrochemical techniques. Nevertheless, decoupling the transport limiting regimes may be obtained from a combination of SECM and transient interrogation.[87] They could be obtained independently or concomitantly by performing at a given tip-substrate separation distance a cyclic voltammogram, a chronoamperometric or impedance measurement in the collection-generation or the feedback modes.

Sub-micrometric probes have also been handled, in the combined AFM-SECM technique, to probe diluted polyethylene glycol (PEG) chains grafted to a gold surface. When the free end of the PEG chains are lablled with a Fc moities, direct electron transfer between the AFM-SECM tip and the terminal Fc is deticated by a current onset. The tip current increases as the tip approaches the substrates. This feedback is attributed to dynamical elastic diffusion of few hundreds of Fc-terminated chain ends between the substrate and the tip. It is then possible to quantify the static and dynamic properties of monolayers of PEG chains.[88] The inverted configuration where the Fc-terminated polymer chain is anchored to the nanometer size tip has also been used as AFM-SECM tip. Such tip allows to probe both topography and electrochemical reactivity of a surface with nanometric resolution[89,90] without being affected by solution diffusion which hampers the SECM resolution.

8.3.5.2 SECM Inspection of Polymer Brushes Tethered to Conducting Surfaces

The SECM characterization of electroactive species transport in the polymeric membrane has been transposed to nanometer thin layers of polymer brushes grown by ATRP on gold electrodes.[52,91] Brushes of polystyrene (PS) and polyglycidylmethacrylate (PGMA) are grown from a nanometer-thin multilayer of the ATRP initiator obtained from the electrografting of the corresponding diazonium salt. This reaction, depicted in Figure 8.12, allows for the covalent anchoring of chemically diverse robust layers of aryl moieties through the electrochemical generation of radicals at different electrode materials.[92,93] However, the electrografting procedure hardly yields monolayers, but rather generates dense and disordered aryl multilayers that are generally several nanometers thick. When the transport of an electroactive species within the polymer brush to the underlying electrode is inspected by SECM, one must take into account the possible contribution of slow limiting transport through this polymerization initiator multilayer.

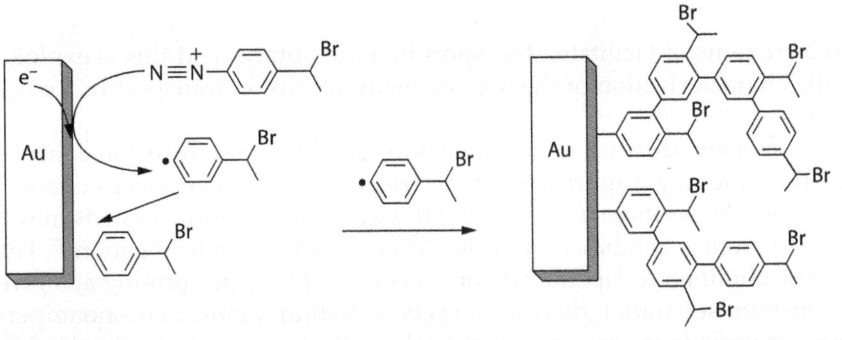

FIGURE 8.12
Immobilization of an ATRP initiator (bromoethylbenzene) by electrografting the corresponding diazonium salt on an (gold) electrode. The electrografting is generally performed by applying during 300 s at the electrode a potential 300 mV more negative than E_p, the peak potential for the reduction of the diazonium ($E_p \sim -0.1$ V vs. saturated calomel electrode SCE). During the course of the electrografting a multilayer is generated.

The transport of different electroactive species in different electrolytic solutions through PS and PGMA brushes has been described. The observed permeabilities corroborate the affinity of the different electroactive species for the different polymer brushes and the compared hydrophobicities of the brush and of the electrolytic solution.

The transport of aqueous FcMeOH in PGMA studied by SECM approach curves is illustrated in Figure 8.11a. It yields permeabilities in agreement with those estimated from cyclic voltammetry, which confirms the potentiality of the SECM approach for transport studies. It is extended to other redox active species or electrolytic solution (dimethylformamide (DMF)). For the highly charged ferro/ferricyanide ions $Fe(CN)_6^{4-/3-}$ the permeability cannot be estimated by SECM, as the approach curve is always representative of the negative feedback response, indicating permeabilities $<10^{-10}$ cm^2 s^{-1}. This is likely due to a poor affinity of these ions in the PGMA film (low value of P), but could also be related to limitation by slow electron transfer often observed with these inner-sphere ions.[94] For outer-sphere redox species such as $Ru(NH_3)_6^{3+}$ and FcMeOH in aqueous electrolytic solution, a significant feedback is measured by SECM, and the effective heterogeneous rate constant, k_{eff}, associated to this redox cycling is estimated from the approach curves. Moreover, as shown in the insert of Figure 8.11a and in Figure 8.11b, species permeation is likely the process under scrutiny as k_{eff} decreases when the brush thickness, e, increases. The measured permeabilities are alike and of the order of 2–5 10^{-10} cm^2 s^{-1} for both species and for the different thicknesses observed. For the thinnest 20 nm thick brushes, however, the process is likely limited by permeation within the electrografted initiator multilayer. When DMF electrolytic solutions are used, higher feedback is observed for the transport of ferrocene (Fc) in PGMA brushes. When measurable, the permeabilities are >100 times faster than in aqueous solution, most likely because of a higher transport ($D_p > 3$ 10^{-9} cm^2 s^{-1}) as a result of the easier transport of Fc within the PGMA brushes, which are better swelled and solvated by DMF than by water.

The transport of redox species in PS brushes has also been investigated, as shown in Figure 8.11c. Whatever their thicknesses, PS brushes are impermeable to any water-soluble redox species, even for FcMeOH, as the negative feedback approach curve is always detected. It demonstrates the efficient collapse of the PS brush when in contact with an aqueous electrolyte. The collapsed PS brush grafted on an electrode creates a very efficient barrier to charge transfer. The transport of species within PS brushes requires the good swelling of the brush by the electrolytic solution. This can be achieved, as for PGMA brushes, by using electroactive species dissolved in a better solvent of the polymer, such as DMF. It could also be achieved by chemically modifying the polymer chain so as to accommodate its swelling by the aqueous solution.

As with PGMA brushes, when the redox species are solvated by DMF, permeation of the PS brushes can be observed by SECM. This is illustrated in Figure 8.11c from the approach curve with the DMF solution of Fc. The feedback detected on the approach curves shows the facilitated transport of Fc in the DMF in PS brushes of different thicknesses (insert of Figure 8.11c) as a result of a good swelling of the chains by the DMF solvent. The permeabilities estimated are also comparable to those observed within PGMA brushes, as PD_p are of the order of 4–10 10^{-9} cm^2 s^{-1}. It indicates that DMF solutions induce similar stretching and swelling of PGMA and PS brushes.[52]

The chemical transformation of PS may be obtained by highly reactive (oxidizing) species. Highly oxidizing and reactive species may be generated at an electrode, where an electrochemical oxidation generates species like Ag^{2+}, NO_2°, or HO°,[47,51,91] or in a plasma. The electrochemical surface transformation resulted in the etching and oxygenation of hydrogenated polymers like PS.[95] The electrochemically assisted surface transformation processes were shown to transform the hydrophobic materials into a more hydrophilic one by introduction of oxygenated functionalities such as hydroxyl, carbonyl, and carboxylic functions in the polymer backbone or by the aromatic ring transformation into phenols or quinones. Such electrochemical treatment is transposed to thin (5 and 25 nm) PS brushes; the oxygenation is demonstrated from X-Ray photoelectron spectroscopy (XPS) and Fourier transform infrared spectroscopy (FT-IR) spectroscopies, but also from the permeation of the oxidized PS structures to electroactive $Fe(CN)_6^{4-/3-}$ or $Ag^{+/2+}$. The oxidized PS brushes easily transport these electroactive ions, and permeabilities of the order of 2 10^{-9} cm^2 s^{-1} are estimated by SECM, in agreement with the transformation of the PS brushes into more hydrophilic chains, which become better swelled by aqueous solutions.[91]

In summary, the SECM is a particularly well-adapted technique to characterize transport processes in polymer brushes. It may offer many other perspective potentialities. Indeed, the advantage of using the SECM as a mechanistic tool resides in its ability to inspect indirectly micrometric domains of the substrate without necessarily biasing it. This is interesting as the SECM inspection solicits the substrate without damaging it, or at least by only soliciting the substrate on a local scale.

Biasing the substrate while inspecting it by SECM might be interesting, as it allows observation indirectly: The impact of an electr(ochem)ical stimulation of the substrate and brush on the change of mass transfer of electroactive species in the brush is revealed from the current response at the nonperturbing tip. This has been illustrated in the investigation of self-assembled monolayers of alkylthiols on gold electrode to decouple electron transfer and mass-transfer-limited processes, like, respectively, the tunneling current through the self-assembled monolayer (SAM) and the current contribution from mass transfer within pinholes or defects present in the SAM.[96] This strategy might be of particular help for characterizing more finely the electrochemical behavior of the barrier formed by collapsed brushes.

8.3.5.3 SECM Inspection of Brushes Tethered to Insulating Surfaces—Transient SECM Analysis

On the other hand, as SECM characterization does not require the biasing of the investigated substrate; the SECM is the single electrochemical technique that permits the characterization of charge or mass transfer processes at polymeric layers immobilized on an insulating surface. This has been illustrated by the observation of the transport of electroactive anions, denoted A⁻ in Figure 8.13a, where A⁻ = ferro- or ferricyanide or iodide ions, during their extraction from nanometer-thin polycationic PMETAC brushes grafted from oxidized silicon surfaces.[74] The extraction of the electroactive ion, A⁻, from the PMETAC brushes is initiated by an external stimulus: the contact of the membrane with an electrolytic solution (water + NaCl), which initiates the exchange of A⁻ anions by Cl⁻ ones in the brush. The composition of the ions, A⁻, in the equilibrated polymer brushes (before and after the ion extraction process) is independently ascertained from spectroscopic XPS and FT-IR analysis. Because the substrate surface is not conducting, it is not possible to determine these absolute compositions of electroactive species A⁻ coordinated in the polymer brush by any electrochemical technique. However, the relative amount of A⁻ in the brush may be determined by SECM. The technique is actually relevant, as it detects the variation of A⁻ expulsed from the brush during the ion exchange process. Combined with the independent

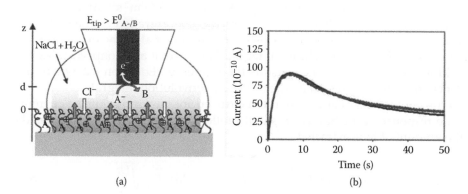

(a)   (b)

FIGURE 8.13
Principle of the transient observation at a SECM tip of the extraction of an electroactive anion, A⁻, from a PMETAC brush by ion exchange with Cl⁻ in the generation-collection mode. (a) Schematic principle of the experiment; the collecting tip is held in the vicinity of the substrate (an ATRP-grown brush from Si/SiO$_2$); the ion exchange in the brush is generated by the deposition of a NaCl droplet on the substrate. (b) Example of chronoamperometric $i_T - t$ curve obtained during the tip detection of Fe(CN)$_6^{4-}$ extraction from a 64 nm (5.3 10^{-8} mol cm^{-2} cationic unit) PMETAC brush by a 5 μm radius SECM tip held at 8 μm from the brush; theoretical black line for the extraction of 3.5 10^{-9} mol cm^{-2} ions limited by their diffusion within the brush with D$_p$ = 8 10^{-12} cm^2 s^{-1}. (Adapted from Combellas, C., et al., *Langmuir*, 25, 5360–5370, 2009; and Combellas, C., et al., *Polymer*, 45, 4669–4675, 2004.)

spectroscopic knowledge of the composition of A⁻ in the brush before the ion exchange process, the SECM is the most pertinent technique for the real-time follow-up of the brush composition during the ion exchange process. To do so, as depicted in Figure 8.13a, the collecting SECM tip is held at micrometric distances, d, from the generating polymer brush. It is biased at a potential that allows the mass-transfer-limited detection of the electroactive ion, A⁻, coordinated by the brush. When a drop of electrolytic solution is deposited on the membrane, the ion exchange between Cl⁻ from the solution and the electroactive ions in the brush proceeds. The electroactive ions A⁻ extracted from the brush diffuse in the solution and reach the tip, where they are detected by the onset of a current. In this substrate (brush) generation–tip collection mode, the noninterfering tip probes the transport of the ions within the brush, and the time-dependent current evolution reflects the kinetics of the extraction process, but convoluted with the dilution of the ion by diffusive transport in the solution. This dilution explains the presence of a peak in the tip current response (Figure 8.13b). It is, however, possible to extract pertinent information about the extraction process at the polymer brush from modeling it.[97] The model proposed assumes the substrate surface is an infinite plane simplifying the transport to a linear 1D diffusion process. Within the thin slab of the polymer brush of thickness e, the ions are transported by a slow diffusion process with a diffusion coefficient D_p; the ions are then ejected (replaced by Cl⁻ anions from the electrolyte) at the brush-solution interface and then diffuse quickly ($D = 7 \cdot 10^{-6}$ cm² s⁻¹) in the solution phase toward the tip. The expression of the concentration of the expulsed ions at any distance z from the brush and any time t after the extraction has begun is given by a convolution integral. The current at a nonperturbing tip held at distance d from the brush is then given by Equation 8.2, where the concentration is expressed by the modeled convolution integral.

One could show that the charge measured by integration of the current-time response at the SECM tip should follow Equation 8.7, and then yields the amount of ions extracted:

$$q(t) = 4nFDa \int_0^t C(d,\tau)d\tau \xrightarrow[t \to \infty]{} 8nFa\Gamma_{A^-}\sqrt{\frac{Dt}{\pi}} \qquad (8.7)$$

where Γ_{A^-} corresponds to the amount of coordinated ions A⁻ in the polymer brush per unit area of the substrate surface. It is found that a 66 nm thick PMETAC brush is coordinated by $3.8 \cdot 10^{15}$ Fe(CN)$_6^{4-}$ ions per cm² (or $5.7 \cdot 10^{13}$ ions per cm² per nm of brush). This result fits with the estimate made by cyclic voltammetry on PMETAC grafted from electrodes,[70] where $5.8 \cdot 10^{14}$ of the same ions were detected per cm² in a 14 nm brush ($4.1 \cdot 10^{13}$ ions per cm² per nm of brush).

Moreover, the fit of the tip current response with the model (Figure 8.13b, black line) also allows the extraction of the time constant of the kinetically limiting process. It is associated with a first-order rate constant of the order 0.2–1 s⁻¹ for brush thicknesses ranging from 64 to 32 nm. It characterizes the slow transport of $Fe(CN)_6^{4-/3-}$ ions within the film of PMETAC brush with an apparent diffusion coefficient D_p ranging between 0.6 and 2 10^{-11} cm^2 s^{-1}. These values are smaller than or comparable to those observed on PMETAC brushes on electrodes[70] but indicate the difference in brush densities.

8.4 Observing and Tuning Reactivity with SECM and Polymer Brushes

8.4.1 Context

Controlling and knowing the chemical reactivity of surfaces and interfaces is important for the design, by bottom-up approach, of chemically nanostructured architectures. The design of a micropatterned polymer structure is actually an important issue to develop smart surfaces, actuators, polymer electronics, or biological microsystems. In this respect, different lithographic techniques have been used for the micro- and nanofabrication of patterned polymer brushes, including electron beam lithography,[98–100] photolithography,[101–105] microcontact printing,[106–108] nanoimprinting,[109,110] Langmuir–Blodgett,[111] or lithographies based on the use of scanning probe microscopies, such as dip-pen[112] lithographies or SECM.[52–54] Recent progresses in the fabrication of nanostructured surfaces with polymer brushes, through different lithographic strategies have been reviewed.[55]

Dense polymer structures with well-controlled compositions and dimensions are preferentially grown from a wide range of surfaces by atom transfer radical polymerization (ATRP).[4–9] Different bromo-terminated molecules ranging from silanes,[4,5] thiols,[104,113–116] or disulfides to aromatic moieties derived from diazonium ions[117–119] have been proposed to initiate the polymer growth.

The combination of surface patterning and polymer brushes also allows an indirect access to chemical reactivity of a surface or interface. In this respect, scanning probe microscopies present indeed an interesting alternative to photolithographic techniques to tune by chemical means the local reactivity of surfaces. However, compared to standard lithographic techniques, the main default of the scanning probe microscopies as lithographic tools relies on the inherent slow writing speed of such pen-writing processes. This default may be circumvented by using, as in microcontact printing, stamp probes made of multiple paralleled probe arrays or of probes having the shape of the pattern that has to be imprinted on the surface. This strategy has been successfully applied from the micrometer to nanometer scale. The scanning electrochemical microscopies are able to handle nanoprobe arrays[120] or band[121] probes for efficient high-throughput surface patterning. At the nanometer scale, the use of multiple parallel arrays of atomic force microscopy (AFM) nanoprobes has also allowed high-throughput paralleled dip-pen nanolithography of surfaces.[122,123]

This section details the uses of patterning strategies, more particularly SECM based, in the context of the controlled growth of polymer brushes. First, a controlled polymerization reaction is used to get insight into the chemical reactivity of organic interfaces. For this purpose the ATRP-controlled polymerization is used to chemically amplify[98,106] the reactivity of molecular moieties immobilized on a surface. Then, a second part explores the SECM strategies proposed to pattern surfaces with polymer brushes.

8.4.2 Polymer Brushes a Mechanistic Tool for the Interrogation of Surface Reactivity

8.4.2.1 Principle

In this approach, a controlled polymerization reaction is used to amplify chemical defects on a surface, and therefore to access indirectly the chemical reactivity of a functional surface. This strategy has been used to interrogate the reactivity of thin assembled layers of different ATRP polymerization initiators. The principle relies on the ability of the selected experimental conditions of ATRP polymerization, to grow polymer brushes of thicknesses proportional to the polymer molecular mass, M_n. Then, for surfaces presenting different densities of polymerization initiator, the growth of polymer brushes of similar mass, M_n, yields polymer thickness, e, proportional to the polymer brush grafting density, σ_p, according to Equation 8.8:[103]

$$\sigma_p = \frac{e}{M_n} d_p N_A \qquad (8.8)$$

where d_p is the density of the polymer and N_A the Avogadro number.

When the thickness of the polymer brushes grown by ATRP also increases proportionally with the initiator surface coverage, Equation 8.8 ensures that the local estimate of the polymer brush thickness is an indirect measurement of the local initiator surface coverage. Such condition is generally demonstrated independently by observing the extent of polymer brush growth from surfaces where the anchored polymerization initiator has been diluted by mixing it with molecules unreactive toward ATRP.[112–116,124] This was also confirmed from the controlled ATRP growth of polymer brushes from substrate where the surface-anchored initiator has been diluted by partial photochemical decomposition[104,125] or electrochemical desorption[126] or decomposition[52–54] of the initiator moieties.

This property of ATRP reaction is particularly appealing, as it gives access indirectly to the polymerization initiator density on a surface. It therefore allows a precise detection of defects sites on a surface (such as presence of polymerization initiator). Then, when photochemical or electrochemical reactions are used to transform chemically and then dilute the initiator seed layer, the polymer brush structure grown on the etched surfaces gives access to the fraction of initiator that has not been transformed during the photochemi-cal or electrochemical activation. This provides indirect access to the electro-chemical or photochemical reactivity of surface-anchored organic moieties.

In this context, the chemical reactivity of ATRP initiators such as bromo-terminated moieties has been illustrated from this chemical amplification scheme of defects by ATRP. The chemical reactivity under investigation corresponds to the reductive debromination of the C-Br bond in bromo-terminated initiators of ATRP immobilized on different surfaces. In organic solvents, the reductive debromination of bromo-derivatives (RBr) has been extensively investigated by homogeneous means (via an electron donor).[127-129] It is also of peculiar interest in the comprehension of ATRP processes as its reaction mechanism is based on an electron transfer reaction initiated and mediated by inner-sphere coordination complexes, like a Cu(I) complex.[7,11]

When an outer-sphere aromatic electron donor, Red in Equation 8.9, is used to mediate homogeneously the RBr reduction, an exchange of two electrons and one proton yields the corresponding debrominated and hydrogenated molecule, RH.

$$RBr + 2\,Red + H^+ \rightarrow RH + 2\,Br^- + 2\,Ox \qquad (8.9)$$

The electron donors able to drive Equation 8.9 are, for example, anion radi-cals, Red = M^-, of aromatic moieties, M, such as the anion radical of naphtha-lene, and can be generated at various electrodes in different organic solvents (DMF, acetonitrile, tetrahydrofuran). The knowledge of the homogeneous reduction kinetics and thermodynamics of bromoalkanes allows one to pre-dict that a surface-anchored bromoalkyl chain might be reduced and there-fore debrominated by reducing radical anions such as that of 2,2′-dipyridyl (E^0 = −2.10 V vs. SCE).

As illustrated in Figure 8.14, this radical anion, Red = M^-, was generated at a microelectrode, radius a = 5 to 25 μm, held in the close vicinity of the substrate surface where a bromo-terminated ATRP initiator has been anchored. Because of the small separation distance between the SECM tip and the substrate, the

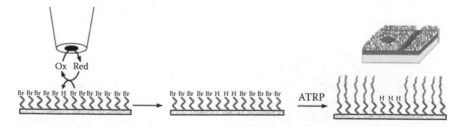

FIGURE 8.14
Schematic principle of the SECM etching of surface-grafted ATRP initiators for surface pat-terning with polymer brushes (or negative transfer of polymer brush on a surface) and local interrogation by chemical amplification of the reactivity of surface-immobilized initiators. (Adapted from Slim, C., et al., *Chem. Mater.*, 20, 6677–6685, 2008.)

electrogenerated reducer M⁻ may reduce locally the substrate surface and debrominate it locally. The tip then acts as a local micrometric eraser of the ATRP initiator seed layer. This micrometric source of electrogenerated etchant is used to erase different ATRP initiators grafted on different substrates. As presented in Figure 8.15, they are bromoisobutyrate alkylchlorosilane covalently grafted as a SAM on Si/SiO₂ or glass substrates,[54] brominated boron-doped diamond,[130] bromoethylbenzene electrografted multilayers obtained from the reduction of the corresponding diazonium at a gold surface (Figure 8.12),[53] or bromo-terminated macromolecular PGMA or PS brushes obtained from the ATRP growth from a Br-terminated initiator on Si/SiO₂ or Au surfaces.[52]

Once locally reduced, the patterned surfaces are submitted to ATRP polymerization of different monomers such as dimethylaminoethylmethacrylate (DMAEMA), styrene, or glycicylmethacrylate (GMA). The controlled poly-merization then chemically amplifies the patterns formed by the SECM tip eraser. The patterns formed on the surface are qualitatively detected by ex situ observation of the preferred condensation of water vapor on the different regions. Figure 8.16a shows such a condensation figure on a patterned and polymerized Si surface; clearly the regions within and outside the patterns have contrasted hydrophobicity and patterns can be efficiently revealed by this technique. The

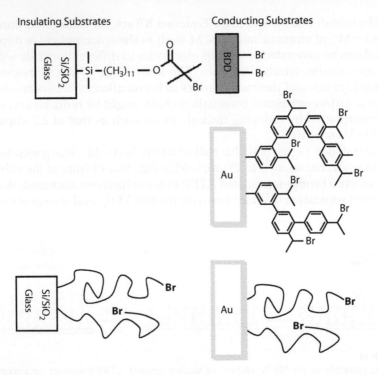

FIGURE 8.15
Nature of the substrates and initiator (mono- to macromolecular from top to bottom of the list) layers used, as in Figure 8.14, for the SECM patterning of surfaces with polymer brushes.

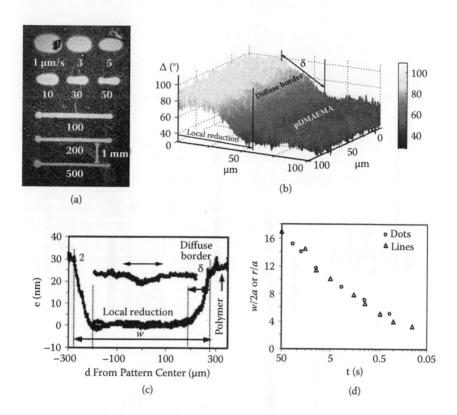

FIGURE 8.16 (See color insert.)
Patterns of PDMAEMA brushes formed by SECM from local erasing of initiator and ATRP from Si/SiO$_2$. (a) Condensation figure showing the patterns formed at different SECM tip writing speed then revealed by ATRP, as in Figure 8.14. (b,c) Local ellipsometry of polymer brush patterns: (b) image of the phase ellipsometric parameter Δ; (c) once reconstructed, Δ gives the evolution of the brush thickness along a pattern made by the tip electrogenerating (1) the 2,2'-dipyridyl ($E^0 = -2.1$ V vs. SCE) or (2) the nitrobenzene ($E^0 = -1.08$ V vs. SCE) anion radicals. (d) Evolution of the dimension of the patterns (width of lines or dots) with the etching time for the kinetic analysis of the ATRP initiator debromination (Equation 8.9). (Adapted from Slim, C., et al., *Chem. Mater.*, 20, 6677–6685, 2008.)

quantification of polymer thickness grown within and outside each pattern is characterized ex situ by different optical microscopic imagery of the surfaces (ellipsometry, interferometry). Such images give qualitative information about the amount of initiator nonchemically etched by the SECM eraser within the pattern. Figure 8.17 summarizes the results obtained in these works.

8.4.2.2 Chemical Reactivity of a Self-Assembled Monolayer

When monolayers of the polymerization initiator are patterned by the tip of the SECM electrogenerating the 2,2'-dipyridyl radical anion, a complete transformation of the initiator is observed, as no polymer grew within the

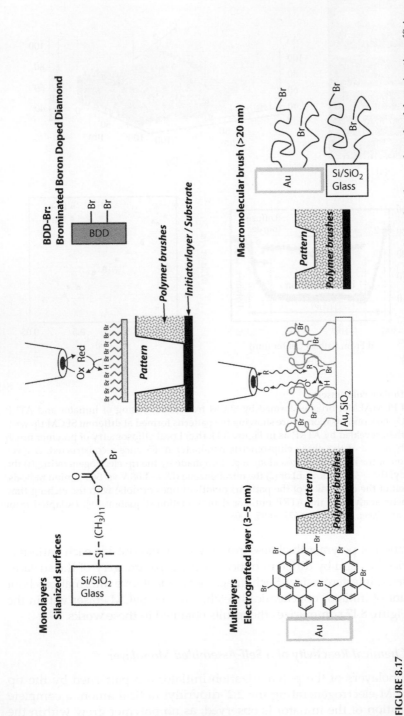

FIGURE 8.17
Schematic illustration of the influence of initiator layer structure on the SECM patterning with polymer brushes and on the debromination efficiency revealed by the ATRP chemical amplification of unreacted initiators.

pattern. This is demonstrated, for example, in Figure 8.16b and c from the local ellipsometric estimate of the polymer thickness profile along the pattern; if 31 nm is grown outside the pattern, no polymer is grafted in the pattern center. The borders of the pattern are regions of gradual change in polymer thickness and therefore of initiator density. The distribution of the unreacted initiator at the substrate surface is paralleled to the distribution of the etchant, Red generated from the tip, in the vicinity of the substrate: the lower the etchant concentration, the lower the amount of initiator etched and, from Equation 8.8, the higher the polymer brush thickness. Then, the kinetic of the transformation of the surface-anchored initiator may be obtained from the evolution of the pattern dimension with the characteristic time of the etching process, as presented in Figure 8.16d. The time evolution of the pattern may be modeled,[57] and the fit of the experimental variations obtained in Figure 8.16d allows an estimate of the initiator reactivity from the rate constant of Equation 8.9, which is about 10^4 s^{-1}.

Changing the etchant reducing strength allows for changing the driving force of the etching reaction (Equation 8.9). Figures 8.16c (trace 2) and 8.18 show the effect of the driving force of Equation 8.9 on the polymer brush grafting density within the pattern. Indeed, for a 1 V less reducing etchant (anion radical of nitrobenzene) the initiator surface can still be patterned, but after ATRP polymerization of DMAEMA, a 18 nm thick polymer brush was grown within the pattern, closer to the 23 nm brush grown outside the pattern. The ATRP growth of polyDMAEMA, PDMAEMA brushes was conducted independently on Si surfaces where the initiator was diluted with alkyl chains that are unreactive toward ATRP. It was then shown that the polymer brush density varies proportionally to the initiator surface coverage.[124] The behavior observed in Figures 8.16c and 8.18 then quantitatively demonstrates the influence of the etching reducing strength on the initiator

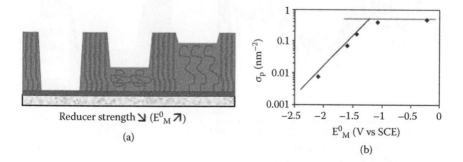

Reducer strength ↘ (E^0_M ↗)

(a)

σ_p (nm^{-2})

E^0_M (V vs SCE)

(b)

FIGURE 8.18

Modulation of the driving force of the debromination efficiency. (a) Schematic representation of the effect of the driving force of Equation 8.9 on the polymer brush conformation within and outside patterns. (b) Experimental variations of the polymer brush grafting density, σ_p, in the patterns, with E^0_M the standard reduction potential of the reducing etchant (driving force of the transformation). (Adapted from Slim, C., et al., *Chem. Mater.*, 20, 6677–6685, 2008.)

transformation efficiency. From Equation 8.8, the polymer brush density can be estimated along the patterned surface from the ellipsometric profile observation of Figure 8.16c (trace 2). It is then suggested that the efficiency of the initiator etching with the anion radical of nitrobenzene is $\rho = 1 - 18/23 = 22\%$. Figure 8.18b summarizes the structure-activity relationship depicting the efficiency of the initiator transformation by different reducing etchants and for an etching time of 1 s. The more negative the E^0 of the electrogenerated etchant, the higher the driving force of the etching reaction and the higher the polymer brush grafting density, or the more complete the transformation of the initiator.

This work is promising and should be valuable for the characterization of the kinetics of surface-initiated polymer brush growth. Indeed, the proposed method could help measure and predict the reactivity of surface-anchored ATRP initiator versus the inner-sphere electron donor coordinating complex generally used in the ATRP reactions. It could also be extended to other controlled polymerization reactions using other initiator functionalities and activation.

Changing the initiator and substrate nature while keeping the reducing etchant, Red, allows for the comparison of the reactivity of initiator-substrate couples. Brominated boron-doped diamond surfaces are also completely etched by the anion radical of 2,2'-dipyridyl.

8.4.2.3 Chemical Reactivity in Thicker and Disordered Layers

When brominated initiators are immobilized as multilayers the transformation efficiency, ρ is lower (Figures 8.17 and 8.19). It is demonstrated from the

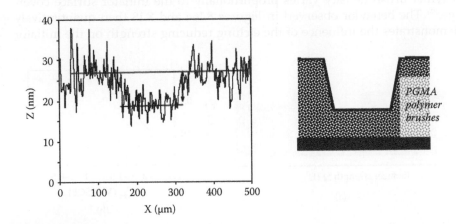

FIGURE 8.19
Interferometric and schematic profile of a diazonium-based ATRP initiator layer on Au surface patterned by SECM. The pattern amplified by ATRP growth of PGMA reveals that unetched ATRP initiators are left within the pattern, and the comparison of polymer brush thickness within (17 nm) and outside (27 nm) the patterns allows the estimate of the ATRP initiator etching efficiency, $\rho = 37\%$. (Adapted from Hauquier, F., et al., *Electrochim. Acta*, 54, 5127–5136, 2009.)

experimental profile of the polymer brush thickness along a pattern formed on 3 nm thick multilayers of electrografted initiator on Au surfaces. Within the pattern, a 17 nm thick PGMA brush was grown, while outside the pattern it was a homogeneous brush 27 nm thick. From these thickness ratios and Equation 8.8, it is shown that 63% of the initiator layer has not been etched by the erasing tip. They correspond to initiator moieties that are buried within the multilayer and are inaccessible to the etchant because of its low permeability during the etching reaction (few seconds). These unetched initiator sites are, however, still accessible for the ATRP reactants mainly because of the much longer reaction time used for the ATRP polymerization (several hours).

The importance of slow transport and permeability of the etchant on the reactivity of the initiator layer is confirmed when studying thicker initiator layers made from (>20 nm) macromolecular brushes. In this case (Figure 8.17), the etching efficiency is comparable to or lower than that for the diazonium-based multilayer. It depends on the nature of the initiator macromolecular brush; for the PS macroinitiating layer the etching efficiency is around ($\rho = 25\%$), which is lower than for the PGMA macroinitiating layer ($\rho = 40\%$). These values corroborate the study of the transport of electroactive species within the whole same polymer brushes, which shows that even in DMF the transport of electroactive species during the course of the etching process (<5 s) in the deepest regions of the brush is limited, even though PGMA brushes are slightly more permeable by DMF solution than PS ones. The partial chain-end etching rate and also its correlation with permeation analysis indicate that the bromo-terminated chain ends are distributed throughout the whole polymer brush, as predicted by theoretical and numerical considerations[131–133] and as demonstrated by neutron reflectivity experiments.[134,135] The comparable etching efficiency of diazonium-based multilayers of initiator and of PS macroinitiator also suggests that both layers are swelled similarly by DMF solution; this confirms the importance of solvent swelling in the electrochemical behavior of diazonium-based multilayers.

8.4.3 Patterning Surfaces with Polymer Brushes by SECM

The different SECM investigations of surface reactivity also demonstrate the possibility of soft lithography strategies based on SECM for the patterning of surfaces with polymer brushes. The SECM may be used to transfer polymer brush structures on different surfaces grafted with layers of polymer brush initiators. This may be achieved by positive or negative transfer of the polymer brush on the initiating surfaces as described, respectively, in Figures 8.20 and 8.14.

8.4.3.1 Positive Transfer of Polymer Brush Patterns by SECM

The ideal lithographic strategy by positive transfer would be to use the SECM tip to drive the ATRP process (Figure 8.20). This is indeed possible

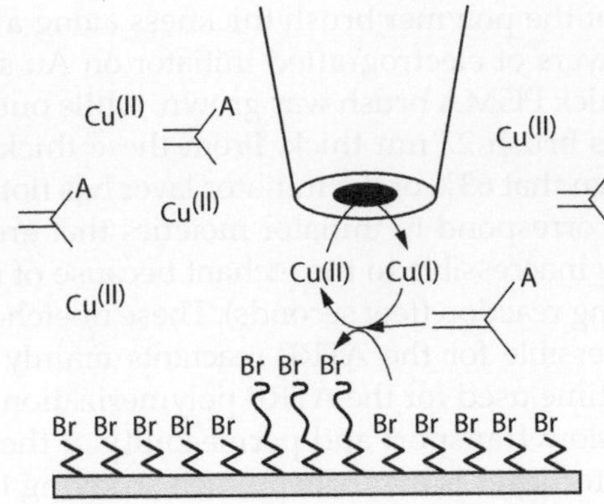

FIGURE 8.20
Schematic concept of positive transfer of polymer brush patterns on ATRP-initiating surfaces by SECM. The strategy detailed in Figure 8.14 corresponds to a negative transfer of polymer brush patterns.

as the ATRP mechanism is based on an electron transfer reaction between the polymer initiator and an electron-donor (reducer) coordinating complex, for example, a Cu(I) complex. One could then hold a SECM tip in the vicinity of an initiator-immobilized surface and use the tip to electrogenerate a Cu(I) reducing species from a Cu(II) solution containing the ingredients for ATRP polymerization. As when ATRP is conducted in the presence of air,[136] the SECM tip is used to control the ratio of propagating/inhibiting chain growth species, the Cu(I)/Cu(II) ratio. This is achieved with GMA as a monomer since the polymerization solvent may be DMF; it allowed the growth of a 5 nm PGMA brush onto a 100 µm radius domain of a Si surface where a Br-terminated silane was self-assembled. Such lithographic strategy works in principle but is not reasonable, as long polymerization times (>3 h) are needed to drive the local ATRP reaction, or for polymerization reactions requiring nondissociating solvents such as those preconized for the growth of PS brushes.

Faster patterning conditions might be obtained with more efficient ATRP coordinating complexes, which should allow faster growths, but also are likely less controlled (higher polydispersity).

Other positive transfer processes could be envisioned from the extension to the local micrometer range of electrochemical routes described in the literature to decorate surfaces with organic moieties. For example, an electrochemical strategy was recently proposed by Advincula's group to anchor an initiator of reversible addition-fragmentation chan transfer polymerization (RAFT) polymerization of a polymer brush on an electrode from the oxidative immobilization of a carabazole-derivatized initiator.[137] This strategy should be easily transposable to surface patterning by using the direct mode of the SECM (Figure 8.5a): the same electrochemical reaction would be performed on the substrate while the SECM tip would be used as a counterelectrode (a cathode), which should allow for the localization of the initiator immobilization on a micrometric domain of the substrate facing the tip.

Similarly, the direct mode of the SECM allowed the local electrografting of a gold substrate with organic moieties by the local generation of radical from an iodonium salt (an onium salt whose reduction behaves similarly to that of diazonium salts; Figure 8.12).[60] Its extension to diazonium salts and to bromoethylbenzenediazonium should allow the patterning of an electrode surface with an ATRP initiator for subsequent positive transfer of patterns of polymer brushes.

8.4.3.2 Negative Transfer of Polymer Brush Patterns by SECM

On the other hand, negative pattern transfers could be envisioned by SECM lithography. In such negative pattern transfers presented in Figure 8.14 and the corresponding section, the SECM tip is used in the feedback mode of the SECM (Figure 8.5b) to erase the immobilized initiator layer. This erasing reaction might be sufficiently fast to allow very efficient and rapid patterning of the initiator layer, as writing speeds as fast as 1 mm/s may be used. Once the initiator surfaces have been patterned, they are submitted to ATRP, the limiting process of the global patterning procedure. The polymerization then reveals micrometric patterns of silanized Si domains surrounded by wide areas of ATRP-grown polymer brushes, as schematized in Figure 8.14 and observed in Figure 8.16. The kinetic analysis of the initiator reactivity or of the pattern evolution with the etching time (Figure 8.16d) allows the precise control of the etching process, and the fast writing speed available permits the fast imprint (<1 min) of positive patterns (micrometric brush domains surrounded by mm^2 silanized Si surfaces).[54] The transposition to nanometer resolution patterning has not been detailed, but it is in principle accessible from the use of different SECM[120] or AFM-SECM[138] nanoprobes.

The procedure was also applied successfully to the patterning of brominated boron-doped diamond. The transposition to such a conducting surface allows the generation of patterns having the resolution of the SECM tip.[130] The strategy is particularly interesting for boron-doped diamond surfaces, which present high roughness (roughness > 2 μm) and are then difficult to pattern, with conventional lithographic procedures preconizing the use of masks or stamps.

From gold surfaces, the same negative transfer lithography is applied when the ATRP initiator layer is obtained from the electrografting of a diazonium salt. As such, the electrografting procedure yields dense multilayers, organic species (etchants) are difficult to transport in these structures, and inaccessible initiators are not etched by the tip-generated etchant. The patterned substrate submitted to ATRP polymerization then yields a pattern of PGMA or PS with different brush thickness (smaller thickness within the pattern)[53] from the local control of the initiator density (Figures 8.17 and 8.19). The complete erasing of the initiating properties within the pattern is also possible for more drastic etching (longer time, higher driving force). In principle, this strategy not only holds to initiator layers obtained from the electrografting of diazonium salts, but could also be applied to SAM of thiols. Actually, a SECM tip has already been used to drive the local reductive desorption of a thiol from micrometric domains of a gold surface.[49] As the local desorption (at the millimeter scale) of SAMs of thiolated ATRP initiator allowed the growth of polymer brush from gold surface with controlled thicknesses,[126] the localization of such thiolated-initiated polymer brush growth should be easily transposed to the micrometer or submicrometer scale by negative pattern transfer SECM-based lithography.

Such lithographic procedure allows for the precise control of the initiator grafting density at the micrometer scale and allows, in principle, the patterning of surfaces with domains presenting controlled polymer brush grafting densities and therefore controlled thicknesses. It is in principle transposable to recently reported ways to electrochemically mediated ATRP polymerization.[12]

The same negative transfer lithographic strategy was also used to pattern surfaces with diblock copolymers A-B when a primer layer of a brush of polymer A grafted from a surface is used as a macroinitiator of the subsequent growth of brushes of a polymer B.[52] Indeed, in principle the proposed SECM-based strategy was used to erase, in part, the macroinitiator layer of brushes of PS or PGMA. The patterned macroinitiator surfaces are then submitted to the ATRP polymerization of GMA or styrene, respectively, to form patterns of brushes of diblocks of PS-PGMA or PGMA-PS on silicon, glass, or gold surfaces (Figure 8.17). However, the macroinitiator debromination is far from complete under the SECM lithographic conditions proposed as a result of the poor permeability of the tip-generated etchant in the macroinitiator primer A. The lithographic procedure then forms surfaces of diblock copolymer brushes A-B with micrometric domains of a less densely grafted polymer chain of B.

8.4.3.3 Local (Electro-)Chemical Functionalization of Polymer Brushes

Finally, it is also possible to use the SECM to etch locally a polymer brush structure grafted from a substrate. This was illustrated by the use of the strong Ag^{2+} oxidant electrogenerated at the SECM tip to etch micrometric domains of a PS brush immobilized on a gold surface. The local etching process allows the local oxidation and oxygenation of the PS chains (chain fragmentation, insertion of hydroxyl, carbonyl, carboxylic functions, oxygenation of the aromatic rings) as depicted in Figure 8.21. It generates micrometric domains of hydrophilic regions surrounded by hydrophobic PS brushes. Such patterned surfaces presenting regions of contrasted hydrophobicity

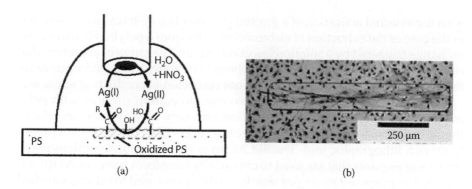

FIGURE 8.21

SECM lithography of PS brushes. (a) Schematic representation of the local oxidative patterning of PS polymer brushes by a SECM tip electrogenerating Ag^{2+} ions in 3M aqueous HNO_3. (b) Selective sorting of myofibroblastic and macrophage cells of heterogeneous population of peritoneal cell from their selective adhesion, respectively, onto oxidized PS pattern and on pristine PS. (From Ktari, N., et al., *Langmuir*, 26, 17348–17356, 2010.)

and chemical functionality also present contrasted affinities for living cells and can be used to discriminate cellular adhesion.[47]

8.4 Outlook and Conclusion

The SECM is a new powerful tool valuable for a wide range of local interrogation, characterization, and chemical or electrochemical transformations at very diverse interfaces. It is then a very promising tool for the inspection, imaging, or patterning of polymer brushes. It has been recently used to decipher precisely the processes associated with the transport or chemical reactivity of electroactive species within polymer brushes grown under ATRP conditions from different insulating or conducting surfaces.

The main advantage of the SECM characterization resides from its ability to interrogate surface properties with a local probe, a tip, which may not perturb the overall surface properties. Different SECM modes and their theoretical descriptions allow a fine mechanistic inspection of mass or charge transfer processes. This was used to observe the permeation of different redox species within polymer brushes. In principle, the strategy depicted in this chapter is applicable to a wide range of chemical species, gases (such as O_2), and also various nonelectroactive ions (K^+), from the use of micropipettes or from the recent development of scanning ion conductance microscopy. The use of a local chemical probe is also particularly interesting, as it allows the in situ and real-time observation of the ion transport associated with the external activation of a grafted polymer brush. It was demonstrated in the case of the extraction of electroactive ions from a polymer brush by ion exchange triggered by a solution injection. Again, the range of actuation that could be followed by a local SECM probe is very wide, and most importantly for an electrochemical technique, it is not restricted to conducting surfaces.

Finally, local SECM probes have been used to pattern surfaces with polymer brushes with micrometer resolution and very high writing speeds. In this section also few examples have been described, but the SECM is a potentially rich lithographic tool. The SECM tip is used to generate micrometric sources of etchants that are used to chemically transform layers of ATRP initiators. The proposed strategies may be easily transposed to other controlled polymerization procedures, as the chemical diversity that can be generated at an electrode, and therefore at a SECM tip, is very large. It is also shown how the lithographic mode of the SECM combined with controlled growth of polymer brushes can elegantly amplify chemical defects to quantify the (photo- electro- and, more generally) chemical reactivity of surface-anchored organic moieties. Usually the electrochemical reactivity of surface-anchored moieties was restricted to electroactive species, which present a reversible and well-defined electrochemical signature; the proposed lithographic method should present a promising alternative route to the chemical reactivity of nonelectrochemically addressable functional surfaces.

References

1. De Gennes, P.G. 1980. Conformations of polymers attached to an interface. *Macromolecules* 13: 1069–1075.
2. Welch, M., Rastogi, A., and Ober, C. 2011. Polymer brushes for electrochemical biosensors. *Soft Matter* 7: 297–302.
3. Senaratne, W., Andruzzi, L., and Ober, C.K. 2005. Self-assembled monolayers and polymer brushes in biotechnology: current applications and future perspectives. *Biomacromolecules* 6: 2427–2448.
4. Matyjaszewski, K., Miller, P.J., Suhkla, N., et al. 1999. Polymers at interfaces: Using atom transfer radical polymerization in the controlled growth of homopolymers and block copolymers from silicon surfaces in the absence of untethered sacrificial initiator. *Macromolecules* 32: 8716–8724.
5. Husseman, M., Malmström, E.E., McNamara, M., et al. 1999. Controlled synthesis of polymer brushes by "living" free radical polymerization techniques. *Macromolecules* 32: 1424–1431.
6. Matyjaszewski, K., and Xia, J. 2001. Atom transfer radical polymerization. *Chem. Rev.* 101: 2921–2990.
7. Kamigaito, M., Ando, T., and Sawamoto, M. 2001. Metal-catalyzed living radical polymerization. *Chem. Rev.* 101: 3689–3746.

8. Braunecker, W., and Matyjaszewski, K. 2007. Controlled/living radical polymerization: features, developments, and perspectives. *Prog. Polym. Sci.* 32: 93–146.

9. Edmonson, S., Osborne, V.L., and Huck, W.T.S. 2004. Polymer brushes via surface-initiated polymerizations. *Chem. Soc. Rev.* 33: 14–22.

10. Barbey, R., Lavanant, L., Paripovic, D., Schuwer, N., Sugnaux, C., Tugulu, S., and Klok, H.-A. 2009. Polymer brushes via surface-initiated controlled radical polymerization: synthesis, characterization, properties, and applications. *Chem. Rev.* 109: 5437–5527.

11. Qiu, J., Matyjaszewski, K., Thouin, L., and Amatore, C. 2000. Cyclic voltammetric studies of copper complexes catalyzing atom transfer radical polymerization. *Macromol. Chem. Phys.* 201: 1625–1631.

12. Magenau, A.J.D., Strandwitz, N.C., Gennaro, A., and Matyjaszewski, K. 2011. Electrochemically mediated atom transfer radical polymerization. *Science* 332: 81–84.

13. Wittstock, G., Burchardt, M., Pust, M.E., Shen, Y., and Zhao, C. 2007. Scanning electrochemical microscopy for direct imaging of reaction rates. *Angew. Chem. Int. Ed.* 46: 1584–1617.

14. Sun, P., Laforge, F.O., and Mirkin, M.V. 2007. Scanning electrochemical microscopy in the 21st century. *Phys. Chem. Chem. Phys.* 9: 802–823.

15. Kanoufi, F. 2007. Microscopie électrochimique; des microelectrodes pour étudier et modifier les interfaces. *Actual Chim.* 311: 36–48.

16. Bard, A.J., and Mirkin, M.V., eds. 2001. *Scanning electrochemical microscopy.* New York: Marcel Dekker.

17. Bard, A.J., and Mirkin, M.V., eds. 2011. *Scanning electrochemical microscopy.* 2nd ed. Boca Raton, FL: Taylor & Francis.

18. Amatore, C., and Fosset, B. 1996. Equivalence between microelectrodes of different shapes: between myth and reality. *Anal. Chem.* 68: 4377–4388.

19. http://www.comsol.com.

20. Wei, C., Bard, A.J., and Mirkin, M.V. 1995. Scanning electrochemical microscopy. 31. Application of SECM to the study of charge transfer processes at the liquid/liquid interface. *J. Phys. Chem.* 99: 16033–16042.

21. Cornut, R., and Lefrou, C. 2008. New analytical approximation of feedback approach curves with a microdisk SECM tip and irreversible kinetic reaction at the substrate. *J. Electroanal. Chem.* 621: 178–184.

22. Engstrom, R.C., and Pharr, C.M. 1989. Scanning electrochemical microscopy. *Anal. Chem.* 61: 1099A–1104A.

23. Scott, E.R., White, H.S., and Phipps, J.B. 1993. Measurement of iontophoretic transport through porous membranes using scanning electrochemical microscopy: application to *in vitro* studies of ion fluxes through skin. *Anal. Chem.* 65: 1537–1545.

24. Scott, E.R., White, H.S., and Phipps, J.B. 1991. Scanning electrochemical microscopy of a porous membrane. *J. Membrane Sci.* 58: 71–87.

25. Bath, B.D., Lee, R.D., White, H.S., and Scott, E.R. 1998. Imaging molecular transport in porous membranes. Observation and analysis of electroosmotic flow in individual pores using the scanning electrochemical microscope. *Anal. Chem.* 70: 1047–1058.

26. Amatore, C., Szunerits, S., Thouin, L., and Warkocz, J.S. 2000. Mapping concentration profiles within the diffusion layer of an electrode. Part III. Steady-state and time-dependent profiles via amperometric measurements with an ultramicroelectrode probe. *Electrochem. Commun.* 2: 353–358.

27. White, R.J., and White, H.S. 2007. Influence of electrophoresis waveforms in determining stochastic nanoparticle capture rates and detection sensitivity. *Anal. Chem.* 79: 6334–6340.

28. Amatore, C., Szunerits, S., Thouin, L., and Warkocz, J.S. 2001. Monitoring concentration profiles *in situ* with an ultramicroelectrode probe. *Electroanalysis* 13: 646–652.

29. Martin, D.R., and Unwin, P.R. 1998. Theory and experiment for the substrate generation/tip collection mode of the scanning electrochemical microscope: application as an approach for measuring the diffusion coefficient ratio of a redox couple. *Anal. Chem.* 70: 276–284.

30. Horrocks, B.R., Mirkin, M.V., Pierce, D.T., Bard A.J., Nagy, G., and Toth, K. 1993. Scanning electrochemical microscopy. 19. Ion-selective potentiometric microscopy. *Anal. Chem.* 65: 1213–1224.

31. Gyurcsanyi, R.E., Pergel, E., Nagy, R., et al. 2001. Direct evidence of ionic fluxes across ion-selective membranes: a scanning electrochemical microscopic and potentiometric study. *Anal. Chem.* 73: 2104–2111.

32. Shim, J.H., Kim, J., Cha, G.S., et al. 2007. Glass nanopore-based ion-selective electrodes. *Anal. Chem.* 79: 3568–3574.

33. Kueng, A., Kranz, C., Lugstein, A., Bertagnolli, E., and Mizaikoff, B. 2005. AFM-tip-integrated amperometric microbiosensors: high-resolution imaging of membrane transport. *Angew. Chem. Int. Ed.* 44: 3419–3422.

34. Kueng, A., Kranz, C., and Mizaikoff, B. 2005. Imaging of ATP membrane transport with dual micro-disk electrodes and scanning electrochemical microscopy. *Biosens. Bioelectron.* 21: 346–353.

35. Chen, C.C., Derylo, M.A., and Baker, L.A. 2009. Measurement of ion currents through porous membranes with scanning ion conductance microscopy. *Anal. Chem.* 81: 4742–4751.

36. Lund, H., and Hammerich, O., eds. 2001. *Organic electrochemistry*. 4th ed. New York: Marcel Dekker.

37. Bard, A.J., Denuault, G., Lee, C., Mandler, D., and Wipf, D.O. 1990. Scanning electrochemical microscopy—a new technique for the characterization and modification of surfaces. *Acc. Chem. Res.* 23: 357–363.

38. Turyan, I., Matsue, T., and Mandler, D. 2000. Patterning and characterization of surfaces with organic and biological molecules by the scanning electrochemical microscope. *Anal. Chem.* 72: 3431–3435.

39. Zhou, J., and Wipf, D.O. 1997. Deposition of conducting polyaniline patterns with the scanning electrochemical microscope. *J. Electrochem. Soc.* 144: 1202–1207.

40. Simeone, F.C., Albonetti, C., and Cavallini, M. 2009. Progress in micro- and nanopatterning via electrochemical lithography. *J. Phys. Chem. C* 113: 18987–18994.

41. Turyan, I., and Mandler, D. 1998. Two-dimensional polyaniline thin film electrodeposited on a self-assembled monolayer. *J. Am. Chem. Soc.* 120: 10733–10742.

42. Kranz, C., Ludwig, M., Gaub, H.E., and Schuhmann, W. 1995. Lateral deposition of polypyrrole lines by means of the scanning electrochemical microscope. *Adv. Mater.* 7: 38–40.

43. Marck, C., Borgwarth, K., and Heinze, J. 2001. Generation of polythiophene micropatterns by scanning electrochemical microscopy. *Chem. Mater.* 13: 747–752.
44. Combellas, C., Ghilane, J., Kanoufi, F., and Mazouzi, D. 2004. Surface modification of halogenated polymers. 7. Local reduction of polytetrafluoroethylene and polychlorotrifluoroethylene by the scanning electrochemical microscope in the feedback mode. *J. Phys. Chem. B* 108: 6391–6397.
45. Combellas, C., Kanoufi, F., and Nunige, S. 2007. Surface modification of halogenated polymers. 10. Redox catalysis induction of the polymerization of vinylic monomers. Application to the localized graft copolymerization of poly(tetrafluoroethylene) surfaces by vinylic monomers. *Chem. Mater.* 19: 3830–3839.
46. Combellas, C., Kanoufi, F., and Mazouzi, D. 2004. Surface modification of halogenated polymers. 8. Local reduction of polytetrafluoroethylene by the scanning electrochemical microscope—transient experiments. *J. Phys. Chem. B* 108: 19260–19268.
47. Ktari, N., Poncet, P., Sénéchal, H., Malaquin, L., Kanoufi, F., and Combellas, C. 2010. Patterning of polystrene by scanning electrochemical microscopy: biological applications to cell adhesion. *Langmuir* 26: 17348–17356.
48. Pust, S.E., Szunerits, S., Boukherroub, R., and Wittstock, G. 2009. Electrooxidative nanopatterning of silane monolayers on boron-doped diamond electrodes. *Nanotechnology* 20: 075302.
49. Wittstock, G., and Schuhmann, W. 1997. Formation and imaging of microscopic enzymatically active spots on an alkanethiolate-covered gold electrode by scanning electrochemical microscopy. *Anal. Chem.* 69: 5059–5066.
50. Zhao, C., Witte, I., and Wittstock, G. 2006. Switching on cell adhesion with microelectrodes. *Angew. Chem. Int. Ed.* 45: 5469–5471.
51. Shiku, H., Takeda, T., Yamada, H., Matsue, T., and Uchida, I. 1995. Microfabrication and characterization of diaphorase-patterned surfaces by scanning electrochemical microscopy. *Anal. Chem.* 67: 312–317.
52. Matrab, T., Hauquier, F., Combellas, C., and Kanoufi, F. 2010. Scanning electrochemical microscopy investigation of molecular transport and reactivity within polymer brushes. *ChemPhysChem* 11: 670–682.
53. Hauquier, F., Matrab, T., Kanoufi, F., and Combellas, C. 2009. Local direct and indirect reduction of electrografted aryldiazonium/gold surfaces for polymer brushes patterning. *Electrochim. Acta* 54: 5127–5136.
54. Slim, C., Tran, Y., Chehimi, M.M., et al. 2008. Microelectrochemical patterning of surfaces with polymer brushes of controlled density grown by atom transfer radical polymerization. *Chem. Mater.* 20: 6677–6685.
55. Orski, S.V., Fries, K.H., Sontag S.K. and J. Locklin. 2011. Fabrication of nanostructures using polymer brushes. *J. Mater. Sci.* 21: 14135–14149.
56. Ktari, N., Nunige, S., Azioune, A., et al. 2010. Managing micrometric sources of solvated electrons: application to the local functionalization of fluorinated self assembled monolayers. *Chem. Mater.* 22: 5725–5731.
57. Hazimeh, H., Nunige, S., Cornut, R., Lefrou, C., Combellas, C., and Kanoufi, F. 2011. Surface reactivity from the electrochemical lithography: illustration in the steady-state reductive etching of perfluorinated surfaces. *Anal. Chem.* 83: 6106–6113.
58. Ku, S.Y., Wong, K.T., and Bard, A.J. 2008. Surface patterning with fluorescent molecules using click chemistry directed by scanning electrochemical microscopy. *J. Am. Chem. Soc.* 130: 2392–2393.

59. Cougnon, C., Gohier, F., Bélanger, D., and Mauzeroll, J. 2009. *In situ* formation of diazonium salts from nitro precursors for scanning electrochemical microscopy patterning of surfaces. *Angew. Chem. Int. Ed.* 48: 4006–4008.
60. Matrab, T., Combellas, C., and Kanoufi, F. 2008. Localized direct electrografting of a gold surface with organic moieties from iodonium salts. *Electrochem. Commun.* 10: 1230–1234.
61. Nunige, S., Cornut, R., Lefrou, C., and Kanoufi, F. 2011. Local etching of copper films by SECM in the feedback mode: a theoretical and experimental investigation. *Electrochim. Acta.* 56: 10701–10707.
62. Azzaroni, O., Brown, A.A., and Huck, W.T.S. 2007. Tunable wettability by clicking into polyelectrolyte brushes. *Adv. Mater.* 19: 151–154.
63. Zhang, N.H., and Ruhe, J. 2005. Swelling of poly(methacrylic acid) brushes: influence of monovalent salts in the environment. *Macromolecules* 38: 4855–4860.
64. Bard, A.-J., Faulkner, L.R. 2001. *Electrochemical methods: principle and applications.* 2nd ed. New York: John Wiley & Sons.
65. Motornov, M., Sheparovych, R., Katz, E., and Minko, S. 2008. Chemical gating with nanostructured responsive polymer brushes: mixed brush versus homopolymer brush. *ACS Nano* 2: 41–52.
66. Bantz, M.R., Brantley, E.L., Weinstein, R.D., Moriarty, R.D., and Jennings, G.K. 2004. Effect of fractional fluorination on the properties of ATRP surface-initiated poly(hydroxyethyl methacrylate) films. *J. Phys. Chem. B* 108: 9787–9794.
67. Tam, T.K., Ornatska, M., Pita, M., Minko, S., and Katz, E. 2008. Polymer brush-modified electrode with switchable and tunable redox activity for bioelectronic applications. *J. Phys. Chem. C* 112: 8438–8445.
68. Zhou, F., Hu, H., Yu, B., Osborne, V.L., Huck, W.T.S., and Liu, W. 2007. Probing the responsive behaviour of polyelectrolyte brushes using electrochemical impedance spectroscopy. *Anal. Chem.* 79: 176–182.
69. Rodriguez Presa, M.J., Gassa, L.M., Azzaroni, O., and Gervasi, C.A. 2009. Estimating diffusion coefficient of probe molecules into polyelectrolyte brushes by electrochemical impedance spectroscopy. *Anal. Chem.* 81: 7936–7943.
70. Spruijt, E., Choi, E.Y., and Huck, W.T.S. 2008. Reversible electrochemical switching of polyelectrolyte brush surface energy using electroactive counterions. *Langmuir* 24: 11253–11260.
71. Choi, E.Y., Azzaroni, O., Cheng, N., Zhou, F., Kelby, T., and Huck, W.T.S. 2007. Electrochemical characteristics of polyelectrolyte brushes with electroactive counterions. *Langmuir* 23: 10389–10394.
72. Tam, T.K., Pita, M., Trotsenko, O., et al. 2010. Reversible "closing" of an electrode interface functionalized with a polymer brush by an electrochemical signal. *Langmuir* 26: 4506–4513.
73. Fulghum, T.M., Estillore, N.C., Vo, C.-D., Armes, S.P., and Advincula, R.C. 2008. Stimuli-responsive polymer ultrathin films with a binary architecture: combined layer-by-layer polyelectrolyte and surface-initiated polymerization approach. *Macromolecules* 41: 429–435.
74. Combellas, C., Kanoufi, F., Sanjuan, S., Slim, C., and Tran, Y. 2009. Electrochemical and spectroscopic investigation of counterions exchange in polyelectrolyte brushes. *Langmuir* 25: 5360–5370.

75. Amatore, C., Savéant, J.M., and Tessier, D. 1983. Charge-transfer at partially blocked surfaces—a model for the case of microscopic active and inactive sites. *J. Electroanal. Chem.* 147: 39–51.

76. Demaille, C., and Moiroux, J. 1999. Sensing a self-assembled protein structure through its permeation by a ferrocene labeled Poly(ethylene glycol) chain. *J. Phys. Chem. B* 103: 9903–9909.

77. Bourdillon, C., Demaille, C., Moiroux, J., and Savéant, J.M. 1995. Catalysis and mass transport in spatially ordered enzyme assemblies on electrodes. *J. Am. Chem. Soc.* 117: 11499–11506.

78. Kwak, J., and Anson, F.C. 1992. Monitoring the ejection and incorporation of ferricyanide $[Fe(CN)_6^{-3}]$ and ferrocyanide $[Fe(CN)_6^{-4}]$ counterions at protonated poly(4-vinylpyridine) coatings on electrodes with the scanning electrochemical microscope. *Anal. Chem.* 64: 250–256.

79. Lee, C., and Anson, F.C. 1992. Use of electrochemical microscopy to examine counterion ejection from Nafion coatings on electrodes. *Anal. Chem.* 64: 528–533.

80. Bertoncello, P., Ciani, I., Li, F., and Unwin, P.R. 2006. Measurement of apparent diffusion coefficients within ultrathin Nafion Langmuir–Schaefer films: comparison of a novel scanning electrochemical microscopy approach with cyclic voltammetry. *Langmuir* 22: 10380–10388.

81. Cornut, R., and Lefrou, C. 2008. Studying permeable films with scanning electrochemical microscopy (SECM): quantitative determination of permeability parameter *J. Electroanal. Chem.* 628: 197–203.

82. Williams, M.E., Stevenson, K.J., Massari A.M., and Hupp, J.T. 2000. Imaging size-selective permeation through micropatterned thin films using scanning electrochemical microscopy. *Anal. Chem.* 72: 3122–3128.

83. Williams, M.E., Benkstein, K.D., Abel, C., Dinolfo, P.H., and Hupp, J.T. 2002. Shape-selective transport through rectangle-based molecular materials: thin-film scanning electrochemical microscopy studies. *Proc. Natl. Acad. Sci. USA* 99: 5171–5177.

84. Guo, J., and Amemiya, S. 2005. Permeability of the nuclear envelope at isolated *Xenopus* oocyte nuclei studied by scanning electrochemical microscopy. *Anal. Chem.* 77: 2147–2156.

85. Kim, E., Xiong, H., Striemer, C.C., et al. 2008. A structure-permeability relationship of ultrathin nanoporous silicon membrane: a comparison with the nuclear envelope. *J. Am. Chem. Soc.* 130: 4230–4231.

86. Kallio, T., Slevin, C., Sundholm, G., Holmlund, P., and Kontturi, K. 2003. Proton transport in radiation-grafted membranes for fuel cells as detected by SECM. *Electrochem. Commun.* 5: 561–565.

87. Rodriguez-Lopez, J., Alpuche-Aviles, M.A., and Bard, A.J. 2008. Interrogation of surfaces for the quantification of adsorbed species on electrodes: oxygen on gold and platinum in neutral media. *J. Am. Chem. Soc.* 130: 16985–16995.

88. Abbou, J., Anne, A., and Demaille, C. 2004. Probing the Structure and Dynamics of End-Grafted Flexible Polymer Chain Layers by Combined Atomic Force-Electrochemical Microscopy. Cyclic Voltammetry within Nanometer-Thick Macromolecular Poly(ethylene glycol) Layers. *J. Am. Chem. Soc.* 126: 10095–10108.

89. Anne, A., Demaille, C., and Goyer, C. 2009. Electrochemical atomic-force microscopy using a tip-attached redox mediator. Proof-of-concept and perspectives for functional probing of nanosystems. *ACS Nano* 3: 349–353.
90. Anne, Cambril, E., Chovin, A., Demaille, C., and Goyer, C. 2009. Electrochemical atomic-force microscopy using a tip-attached redox mediator for topographic and functional imaging of nanosystems. *ACS Nano* 3: 2927–2940.
91. Ktari, N., Combellas, C., and Kanoufi, F. 2011. Local oxidation of polystyrene by scanning electrochemical microscopy. *J. Phys. Chem. C* 115: 17891–17897.
92. Pinson, J., and Podvorica, F.I. 2005. Attachment of organic layers to conductive or semiconductive surfaces by reduction of diazonium salts. *Chem. Soc. Rev.* 34: 429–439.
93. Bélanger, D., and Pinson, J. 2011. Electrografting: a powerful method for surface modification. *Chem. Soc. Rev.* 40: 3995–4048.
94. McCreery, R.L. 2008. Advanced carbon electrode materials for molecular electrochemistry. *Chem. Rev.* 108: 2646–2687.
95. Brewis, D.M., Dahm, R.H., and Mathieson, I. 2000. Electrochemical pretreatment of polymers with dilute nitric acid either alone or in the presence of silver ions. *J. Adhes.* 72: 373–386.
96. Liu, B., Bard, A.J., Mirkin, M.V., and Creager, S.E. 2004. Electron transfer at self-assembled monolayers measured by scanning electrochemical microscopy. *J. Am. Chem. Soc.* 126: 1485–1492.
97. Combellas, C., Fuchs, A., Kanoufi, F., Mazouzi, D., and Nunige, S. 2004. Surface modification of halogenated polymers. 6. Graft copolymerization of poly(tetrafluoroethylene) surfaces by polyacrylic acid. *Polymer* 45: 4669–4675.
98. Schmelmer, U., Jordan, R., Geyer, W., et al. 2003. Surface-initiated polymerization on self-assembled monolayers: amplification of patterns on the micrometer and nanometer scale. *Angew. Chem. Int. Ed.* 42: 559–560.
99. Ahn, S.J., Kaholek, M., Lee, W.-K., LaMattina, B., LaBean, T.H., Zauscher, S. 2004. Surface-initiated polymerization on nanopatterns fabricated by electron-beam lithography. *Adv. Mater.* 16: 2141–2142.
100. He, Q., Küller, A., Grunze, M., Li, J. 2007. Fabrication of thermosensitive polymer nanopatterns through chemical lithography and atom transfer radical polymerization. *Langmuir* 23: 3981–3987.
101. Husemann, M., Morrison, M., Benoit, D., et al. 2000. Manipulation of surface properties by patterning of covalently bound polymer brushes. *J. Am. Chem. Soc.* 122: 1844–1845.
102. Iwata, P., Suk-In, P., Hoven, V.P., Takahara, A., Akiyoshi, K., and Iwasaki, Y. 2004. Control of nanobiointerfaces generated from well-defined biomimetic polymer brushes for protein and cell manipulations. *Biomacromolecules* 5: 2308–2314.
103. Konradi, R., and Ruhe, J. 2006. Fabrication of chemically microstructured polymer brushes. *Langmuir* 22: 8571–8575.
104. Khire, V.S., Harant, A.W., Watkins, A.W., Anseth, K.S., Bowman, C.N. 2006. Ultrathin patterned polymer films on surfaces using thiol–ene polymerizations. *Macromolecules* 39: 5081–5086.
105. Dong, R., Krishnan, S., Baird, B.A., Lindau, M., and Ober, C.K. 2007. Patterned biofunctional poly(acrylic acid) brushes on silicon surfaces. *Biomacromolecules* 8: 3082–3092.

106. Husemann, M., Mecerreyes, D., Hawker, C.J., Hedrick, J.L., Shah, R., and Abbott, N.L. 1999. Surface-initiated polymerization for amplification of self-assembled monolayers patterned by microcontact printing. *Angew. Chem. Int. Ed.* 38: 647–649.

107. Tu, H., Heitzman, C.E., and Braun, P.V. 2004. Patterned poly(N-isopropylacrylamide) brushes on silica surfaces by microcontact printing followed by surface-initiated polymerization. *Langmuir* 20: 8313–8320.

108. Edmondson, S., Vo, C.D., Armes, S.P., and Unali, G.F. 2007. Surface polymerization from planar surfaces by atom transfer radical polymerization using polyelectrolytic macroinitiators. *Macromolecules* 40: 5271–5278.

109. Beinhoff, M., Appapillai, A.T., Underwood, L.D., Frommer, J.E., and Carter, K.R. 2006. Patterned polyfluorene surfaces by functionalization of nanoimprinted polymeric features. *Langmuir* 22: 2411–2414.

110. Genua, A., Alduncín, J.A., Pomposo, J.A., Grande, H., Kehagias, N., Reboud, V., Sotomayor, C., Mondragon, I., and Mecerreyes, D. 2007. Functional patterns obtained by nanoimprinting lithography and subsequent growth of polymer brushes. *Nanotechnology* 18: 215301.

111. Brinks, M.K., Hirtz, M., Chi, L., Fuchs, H., and Studer, A. 2007. Site-selective surface-initiated polymerization by Langmuir-Blodgett lithography. *Angew. Chem. Int. Ed.* 46: 5231–5233.

112. Hou, S., Li, Q., and Liu, Z. 2004. Poly(methyl methacrylate) nanobrushes on silicon based on localized surface-initiated polymerization. *Appl. Surf. Sci.* 222: 338–345.

113. Wu, T., Efimenko, K., and Genzer, J. 2002. Combinatorial study of the mushroom-to-brush crossover in surface anchored polyacrylamide. *J. Am. Chem. Soc.* 124: 9394–9395.

114. Jones, D.M., Brown, A.A., and Huck, W.T.S. 2002. Surface-initiated polymerizations in aqueous media: effect of initiator density. *Langmuir* 18: 1265–1269.

115. Wu, T., Efimenko, K., Vlcek, P., Subr, V., and Genzer, J. 2003. Formation and properties of anchored polymers with a gradual variation of grafting densities on flat substrates. *Macromolecules* 36: 2448–2453.

116. Khire, V.S., Lee, T.Y., and Bowman, C.N. 2007. Surface modification using thiol–acrylate conjugate addition reactions. *Macromolecules* 40: 5669–5677.

117. Matrab, T., Chehimi, M.M., Perruchot, C., et al. 2005. Novel approach for metallic surface-initiated atom transfer radical polymerization using electrografted initiators based on aryl diazonium salts. *Langmuir* 21: 4686–4694.

118. Matrab, T., Chancolon, J., Mayne L'Hermite, M., et al. 2006. Atom transfer radical polymerization (ATRP) initiated by aryl diazonium salts: a new route for surface modification of multiwalled carbon nanotubes by tethered polymer chains. *Colloids Surf. A* 287: 217–221.

119. Matrab, T., Save, M., Charleux, B., et al. 2007. Grafting densely-packed poly(n-butyl methacrylate) chains from an iron substrate by aryl diazonium surface-initiated ATRP: XPS monitoring. *Surf. Sci.* 601: 2357–2366.

120. Deiss, F., Combellas, C., Sojic, N., and Kanoufi, F. 2010. Lithography by scanning electrochemical microscopy with a multiscaled electrode. *Anal. Chem.* 82: 5169–5175.

121. Combellas, C., Fuchs, A., and Kanoufi, F. 2004. Scanning electrochemical microscopy with a band microelectrode. *Anal. Chem.* 76: 3612–3618.

122. Salaita, K., Wang, Y., Fragala, J., Vega, R.A., Liu, C., and Mirkin, C.A. 2006. Massively parallel dip-pen nanolithography with 55000-pen two-dimensional arrays. *Angew. Chem. Int. Ed.* 45: 7220–7223.
123. Mirkin, C.A. 2007. The power of the pen: development of massively parallel dip-pen nanolithography. *ACS Nano* 1: 79–83.
124. Sanjuan, S., Perrin, P., Pantoustier, N., and Tran, Y. 2007. Synthesis and swelling behavior of pH-responsive polybase brushes. *Langmuir* 23: 5769–5778.
125. Yamamoto, S., Ejaz, M., Tsuji, Y., Matsumoto, M., and Fukuda, T. 2000. Surface Interaction forces of well-defined, high-density polymer brushes studied by atomic force microscopy. 2. Effect of graft density. *Macromolecules* 33: 5608–5612.
126. Wang, X., Tu, H., Braun, P.V., and Bohn, P.W. 2006. Length scale heterogeneity in lateral gradients of poly(N-isopropylacrylamide) polymer brushes prepared by surface-initiated atom transfer radical polymerization coupled with in-plane electrochemical potential gradients. *Langmuir* 22: 817–823.
127. Kjærsbo, T., Daasbjerg, K., and Pedersen, S.U. 2003. Study of the coupling reactions between electrochemically generated aromatic radical anions and methyl, alkyl and benzyl radicals. *Electrochim. Acta* 48: 1807–1816, and references therein.
128. Costentin, C., Robert, M., and Savéant, J.M. 2006. Electron transfer and bond breaking: recent advances. *Chem. Phys.* 324: 40–56.
129. Houmam, A. 2008. Electron transfer initiated reactions: bond formation and bond dissociation. *Chem. Rev.* 108: 2180–2237.
130. Szunerits, S., Kanoufi, F. Unpublished Manuscript.
131. Millner, S.T., Witten, T.A., and Cates, M.E. 1989. Effects of polydispersity in the end-grafted polymer brush. *Macromolecules* 22: 853–861.
132. Murat, M., and Grest, G.S. 1993. Structure of grafted polymeric brushes in solvents of varying quality: a molecular dynamics study. *Macromolecules* 26: 3108–3117.
133. Laradji, M., Guo, H., and Zuckermann, M.J. 1994. Off-lattice Monte-Carlo simulation of polymer brushes in good solvents. *Phys. Rev. E* 49: 3199–3206.
134. Sirard, S.M., Gupta, R.R., Russell, T.R., Watkins, J.J., Green, P.F., and Johnston, K.P. 2003. Structure of end-grafted polymer brushes in liquid and supercritical carbon dioxide: a neutron reflectivity study. *Macromolecules* 36: 3365–3373.
135. Auroy, P., Mir, Y., and Auvray, L. 1992. Local-structure and density profile of polymer brushes. *Phys. Rev. Lett.* 69: 93–95.
136. Matyjaszewski, K., Dong, H., Jakubowski, W., Pietrasik, J., and Kusumo, A. 2007. Grafting from surfaces for "everyone": ARGET ATRP in the presence of air. *Langmuir* 23: 4528–4531.
137. Tria, M.C.R., Grande, C.D.T., Ponnapati, R.R., and Advincula, R.C. 2010. Electrochemical deposition and surface-initiated RAFT polymerization: protein and cell-resistant PPEGMEMA polymer brushes. *Biomacromolecules* 11: 3422–3431.
138. Ghorbal, A., Grisotto, F., Charlier, J., Palacin, S., Goyer, C., and Demaille, C. 2009. Localized electrografting of vinylic monomers on a conducting substrate by means of an integrated electrochemical AFM probe. *ChemPhysChem* 10: 1053–1057.

9

Comparison of Surface-Confined ATRP and SET-LRP Syntheses of Polymer Brushes

Keisha B. Walters

Dave C. Swalm School of Chemical Engineering, Mississippi State University, Mississippi State, Mississippi

Shijie Ding

College of Life Science and Chemical Engineering, Jiangsu Provincial Key Laboratory of Palygorskite Science and Applied Technology, Huaiyin Institute of Technology, Huaian, People's Republic of China

CONTENTS

In the last two decades, extensive development of new polymer synthesis methods has allowed for novel polymer chemistries and architectures. Specifically, the emergence of a variety of controlled/living radical polymerizations (CRP/LRP) has provided techniques allowing for more precise design of the composition, size, topology, functionality, and architecture of macromolecules—the dream of many polymer chemists. These strategies include nitroxide-mediated living radical polymerization (NMP), transition metal complex-mediated living radical polymerization (ATRP), and living radical polymerization via reversible addition fragmentation transfer polymerization (RAFT).

9.1 Atom Transfer Radical Polymerization (ATRP) and Single-Electron Transfer–Living Radical Polymerization (SET-LRP)

ATRP was first reported in 1995 [1–3] and is of special interest due to its versatility, robustness, controllability, and the living nature of the polymerization. The general mechanism of ATRP is shown in Scheme 9.1. By employing a transition metal complex ($Y - M_t^x$/ligand) as catalysts in a redox reaction, ATRP establishes a rapid dynamic equilibrium between a minute amount of growing free radicals ($\sim P_m \bullet$) and a large majority of the dormant species ($\sim P_n - X$). The radical propagates monomers with an activity similar to that of a conventional free radical. Then the radical is very quickly deactivated to its dormant state—the polymer chain terminally capped with a halide group ($\sim P_n - X$). Since the deactivation rate constant is substantially higher than that of the activation reaction ($K_{eq} = k_{act}/k_{deact} \sim 10^{-7}$), each polymer chain is protected by spending most of the time in the dormant state, thereby substantially reducing the likelihood of permanent termination via radical coupling and disproportionation. In a well-controlled ATRP, only several percent of the chain ends die via termination [4].

ATRP has been utilized for the preparation of a large variety of (co)polymers, with monomers including (meth)acrylates, styrenes, (meth)acrylamides, acrylonitrile, vinyl acetate, etc., characterized by well-controlled molecular weights and structures. ATRP can be performed in bulk, organic/aqueous solvents, emulsion, and supercritical carbon dioxide polymerization conditions [4,5]. Among various transition metal complexes—such as Mo, Ru, Fe, Co, Ni, Pa, and Rh—the most versatile ATRP catalysts are copper halides, which generally complex with nitrogen-containing ligands (e.g., N,N,N',N',N''-pentamethyldiethylenetriamine (PMDETA), 1,1,4,7,10,10-hexa-methyl-triethylenetetramine (HMTETA), 1,4,8,11-tetramethyl-1,4,8,11-tetraazacyclotetradecane (Me$_4$Cyclam), tris[2-(dimethylamino)ethyl]amine (Me$_6$TREN), and 2,2'-bipyridine (bpy)). Parameters such as the ligand type, catalyst-to-ligand ratio, solvent(s), ratios of Cu(I) and Cu(II), and initiator all have critical effects on the performance of a particular ATRP reaction. Notable, it has been found that polymerization rates can be dramatically increased with the addition of a small amount of Cu(0) [6,7]. The role of Cu(0) is to reduce Cu(II)/L through comproportionation, and increase the formation of Cu(I)/L (Scheme 9.2).

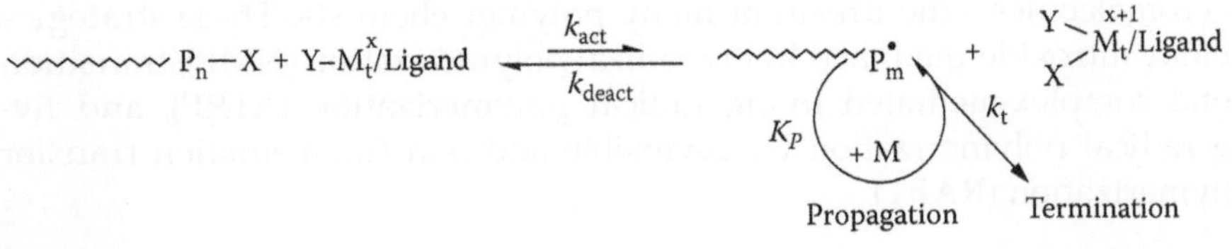

$$\sim\sim\sim P_n{-}X + Y{-}M_t^x/Ligand \underset{k_{deact}}{\overset{k_{act}}{\rightleftharpoons}} \sim\sim\sim P_m^\bullet + \underset{X}{\overset{Y\diagdown\ \ ^{x+1}}{M_t/Ligand}}$$

$$K_P \diagdown +\ M \qquad k_t$$

Propagation　　　Termination

SCHEME 9.1
The general mechanism of ATRP. (From Shen, Y., et al., *Prog. Polym. Sci.*, 29, 1053–1078, 2004.)

This methodology has been extended to the use of organic reducing agents, including sugars, ascorbic acid, hydrazine, amines, and phenols, and formed a new technique known as activators regenerated by electron transfer (ARGET) ATRP [8]. Most recently, ARGET ATRP of methyl acrylate was performed with "inexpensive ligands" such as diethylenetriamine (DETA), PMDETA, and tris(2-aminoethyl)amine (TREN), and ppm levels of Cu(II) Br_2/L in the presence of a zero-valent copper metal (i.e., copper powder or wire) at 25°C [9]. With this minuscule amount of catalyst, the reaction solution was almost colorless, thus eliminating the need for purification of the polymer product.

In 2006, a new mechanism named low-activation energy outer-sphere single-electron transfer (SET) was proposed for the metal-catalyzed living radical polymerization using Cu(0), Cu_2Se, Cu_2Te, Cu_2S, or Cu_2O species as the catalyst instead of Cu(I)X. SET polymerizations (namely, SET-LRP) are carried out at room temperature in polar solvents, including water, alcohol, and dipolar aprotic solvents, and result in ultrafast polymerization rates and ultrahigh molecular weights for polymers of methyl acrylate (MA), methyl methacrylate (MMA), ethyl acrylate (EA), n-butyl acrylate (BA), and vinyl chloride (VC), monomers with electron-withdrawing groups [10]. In SET-LRP, Cu(0) species act as electron donors, and the initiator and dormant propagating species act as electron acceptors. As shown in Scheme 9.3, the active Cu(0) species activates the R-X group to generate R• and a Cu(I)X species. The inactive Cu(I)X species generated instantaneously disproportionates into Cu(II)X and Cu(0), whereas deactivation of propagating macroradicals is mediated by Cu(II)X complex. By this mechanism, the inactive Cu(I)X species are spontaneously consumed and the reactive Cu(0) species are continually produced [10].

Essential to the SET-LRP mechanism is the establishment of an appropriate balance of Cu(0) and Cu(II) species by the disproportionation of the Cu(I)X generated in situ via activation and deactivation. Disproportionation

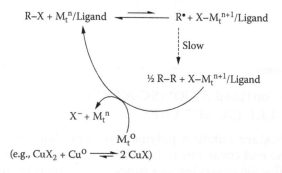

SCHEME 9.2
Removal of excess high oxidation state metal halide by reaction with a zero-valent metal. (From Matyjaszewski, K., et al., *Macromolecules*, 30, 7348–7350, 1997.)

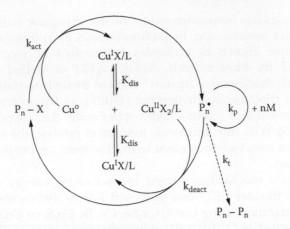

SCHEME 9.3
Mechanism of SET-LRP. (From Percec, V., et al., *J. Am. Chem. Soc.*, 128, 14156–14165, 2006.)

is not rapid or extensive under most conditions; rather, it requires a combination of an appropriate solvent (Table 9.1) and an appropriate N-ligand, such as Me₆TREN, TREN, or PMDETA. Use of solvents that do not support disproportionation, such as acetone, MeCN, and toluene, result in non-first-order kinetics and nonliving polymerizations. Addition of polar alcohol or phenol additives to toluene or addition of water to organic solvents gives a solvent mixture capable of mediating disproportion of Cu(I)X and supporting a living SET-LRP [11–14]. It was also found that the ligand concentration and the particle size of Cu(0) powder play strong roles in the SET-LRP kinetics [15,16]. Recently, Cu(0) wire was reported to exhibit significantly greater control of molecular weight distribution than Cu(0) powder [17]. Control over molecular weight distribution combined with the advantages of easier catalyst preparation, handling, predictability, tunability, and simple recovery/recycling have made Cu(0) wire an attractive alternative to powders [17–19].

9.2 Surface-Confined ATRP (SC-ATRP) and SET-LRP (SC-SET-LRP)

Polymer brushes are ultrathin polymer coatings that consist of polymer chains with one end covalently tethered to a substrate, i.e., a surface either needing modification or serving as a support [20]. There are two methods to synthesize polymer brushes: grafting to and grafting from. Grafting to techniques employ prefabricated, end-functionalized polymer chains that are

TABLE 9.1

Equilibrium Constants for the
Disproportionation of Cu(I)X,
K_{disp}, in Various Solvents

Solvent	K_{disp}
Acetone	0.03
DMF	1.82×10^4
DMSO	1.5–4.4
EtOH	3.6
H_2O	0.89×10^6–5.8×10^7
MeCN	6.3×10^{-21}
MeOH	4–6.3×10^3

Source: Rosen, B.M., et al., *J. Polym. Sci. A Polym. Chem.*, 47, 5606–5628, 2009.

attached to the substrate surface. The primary limitation of the grafting to strategy is obvious: previously immobilized polymer chains (with coiled or globular conformations) hinder the positioning and attachment of additional chains. This steric repulsion thwarts the diffusion of approaching chains to the reactive sites at the surface, resulting in low graft density. In contrast, the grafting from method can produce thick, highly dense polymer brushes by directly initiating polymerization from the initiator-immobilized surface [20]; thus, these polymerizations are named surface-confined (SC) or surface-initiated (SI) polymerizations.

Currently, most polymer brushes are prepared via surface confined controlled/living radical polymerizations. SC-ATRP has been the most extensively used SC-LRP technique, mainly due to the extreme versatility and robustness of the chemistry, such as compatibility with a large variety of functional groups, mild reaction conditions, relatively easy synthesis, and high tolerance for impurities [20]. Since the origin of the first SC-ATRP reactions in 1997, SC-ATRP has been applied for the preparation of innumerable polymer brushes on various substrates using a wide range of catalyst systems and solvents. Through these investigations, one important finding is that there is a significant increase in the polymerization rate for SC-ATRP performed in polar/aqueous media [20,21–22].

While SC-ATRP flourished and is well established in terms of being used widely [23], SC-SET-LRP is just beginning to emerge as a valuable technique. This is mainly due to SET-LRP being 10 years younger than ATRP. While there are a very limited number of reports on SC-SET-LRP, the objective of this paper is to compare the SC-ATRP and SC-SET-LRP synthesis techniques using the published data and to aid researchers in further developing these techniques, as well as novel polymer brush systems.

9.2.1 SC-SET-LRP and SC-ATRP of a Series of Amino (Meth)Acrylate Polymer Brushes

The first SC-SET-LRP was reported on the syntheses of a series of amino (meth)acrylate polymer brushes, such as poly(2-(dimethylamino)ethyl methacrylate) (PDMAEMA), poly(2-(diethylamino)ethyl methacrylate) (PDEAEMA), poly(2-(dimethylamino)ethyl acrylate) (PDMAEA), and poly(2-(*tert*-butylamino)ethyl methacrylate) (PTBAEMA), by the authors [24]. (*Tert*-)amino (meth)acrylate polymers are one example of a weakly basic polymer that has attracted considerable interest in the past decade, mainly due to their pH stimuli-responsive behavior favorable to many applications, especially in biomedical fields [20,25]. The most extensively studied amino (meth)acrylate polymer brush is PDMAEMA [20]. SC-ATRP of 2-(dimethylamino)ethyl methacrylate (DMAEMA) has widely been reported on different substrates [26–42] using a variety of catalysts [43–46] and solvents [47–51]. Therefore, prior to attempting SC-SET-LRP, SC-ATRP of DMAEMA was tried with $CuBr/CuBr_2$/bipyridine as the catalyst system and isopropyl alcohol-water solution as the solvent. This polymerization was carried out at a relatively low temperature (40°C), and the dry thickness of the polymer layer, as measured by ellipsometry, reached ~30 nm after polymerization for 48 h. In order to gain thicker polymer brushes and reduce the polymerization time, several modifications to the reaction methods were attempted, including increased catalyst concentrations, higher ligand/catalyst ratios, and use of a N_2-purged (dry) glove box to avoid catalyst oxygenation, but none of these alternate methods showed measurable increased dry layer thicknesses.

Compared to DMAEMA, there are only a few reports on the ATRP of DEAEMA polymer brushes [20,52,53]. Our efforts involved the SC-ATRP of PDEAEMA from silanized silicon (Si) wafer substrates, and different catalyst (CuBr, CuCl), initiator (Si-Br, Si-Cl), solvent (IPA/H_2O, methanol, toluene, THF, anisole), and reaction temperature (25°C–90°C) combinations were examined. The thickest PDEAEMA brushes were obtained using CuCl/$CuCl_2$/HMTETA as the catalyst system with a 2-bromoisobutyryl-based initiator, MeOH as the solvent, and a reaction temperature of 60°C. Under these conditions, the maximum PDEAEMA dry layer thickness obtained was 19 nm after polymerization for 72 h.

Currently, there has been only one report in the literature concerning ATRP of 2-(dimethylamino)ethyl acrylate (DMAEA) which involved homogeneous (bulk) polymerization (instead of polymer brush) formation [54]. To the authors' knowledge, the only LRP technique applied in the synthesis of PDMAEA polymer brushes is nitroxide-mediated radical polymerization [55–57]. The first synthesis of PDMAEA brushes from Si substrates by SC-ATRP used a CuCl/$CuCl_2$/Me_6TREN/2-bromopropionyl type initiator catalyst system and reaction temperature of 70°C. PDMAEA dry layer thicknesses reached 34 ± 5.9 nm after 72 h.

Poly(2-(*tert*-butylamino)ethyl methacrylate) (PTBAEMA) is a neutral polymer, which has attracted recent interest mainly due to potential applications in antifouling paints and coatings as a water-insoluble biocide [58]. PTBAEMA has been grafted from stainless steel surfaces via ATRP using Grubbs' catalyst or grafted onto polypropylene by reactive blending [59–60]. In order to compare the impact of secondary amine structure on LRP efficacy (and in subsequent work, its pH and temperature-responsive behavior), along with the previously described poly(tertiary amines), PTBAEMA was grafted from Si via SC-ATRP using CuCl/CuCl₂/HMTETA as the catalyst system, 2-bromoisobutyryl bromide as the initiator, and a reaction temperature of 90°C. From ellipsometry measurements, the maximum dry thickness of PTBAEMA obtained under these SC-ATRP conditions was 20 ± 0.73 nm.

Table 9.2 presents a summary of the surface-confined (SC) ATRP reaction conditions and corresponding ellipsometry results for the poly(amino (meth)acrylate)s discussed above, as well as analogous systems synthesized with SC-SET-LRP. Due to the attraction of ultrafast polymerization rates for SET-LRP, SC-SET-LRP polymerizations of amino (meth)acrylates were conducted from Si substrates. The same type of SAM substrate, 3-aminopropyltriethoxysilane (APTES)-modified Si, was used in all samples as those in SC-ATRP, but for SC-SET-LRP polymerizations the catalyst was Cu(0) powder instead of copper halides and the ligand selections were guided by the aforementioned SC-ATRP investigations. For the SET-LRP reactions, the polymerization mixture was sonicated or stirred during the reaction to disperse the

TABLE 9.2

Reaction Conditions and Average Dry Polymer Brush Thicknesses (with 95% Confidence Intervals) for SC-LRP of Poly(Amino (Meth)Acrylate) Brushes from Si-SAM Substrates Using ATRP and SET-LRP Methods

	Sample	Dry Brush Thickness (nm)	Catalyst(s)	Ligand	Solvent	Temp (°C)	Reaction Time (h)
1	Si-PDMAEMA	29 ± 7.4	CuBr/CuBr₂	Bipyridine	IPA/water	40	48
2	Si-PDEAEMA	17 ± 2.2	CuCl/CuCl₂	HMTETA	MeOH	60	72
3	Si-PDMAEA	34 ± 5.9	CuCl/CuCl₂	Me₆TREN	Bulk	90	72
4	Si-PTBAEMA	20 ± 0.73	CuCl/CuCl₂	HMTETA	Bulk	90	72
5	Si-PDMAEMA	54 ± 5.8	Cu (0)	Bipyridine	Bulk	40	10
6	Si-PDEAEMA	21 ± 1.9	Cu (0)	HMTETA	Bulk	40	4
7	Si-PDMAEA	6 ± 0.61	Cu (0)	Me₆TREN	DMSO	60	72
8	Si-PTBAEMA	17 ± 6.4	Cu (0)	HMTETA	DMSO	75	96

Source: Ding, S., et al., *J. Polym. Sci. A Polym. Chem.,* 47(23), 6552–6560, 2009.

copper powder uniformly in solution. High molar ratios of ligand to Cu(0) catalyst were used since both the monomers and the solvent (DMSO) were reported to be able to coordinate with copper catalyst and thus compete with the ligand and change the catalysts' nature [50,61]. Polymerization conditions and results for the SC-SET-LRP experiments are also listed in Table 9.2. While ATRP and SET-LRP of a given polymer involved changes in more than one method variable, it can be generally seen that the SET-LRP polymerizations proceeded faster or at relatively lower temperatures vs. the ATRP polymerizations. An obvious improvement was achieved for the SC-SET-LRP of Si-PDMAEMA, which showed significantly greater growth (as demonstrated by polymer brush thicknesses) in a much shorter time vs. SC-ATRP. Further investigation of the polymerization kinetics was conducted and the results are shown in Figure 9.1. There is an approximately linear increase in the dry layer thickness of the grafted PDMAEMA brushes with polymerization time, indicating the chain growth from the Si-SAM substrates is consistent with a controlled or living process.

9.2.2 Poly(*N*-Isopropylacrylamide) (PNIPAM) Brushes

Surface-confined PNIPAM has been one of the most studied polymer brushes due to its thermoresponsive behavior, and thus its application in medicine, membrane technologies, and catalysis [20]. PNIPAM shows lower critical solution temperature (LCST) in aqueous solutions at ~ 32°C. At temperatures below the LCST, PNIPAM brushes are hydrophilic and hydrogen bonds with

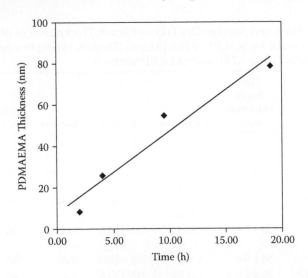

FIGURE 9.1
Dependence of Si-PDMAEMA polymer brush thickness on time via SC-SET-LRP method; experimental conditions: DMAEMA = 3.0 ml, [DMAEMA]/[Cu(0)]/[bipyridine] = 100:1:4, stirred at 40°C. (From Ding, S., et al., *J. Polym. Sci. A Polym. Chem.*, 47(23), 6552–6560, 2009.)

water, whereas at temperatures higher than the LCST, the brushes become hydrophobic and collapse in water [62].

Among the pioneering work on SC-ATRP of PNIPAM brushes, the synthesis system of Huck's group should be noted. The catalyst they used was CuBr/ PMDETA (1:3 molar ratio) with MeOH/H$_2$O (1:1 volume ratio) as solvent. The high molar ratio of catalyst to ligand was perhaps used to reduce the coordination of the copper catalyst to the formed polymer [62]. Polymerization times from 5 to 100 min at ambient temperature resulted in PNIPAM brushes with thicknesses from 13 to >100 nm [62]. After the report by Huck and coworkers, other investigations have employed this synthetic system to prepare PNIPAM brushes on various substrates (Table 9.3) [63–80]. Variations in the volume ratio of MeOH to H$_2$O has been studied (from 1:1 to 3:1, 7:3, 1:2, 1:38, and 1.5:40), along with pure H$_2$O, and it was found that increasing the water content accelerates the ATRP reaction rate [71]. Other researchers have replaced the CuBr catalyst with CuCl, though still with excess amount of ligand (catalyst/ligand molar ratio ~1:5) [76], or adding CuBr$_2$ to improve the fast, yet nonlinear (less controlled) aqueous SC-ATRP of PNIPAM [77].

It has been reported that tetradendate HMTETA produces more reactive complexes than PMDETA [81]. Perhaps due to this reason, some groups chose HMTETA instead of PMDETA [82,83]. Among them, Brooks' group investigated grafting PNIPAM onto functionalized polystyrene latex using two different ligand–copper chloride combinations, PMDETA/CuCl and HMTETA/ CuCl. They found that HMTETA/CuCl produced a higher molecular weight than PMDETA/CuCl under identical reaction conditions. The difference in reactivity of PMDETA/CuCl compared to that of HMTETA/CuCl may be the reason for the lower molecular weight of the grafted polymers produced in the PMDETA system. However, PMDETA/CuCl generally gives higher grafting densities than HMTETA/CuCl, showing that more chains are initiated for this catalyst under identical reaction conditions. They also found that besides the nature of the surface, surface initiator density, surface charge density, catalyst, and monomer concentrations—all of which influence surface-initiated aqueous ATRP reactions—an additional factor, the nature of the monomer, also influences polymerization from surfaces. Therefore, for successful SC polymerization a unique set of reaction conditions is required for a particular monomer-catalyst pair [84–86]. A noteworthy point is that Cu(0) powder had been used together with Cu(I)Cl and Cu(II)Cl$_2$ (Cu(0):Cu(I) Cl:Cu(II)Cl$_2$ molar ratio = 16:15:3) as catalysts, but the effect of Cu(0) on the polymerizations was not explained.

Some groups have used 2,2′-bipyridine (bpy) as the ligand [87–89]. For example, PNIPAM brushes have been prepared in DMF with CuCl/bpy as the catalyst at a reaction temperature of 130°C and time of 40 h [89]. It has been shown that CuX/bpy is not a suitable catalytic system for the ATRP of acrylamido monomers, and NIPAM has not been controllably polymerized by ATRP when bpy was employed as the ligand [90]. Therefore, a more active ligand, Me$_6$TREN, was employed [90,91]. Cu(II)Cl$_2$ has been added in order to

TABLE 9.3

Summary of the SC-ATRP and SC-SET-LRP of PNIPAM Reported in the Literature

Method	Substrate	Catalyst	Ligand	Solvent (vol. ratio)	Temp. (°C)	Ref.
SC-ATRP	Gold	CuBr	PMDETA	MeOH/H$_2$O (1:1)	r.t.	62
	Silicon	CuBr	PMDETA	MeOH/H$_2$O (1:1)	r.t.	63
	Aligned carbon nanotubes (CNTs)	CuBr	PMDETA	MeOH/H$_2$O (1:1)	r.t.	64
	Au/silicon Au/SF10 glass	CuBr	PMDETA	MeOH/H$_2$O (1:1)	r.t.	65
	Gold-coated silicon	CuBr	PMDETA	MeOH/H$_2$O (1:1)	r.t.	66
	Oxidized silicon wafer and glass	CuBr	PMDETA	MeOH/H$_2$O (1:1)	r.t.	67
	Gold	CuBr	PMDETA	MeOH/H$_2$O (1:1)	r.t.	68
	Gold	CuBr	PMDETA	MeOH/H$_2$O (1:1)	60	69
	PAA/PAH	CuBr	PMDETA	MeOH/H$_2$O (1:1)	65	70
	Au-coated PC membrane	CuBr	PMDETA	MeOH/H$_2$O (1:1 or 3:1)	r.t.	71
	Si wafer	CuBr	PMDETA	MeOH/H$_2$O (1.5:40)	r.t.	72
	Au-patterned silicon	CuBr	PMDETA	MeOH/H$_2$O (1:38)	r.t.	73
	Gold	CuBr	PMDETA	MeOH/H$_2$O (7:3)	r.t.	74
	PET	CuBr	PMDETA	MeOH/H$_2$O (7:3)	r.t.	75
	Silica sphere	CuCl	PMDETA	MeOH/H$_2$O (7:3)	r.t.	76
	Polymer coated-gold or silicon	CuBr/CuBr$_2$	PMDETA	MeOH/H$_2$O (1:2)	r.t.	77
	Gold nanoparticle	CuBr	PMDETA	THF/H$_2$O (1:1)	r.t.	78
	Multiwall carbon nanotubes (MWNTs)	CuBr	PMDETA	H$_2$O	r.t.	79

TABLE 9.3 (CONTINUED)

Summary of the SC-ATRP and SC-SET-LRP of PNIPAM Reported in the Literature

Method	Substrate	Catalyst	Ligand	Solvent (vol. ratio)	Temp. (°C)	Ref.
	CNT-PS	CuBr	PMDETA	H_2O	80	80
	Gold	CuBr	Me$_4$Cyclam	DMF	r.t.	95,96
	Cross-linked dextran (CLD) microsphere	CuBr	2,2'-Bipyridine	H_2O	r.t.	87
	PEG-coated cross-linked PS bead	CuBr	2,2'-Bipyridine	H_2O	20	88
	Silica particle	CuCl	2,2'-Bipyridine	DMF	130	89
	Silica nanoparticle (NP)	CuCl/ CuCl$_2$	Me$_6$TREN	2-Propanol	25	91
	Silica bead	CuCl/ CuCl$_2$	Me$_6$TREN	2-Propanol	25	93
	Silica bead	CuCl	Me$_6$TREN	DMF	25	94
	Glass	CuCl/ CuCl$_2$	Me$_6$TREN	H_2O	25	90
	Filter paper	CuCl/ CuCl$_2$	Me$_6$TREN	MeOH/ H_2O (2:1)	r.t.	92
	Magnetic NP	CuBr	HMTETA	MeOH/ H_2O (1:3)	r.t.	82
	PS Latex Particle	Cu/ CuCl/ CuCl$_2$	HMTETA, PMDETA	H_2O	r.t. (22)	84–86
	Si(100)	CuBr/ CuBr$_2$	HMTETA	DMSO	40	83
SC-SET-LRP	Cellulose Nanocrystal	CuBr	PMDETA	MeOH/ H_2O	r.t.	97
	Silicon Wafer	CuBr/ CuBr$_2$	2,2'-Bipyridine	DMF	90	98

ensure an efficient exchange between the dormant and active species [91,92]. Idota et al. first used an aqueous SC-ATRP technique, but which was proven to suffer from disadvantages such as high concentration of polymeric radicals and unavoidable chain transfer or termination [90]. Then they changed the reaction solvent from water to DMF or 2-propanol to better control the polymerization rates [93,94]. Other groups have used CuBr and Me$_4$Cyclam to prepare PNIPAM brushes, and after polymerization for 1 h a grafted polymer layer with a dry thickness of 50 nm was obtained [95,96].

In total, there are, to date, two reports on the preparation of PNIPAM brushes via SC-SET-LRP [97,98]. Zoppe et al. [97] have grafted PNIPAM brushes from cellulose nanocrystals at room temperature. Although a traditional ATRP system (Cu(I)Br/PMDETA, catalyst; MeOH/H$_2$O (variable ratios), solvent) was applied, they deemed their polymerization mechanism to be SET instead of ATRP. This conclusion was mainly based on Feng's work [99], which used CuCl/Me$_6$TREN as the catalyst and DMF/H$_2$O (1:1 volume ratio) as the solvent, and found Cu(0) precipitation at the bottom of the flask formed by the disporpotionation of Cu(I) in DMF/H$_2$O in the presence of Me$_6$TREN in the control experiment. Similarly, Turan and Caykara also prepared PNIPAM brushes from Si wafers with CuBr/CuBr$_2$/bpy in DMF at 90°C. By properly selecting the ratio of [CuBr]/[CuBr$_2$], SC-SET-LRP was well controlled [98]. However, caution should be applied when reaching conclusions that can be generalized to other reaction systems where different ligand or solvent is utilized.

9.3 Conclusions

In this chapter SC-ATRP and SC-SET living radical polymerizations were compared and contrasted by reviewing prior methods used to prepare (meth) acrylate polymer brushes—including a series of poly(amino (meth)acrylate)s and poly(N-isopropylacrylamide) (PNIPAM) brushes—from various substrates. A wide range of catalytic systems have been discussed, including various copper catalyst, ligand, and solvent combinations. Further investigation is still needed to explore the detailed mechanism present in certain polymerizations, but SC-SET-LRP is expected to allow for the efficient synthesis of a wider range of polymer brushes.

References

1. Wang, J.-S., Matyjaszewski, K. 1995. Controlled "Living" Radical Polymerization. Atom Transfer Radical Polymerization in the Presence of Transition-Metal Complexes. *J. Am. Chem. Soc.* 117:5614–5615.
2. Kato, M., Kamigaito, M., Sawamoto, M., Higashimura, T. 1995. Polymerization of Methyl Methacrylate with the Carbon Tetrachloride/Dichlorotris-(Triphenylphosphine) Ruthedum(II)/Methylaluminum Bis(2,6-Di-Tert-Butylphenoxidei) Initiating System: Possibility of Living Radical Polymerization. *Macromolecules* 28:1721–1723.
3. Percec, V., Barboiu, B. 1995. "Living" Radical Polymerization of Styrene Initiated by Arenesulfonyl Chlorides and Cu^I(bpy)$_n$Cl. *Macromolecules* 28:7970–7972.
4. Matyjaszewski, K., Xia J. 2001. Atom Transfer Radical Polymerization. *Chem. Rev.* 101(12):2921–2990.

5. Shen, Y., Tang, H., Ding, S. 2004. Catalyst Separation in Atom Transfer Radical Polymerization. *Prog. Polym. Sci.* 29:1053–1078.

6. Matyjaszewski, K., Coca, S., Gaynor, S.G., Wei, M., Woodworth, B.E. 1997. Zerovalent Metals in Controlled/"Living" Radical Polymerization. *Macromolecules* 30:7348–7350.

7. Queffelec, J., Gaynor, S.G., Matyjaszewski, K. 2000. Optimization of Atom Transfer Radical Polymerization Using Cu(I)/Tris(2-(Dimethylamino)Ethyl) Amine as a Catalyst. *Macromolecules* 33:8629–8639.

8. Matyjaszewski, K,. Tsarevsky, N.V., Braunecker, W.A., Dong, H., Huang, J., Jakubowski, W., Kwak, Y., Nicolay, R., Tang, W., Yoon, J.A. 2007. Role of Cu^0 in Controlled/"Living" Radical Polymerization. *Macromolecules* 40:7795–7806.

9. Kwak, Y., Magenau, A.J.D., Matyjaszewski K. 2011. ARGET ATRP of Methyl Acrylate with Inexpensive Ligands and ppm Concentrations of Catalyst. *Macromolecules* 44:811–819.

10. Percec, V., Guliashvili, T., Ladislaw, J.S., Wistrand, A., Stjerndahl, A., Sienkowska, M.J., Monteiro, M.J., Sahoo, S. 2006. Ultrafast Synthesis of Ultrahigh Molar Mass Polymers by Metal-Catalyzed Living Radical Polymerization of Acrylates, Methacrylates, and Vinyl Chloride Mediated by SET at 25°C. *J. Am. Chem. Soc.* 128:14156–14165.

11. Rosen, B.M., Jiang, X., Wilson, C.J., Nguyen, N.H., Monteiro, M.J., Percec, V. 2009. The Disproportionation of Cu(I)X Mediated by Ligand and Solvent into Cu(0) and Cu(II)X_2 and Its Implications for SET-LRP. *J. Polym. Sci. A Polym. Chem.* 47:5606–5628.

12. Wright, P.M., Mantovani, G., Haddleton, D.M. 2008. Polymerization of Methyl Acrylate Mediated by Copper(0)/Me$_6$-TREN in Hydrophobic Media Enhanced by Phenols; Single Electron Transfer-Living Radical Polymerization. *J. Polym. Sci. A Polym. Chem.* 46:7376 7385.

13. Wang, W., Zhang, Z., Zhu, J., Zhou, N., Zhu, X. 2009. Single Electron Transfer-Living Radical Polymerization of Methyl Methacrylate in Fluoroalcohol: Dual Control Over Molecular Weight and Tacticity. *J. Polym. Sci. A Polym. Chem.* 47:6316–6327.

14. Levere, M.E., Willoughby, I., Wright, P.M., Grice, A.J., FIdge, C., Becer, C.R., Haddleton, D.M. 2011. Cu(O) Mediated Polymerization in Toluene Using Online Rapid GPC Monitoring. *J. Polym. Sci. A Polym. Chem.* 49:1756–1763.

15. Nguyen, N.H., Jiang, X., Fleischmann, S., Rosen, B.M., Percec, V. 2009. The Effect of Ligand on the Rate of Propagation of Cu(0)-Wire Catalyzed SET-LRP of MA in DMSO at 25°C. *J. Polym. Sci. A Polym. Chem.* 47:5629–5638.

16. Lligadas, G., Rosen, B.M., Bell, C.A., Monteiro, M.J., and Percec, V. 2008. Effect of Cu(0) Particle Size on the Kinetics of SET-LRP in DMSO and Cu-Mediated Radical Polymerization in MeCN at 25°C. *Macromolecules* 41:8365–8371.

17. Nguyen, N.H., Rosen, B.M., Lligadas, G., and Percec, V. 2009. Surface-Dependent Kinetics of Cu(0)-Wire-Catalyzed Single-Electron Transfer Living Radical Polymerization of Methyl Acrylate in DMSO at 25°C. *Macromolecules* 42:2379–2386.

18. Levere, M.E., Willoughby, I., O'Donohue, S., de Cuendias, A., Grice, A.J., Fidge, C., Becer, C.R., Haddleton, D.M. 2010. Assessment of SET-LRP in DMSO using online monitoring and Rapid GPC. *Polym. Chem.* 1:1086–1094.

19. Fleischmann, S., Ropsen, B.M., Percec, V. SET_LRP of Acrylates in Air. *J. Polym. Sci. A Polym. Chem.* 48:1190–1196.

20. Barbey, R., Lavanant, L., Paripovic, D., Schöwer, N., Sugnaux, C., Tugulu,S., Klok, H. 2009. Polymer Brushes via Surface-Initiated Controlled Radical Polymerization: Synthesis, Characterization, Properties, and Applications. *Chem. Rev.* 109:5437–5527.

21. Nanda, A.K., Matyjaszewski K. 2003. Effect of [bpy]/[Cu(I)] Ratio, Solvent, Counterion, and Alkyl Bromides on the Activation Rate Constants in Atom Transfer Radical Polymerization. *Macromolecules* 36:599–604.

22. Wang, X.-S., Armes, S.P. 2000. Facile Atom Transfer Radical Polymerization of Methoxy-Capped Oligo(Ethylene Glycol) Methacrylate in Aqueous Media at Ambient Temperature. *Macromolecules* 33:6640–6647.

23. Ouchi, M., Terashima, T., Sawamoto, M. 2009. Transition Metal-Catalyzed Living Radical Polymerization: Toward Perfection in Catalysis and Precision Polymer Synthesis. *Chem. Rev.* 109:4963–5050.

24. Ding, S., Floyd, J.A., and Walters, K.B. 2009. Comparison of Surface Confined ATRP and SET-LRP Syntheses for a Series of Amino (Meth)Acrylate Polymer Brushes on Silicon Substrates. *J. Polym. Sci. A Polym. Chem.* 47(23):6552–6560.

25. Walters, K.B. Surface-Grafting of pH-Responsive Polymer Layers. PhD dissertation, Clemson University, Clemson, SC, August 2005.

26. Shah, R.R., Merreceyes, D., Husemann, M., Rees, I., Abbott, N.L., Hawker, C.J. and Hedrick, J.L. 2000. Using Atom Transfer Radical Polymerization to Amplify Monolayers of Initiators Patterned by Microcontact Printing into Polymer Brushes for Pattern Transfer. *Macromolecules* 33:597–605.

27. Zhao, B., Brittain, W.J. 2000. Synthesis, Characterization, and Properties of Tethered Polystyrene-*b*-Polyacrylate Brushes on Flat Silicate Substrates. *Macromolecules* 33:8813–8820.

28. Zheng, G., Stover, H.D.H. 2002. Grafting of Poly(Alkyl (Meth)Acrylates) from Swellable Poly(DVB80-co-HEMA) Microspheres by Atom Transfer Radical Polymerization. *Macromolecules* 35:7612–7619.

29. Yu, W.H., Kang, E.T., Neoh, K.G. and Zhu, S. 2003. Controlled Grafting of Well-Defined Polymers on Hydrogen-Terminated Silicon Substrates by Surface-Initiated Atom Transfer Radical Polymerization. *J. Phys. Chem. B* 107:10198–10205.

30. Xu, F.J., Cai, Q.J., Kang, E.T., and Neoh, K.G. 2005. Surface-Initiated Atom Transfer Radical Polymerization from Halogen-Terminated Si(111) (Si-X, X) Cl, Br) Surfaces for the Preparation of Well-Defined Polymer-Si Hybrids. *Langmuir* 21:3221–3225.

31. Duan, H., Kuang, M., Wang, D., Kurth, D.G., and Mohwald, H. 2005. Colloidal Stable Amphibious Nanocrystals Derived from Poly[(2-Dimethylamino)Ethyl Methacrylate] Capping. *Angew. Chem. Int. Ed.* 44:1717–1720.

32. Lee, S., Koepsel, R.R., Morley, S.W., Matyjaszewski, K., Sun, Y., and Russell, A.J. 2004. Permanent, Nonleaching Antibacterial Surfaces. 1. Synthesis by Atom Transfer Radical Polymerization. *Biomacromolecules* 5:877–882.

33. Chen, Z., Zhu, X., Shi, Z.L., Neoh, K.G., Kang, E.T. 2005. Polymer Microspheres with Permanent Antibacterial Surface from Surface-Initiated Atom Transfer Radical Polymerization. *Ind. Eng. Chem. Res.* 44:7098–7104.

34. Liu, T., Jia, S., Kowalewski, T., and Matyjaszewski, K. 2006. Water-Dispersible Carbon Black Nanocomposites Prepared by Surface-Initiated Atom Transfer Radical Polymerization in Protic Media. *Macromolecules* 39:548–556.

35. Xu, C., Wu, T., Drain, C.M., Batteas, J.D., Fasolka, M.J., Beers, K.L. 2006. Effect of Block Length on Solvent Response of Block Copolymer Brushes: Combinatorial Study with Block Copolymer Brush Gradients. *Macromolecules* 39:3359–3364.

36. Chen, X., Randall, D.P., Perruchot, C., Watts, J.F., Patten, T.E., Werne, T., Armes, S.P. 2003. Synthesis and Aqueous Solution Properties of Polyelectrolyte-Grafted Silica Particles Prepared by Surface-Initiated Atom Transfer Radical Polymerization. *J. Colloid Interface Sci.* 257:56–64.

37. Zhang, M., Liu, L., Zhao, H., Yang, Y., Fu, G., He, B. 2006. Double-Responsive Polymer Brushes on the Surface of Colloid Particles. *J. Colloid Interface Sci.* 301:85–91.

38. Zhang, F., Xu, F.J., Kang, E.T., Neoh, K.G. 2006. Modification of Titanium via Surface-Initiated Atom Transfer Radical Polymerization (ATRP). *Ind. Eng. Chem. Res.* 45:3067–3073.

39. Kusumo, A., Bombalski, L., Lin, Q., Matyjaszewski, K., Schneider, J.W., Tilton, R.D. 2007. High Capacity, Charge-Selective Protein Uptake by Polyelectrolyte Brushes. *Langmuir* 23:4448–4454.

40. Zhou, F., Hu, H., Yu, B., Osborne, V.L., Huck, W.T.S., Liu, W. 2007. Probing the Responsive Behavior of Polyelectrolyte Brushes Using Electrochemical Impedance Spectroscopy. *Anal. Chem.* 79:176–182.

41. Huang, J., Murata, H., Koepsel, R.R., Russell, A.J., Matyjaszewski, K. 2007. Antibacterial Polypropylene via Surface-Initiated Atom Transfer Radical Polymerization. *Biomacromolecules* 8:1396–1399.

42. Sanjuan, S., Perrin, P., Pantoustier, N., Tran, Y. 2007. Synthesis and Swelling Behavior of pH-Responsive Polybase Brushes. *Langmuir* 23:5769–5778.

43. Zhang, X., Xia, J., Matyjaszewski, K. 1998. Controlled/"Living" Radical Polymerization of 2-(Dimethylamino)ethyl Methacrylate. *Macromolecules* 31:5167–5169.

44. Zhang, X., Xia, J., Matyjaszewski, K. 1999. Synthesis of Well-Defined Amphiphilic Block Copolymers with 2-(Dimethylamino)Ethyl Methacrylate by Controlled Radical Polymerization. *Macromolecules* 32:1763–1766.

45. Zeng, F., Shen, Y., Zhu, S., Pelton, R. 2000. Synthesis and Characterization of Comb-Branched Polyelectrolytes. 1. Preparation of Cationic Macromonomer of 2-(Dimethylamino)Ethyl Methacrylate by Atom Transfer Radical Polymerization *Macromolecules* 33:1628–1635.

46. Mao, B., Gan, L., Gan, Y., Li, X., Ravi, P., Tam, K. 2004. Controlled Polymerizations of 2-(Dialkylamino)ethyl Methacrylates and Their Block Copolymers in Protic Solvents at Ambient Temperature via ATRP. *J. Polym. Sci.: Part A: Polym. Chem.* 42:5161–5169.

47. Zeng, F., Shen, Y., Zhu, S., Pelton, R. 2000. Atom Transfer Radical Polymerization of 2-(Dimethylamino)ethyl Methacrylate in Aqueous Media. *J. Polym. Sci. A Polym. Chem.* 38:3821–3827.

48. Butun, V., Wang, X.-S., de Paz Banez, M.V., Robinson, K.L., Billingham, N.C., Armes, S.P., Tuzar, Z. 2000. Synthesis of Shell Cross-Linked Micelles at High Solids in Aqueous Media. *Macromolecules* 33:1–3.

49. Bories-Azeau, X., Armes, S.P. 2002. Unexpected Transesterification of Tertiary Amine Methacrylates during Methanolic ATRP at Ambient Temperature: A Cautionary Tale. *Macromolecules* 35:10241–10243.

50. Monge, S., Darcos, V., Haddleton, D.M. 2004. Effect of DMSO Used as Solvent in Copper Mediated Living Radical Polymerization. *J. Polym. Sci. A Polym. Chem.* 42:6299–6308.
51. Lee, S., Russell, A.J., Matyjaszewski, K. 2003. ATRP Synthesis of Amphiphilic Random, Gradient, and Block Copolymers of 2-(Dimethylamino)ethyl Methacrylate and n-Butyl Methacrylate in Aqueous Media. *Biomacromolecules* 4:1386–1393.
52. Ryan, A.J., Crook, C.J., Howse, J.R., Topham, P., Jones, R.A.L., Geoghegan, M., Parnell, A.J., Ruiz-Perez, L., Martin, S.J., Cadby, A., Menelle, A., Webster, J.R.P., Gleeson, A.J., Bras, W. 2005. Responsive Brushes and Gels as Components of Soft Nanotechnology. *Faraday Discuss.* 128:55–74.
53. Topham, P., Howse, J.R., Crook, C.J., Parnell, A.J., Geoghegan, M., Jones, R.A.L., Ryan, A.J. 2006. Controlled Growth of Poly(2-(Diethylamino)Ethyl Methacrylate) Brushes via Atom Transfer Radical Polymerisation on Planar Silicon Surfaces. *Polym. Int.* 55:808–815.
54. Zeng, F., Shen, Y., Zhu, S. 2002. Atom Transfer-Radical Polymerization of 2-(N,N-Dimethylamino)Ethyl Acrylate. *Macomol. Rapid Commun.* 23:1113–1117.
55. Ignatova, M., Voccia, S., Gilbert, B., Markova, N., Mercuri, P.S., Galleni, M., Sciannamea, V., Lenoir, S., Cossement, D., Gouttebaron, R., Jerome, R., Jerome, C. 2004. Synthesis of Copolymer Brushes Endowed with Adhesion to Stainless Steel Surfaces and Antibacterial Properties by Controlled Nitroxide-Mediated Radical Polymerization. *Langmuir* 20:10718–10726.
56. Voccia, S., Ignatova, M., Jerome, R., Jerome, C. 2006. Design of Antibacterial Surfaces by a Combination of Electrochemistry and Controlled Radical Polymerization. *Lagmuir*, 22:8607–8613.
57. Bian, K., Cunningham, M.F. 2006. Surface-Initiated Nitroxide-Mediated Radical Polymerization of 2-(Dimethylamino)Ethyl Acrylate on Polymeric Microspheres. *Polymer* 47:5744–5753.
58. Lenoir, S., Pagnoulle, C., Galleni, M., Compere, P., Jerome, R., Detrembleur, C. 2006. Polyolefin Matrixes with Permanent Antibacterial Activity: Preparation, Antibacterial Activity, and Action Mode of the Active Species. *Biomacromolecules* 7:2291–2296.
59. Ignatova, M., Voccia, S., Gilbert, B., Markova, N., Cossement, D., Gouttebaron, R., Jerome, R., Jerome, C. 2006. Combination of Electrografting and Atom-Transfer Radical Polymerization for Making the Stainless Steel Surface Antibacterial and Protein Antiadhesive. *Langmuir* 22:255–262.
60. Thomassin, J.-M., Lenoir, S., Riga, J., Jerome, R., Detrembleur, C. 2007. Grafting of Poly[2-(Tert-Butylamino)Ethyl Methacrylate] onto Polypropylene by Reactive Blending and Antibacterial Activity of the Copolymer. *Biomacromolecules* 8:1171–1177.
61. Percec, V., Guliashvili, T., Popov, A.V., Ramirez-Castillo, E. 2005. Catalytic Effect of Dimethyl Sulfoxide in the Cu(0)/Tris(2-Dimethylaminoethyl)Amine-Catalyzed Living Radical Polymerization of Methyl Methacrylate at 0–90 °C Initiated with CH_3CHCII as a Model Compound for α,ω-Di(Iodo)Poly(Vinyl Chloride) Chain Ends. *J. Polym. Sci. A Polym. Chem.* 43:1935–1947.
62. Jones, D.M., Smith, J.R., Huck, W.T.S., Alexander, C. 2002. Variable Adhesion of Micropatterned Thermoresponsive Polymer Brushes: AFM Investigations of Poly(N-Isopropylacrylamide) Brushes Prepared by Surface-Initiated Polymerizations. *Adv. Mater.* 14:1130–1134.

63. Sun, T.L., Wang, G.J., Feng, L., Liu, B.Q., Ma, Y.M., Jiang, L., Zhu, D.B. 2004. Reversible Switching between Superhydrophilicity and Superhydrophobicity. *Angew. Chem. Int. Ed.* 43:357–360.

64. Sun, T.L., Liu, H.A., Song, W.L., Wang, X., Jiang, L., Li, L., Zhu, D.B. 2004. Responsive Aligned Carbon Nanotubes. *Angew. Chem. Int. Ed.* 43:4763–4766.

65. Wang, X.J., Tu, H.L., Braun, P.V., Bohn, P.W. 2006. Length Scale Heterogeneity in Lateral Gradients of Poly(*N*-Isopropylacrylamide) Polymer Brushes Prepared by Surface-Initiated Atom Transfer Radical Polymerization Coupled with In-Plane Electrochemical Potential Gradients. *Langmuir* 22:817–823.

66. Kaholek, M., Lee, W.-K., LaMattina, B., Caster, K.C., Zauscher, S. 2004. Fabrication of Stimulus-Responsive Nanopatterned Polymer Brushes by Scanning-Probe Lithography. *Nano Lett.* 4:373–376.

67. Tu, H., Heitzman, C.E., Braun, P.V. 2004. Patterned Poly(*N*-Isopropyl-Acrylamide) Brushes on Silica Surfaces by Microcontact Printing Followed by Surface-Initiated Polymerization. *Langmuir* 20:8313–8320.

68. He, Q., Küller, A., Grunze, M., Li, J.B. 2007. Fabrication of Thermosensitive Polymer Nanopatterns through Chemical Lithography and Atom Transfer Radical Polymerization. *Langmuir* 23:3981–3987.

69. Zhou, F., Zheng, Z.J., Yu, B., Liu, W.M., Huck, W.T.S. 2006. Multicomponent Polymer Brushes. *J. Am. Chem. Soc.* 128:16253–16258.

70. Fulghum, T.M., Estillore, N.C., Vo, C.-D., Armes, S.P., Advincula, R.C. 2008. Stimuli-Responsive Polymer Ultrathin Films with a Binary Architecture: Combined Layer-by-Layer Polyelectrolyte and Surface-Initiated Polymerization Approach. *Macromolecules* 41:429–435.

71. Lokuge, I., Wang, X., Bohn, P.W. 2007. Temperature-Controlled Flow Switching in Nanocapillary Array Membranes Mediated by Poly(*N*-Isopropylacrylamide) Polymer Brushes Grafted by Atom Transfer Radical Polymerization. *Langmuir* 23:305–311.

72. Liu, Y., Klep, V., Luzinov, I. 2006. To Patterned Binary Polymer Brushes via Capillary Force Lithography and Surface-Initiated Polymerization. *J. Am. Chem. Soc.* 128:8106–8107.

73. Ahn, S.J., Kaholek, M., Lee, W.-K., LaMattina, B., LaBean, T.H., Zauscher, S. 2004. Surface-Initiated Polymerization on Nanopatterns Fabricated by Electron-Beam Lithography. *Adv. Mater.* 16:2141–2145.

74. Plunkett, K.N., Zhu, X., Moore, J.S., Leckband, D.E. 2006. PNIPAM Chain Collapse Depends on the Molecular Weight and Grafting Density. *Langmuir* 22:4259–4266.

75. Friebe, A., Ulbricht, M. 2007. Controlled Pore Functionalization of Poly(Ethylene Terephthalate) Track-Etched Membranes via Surface-Initiated Atom Transfer Radical Polymerization. *Langmuir* 23:10316–10322.

76. Schepelina, O., Zharov, I. 2007. PNIPAAM-Modified Nanoporous Colloidal Films with Positive and Negative Temperature Gating. *Langmuir* 23:12704–12709.

77. Teare, D.O.H., Barwick, D.C., Schofield, W.C.E., Garrod, R.P., Ward, L.J., Badyal, J.P.S. 2005. Substrate-Independent Approach for Polymer Brush Growth by Surface Atom Transfer Radical Polymerization. *Langmuir* 21:11425–11430.

78. Roth, P.J., Theato, P. 2008. Versatile Synthesis of Functional Gold Nanoparticles: Grafting Polymers From and Onto. *Chem. Mater.* 20:1614–1621.

79. Kong, H., Li, W.W., Gao, C., Yan, D.Y., Jin, Y.Z., Walton, D.R.M., Kroto, H.W. 2004. Poly(*N*-Isopropylacrylamide)-Coated Carbon Nanotubes: Temperature-Sensitive Molecular Nanohybrids in Water. *Macromolecules* 37:6683–6686.

80. Liu, Y.-L., Chen, W.-H. 2007. Modification of Multiwall Carbon Nanotubes with Initiators and Macroinitiators of Atom Transfer Radical Polymerization. *Macromolecules* 40:8881–8886.
81. Xia, J.H., Matyjaszewski, K. 1997. Controlled/"Living" Radical Polymerization. Atom Transfer Radical Polymerization Using Multidentate Amine Ligands. *Macromolecules* 30:7697–7700.
82. Lattuada, M., Hatton, T.A. 2007. Functionalization of Monodisperse Magnetic Nanoparticles. *Langmuir* 23:2158–2168.
83. Xu, F.J., Zhong, S.P., Yung, L.Y.L., Kang, E.T., Neoh, K.G. 2004. Surface-Active and Stimuli-Responsive Polymer-Si(100) Hybrids from Surface-Initiated Atom Transfer Radical Polymerization for Control of Cell Adhesion. *Biomacromolecules* 5:2392–2403.
84. Kizhakkedathu, J.N., Norris-Jones, R., Brooks, D.E. 2004. Synthesis of Well-Defined Environmentally Responsive Polymer Brushes by Aqueous ATRP. *Macromolecules* 37:734–743.
85. Goodman, D., Kizhakkedathu, J.N., Brooks, D.E. 2004. Attractive Bridging Interactions in Dense Polymer Brushes in Good Solvent Measured by Atomic Force Microscopy. *Langmuir* 20:2333–2340.
86. Goodman, D., Kizhakkedathu, J.N., Brooks, D.E. 2004. Molecular Weight and Polydispersity Estimation of Adsorbing Polymer Brushes by Atomic Force Microscopy. *Langmuir* 20:3297–3303.
87. Kim, D.J., Heo, J.-Y., Kim, K.S., Choi, I.S. 2003. Formation of Thermoresponsive Poly(Nisopropylacrylamide)/Dextran Particles by Atom Transfer Radical Polymerization. *Macromol. Rapid Commun.* 24:517–521.
88. Bontempo, D., Tirelli, N., Masci, G., Crescenzi, V., Hubbell, J.A. 2002. Thick Coating and Functionalization of Organic Surfaces via ATRP in Water. *Macromol. Rapid Commun.* 23:417–422.
89. Fu, Q., Rama Rao, G.V., Basame, S.B., Keller, D.J., Artyushkova, K., Fulghum, J.E., López, G.P. 2004. Reversible Control of Free Energy and Topography of Nanostructured Surfaces. *J. Am. Chem. Soc.*126:8904–8905.
90. Idota, N., Kikuchi, A., Kobayashi, J., Akiyama, Y., Sakai, K., Okano, T. 2006. Thermal Modulated Interaction of Aqueous Steroids Using Polymer-Grafted Capillaries. *Langmuir* 22:425–430.
91 Wu, T., Zhang, Y.F., Wang, X.F., Liu, S.Y. 2008. Fabrication of Hybrid Silica Nanoparticles Densely Grafted with Thermoresponsive Poly(*N*-Isopropylacrylamide) Brushes of Controlled Thickness via Surface-Initiated Atom Transfer Radical Polymerization. *Chem. Mater.* 20:101–109.
92. Lindqvist, J., Nystrom, D., Ostmark, E., Antoni, P., Carlmark, A., Johansson, M., Hult, A., Malmstrom, E. 2008. Intelligent Dual-Responsive Cellulose Surfaces via Surface-Initiated ATRP. *Biomacromolecules* 9:2139–2145.
93. Nagase, K., Kobayashi, J., Kikuchi, A.I., Akiyama, Y., Kanazawa, H., Okano, T. 2008. Effects of Graft Densities and Chain Lengths on Separation of Bioactive Compounds by Nanolayered Thermoresponsive Polymer Brush Surfaces. *Langmuir* 24:511–517.
94. Nagase, K., Kobayashi, J., Kikuchi, A., Akiyama, Y., Kanazaw, H., Okano, T. 2007. Interfacial Property Modulation of Thermoresponsive Polymer Brush Surfaces and Their Interaction with Biomolecules. *Langmuir* 23:9409–9415.

95. Balamurugan, S., Mendez, S., Balamurugan, S.S., O'Brien, M.J., López, G.P. 2003. Thermal Response of Poly(*N*-Isopropylacrylamide) Brushes Probed by Surface Plasmon Resonance. *Langmuir* 19:2545–2549.
96. Yim, H., Kent, M.S., Mendez, S., Balamurugan, S.S., Balamurugan, S., Lopez, G.P., Satija, S. 2004. Temperature-Dependent Conformational Change of PNIPAM Grafted Chains at High Surface Density in Water. *Macromolecules* 37:1994–1997.
97. Zoppe, J., Habibi, Y., Rojas, O.J., Venditti, R.A., Johansson, L.-S., Efimenko, K., Osterberg, M., Laine, J. 2010. Poly(*N*-Isopropylacrylamide) Brushes Grafted from Cellulose Nanocrystals via Surface-Initiated Single-Electron Transfer Living Radical Polymerization. *Biomacromolecules* 11:2683–2691.
98. Turan, E., Caykara, T. 2010. Kinetic Analysis of Surface-Initiated SET-LRP of Poly(N-Isopropylacrylamide). *J. Polym. Sci. A Polym. Chem.* 48:5842–5847.
99. Feng, C., Shen, Z., Li, Y., Gu, L., Zhang, Y., Lu, G., Huang, X. 2009. PNIPAM-b-(PEA-g-PDMAEA) Double-Hydrophilic Graft Copolymer: Synthesis and Its Application for Preparation of Gold Nanoparticles in Aqueous Media. *J. Polym. Sci. A Polym. Chem.* 47:1811–1824.

10

Stimulus-Responsive Polymer Brushes on Polymer Particles' Surfaces and Applications

V. Mittal

Chemical Engineering Department, Petroleum Institute,
Abu Dhabi, United Arab Emirates

CONTENTS

10.1 Introduction

Surface functionalization is an extremely important field of science as the surface properties significantly affect the application of solid materials. Development on specific surface characteristics is thus always of importance and need, and has seen growing interest in recent years. One such surface property of recent interest is the reversible response of the materials to stimulants like temperature, pH, salt concentration changes, etc. Generation of this property on the surface of the materials leads to functionalized materials in which their nature, e.g., hydrophobicity or hydrophilicity, can be finely controlled by environmental stimulants like temperature, salt, pH, etc. This property translates the use of these materials in a number of applications, like temperature-controlled adsorption and desorption of biomolecules, temperature-controlled drug delivery processes, etc.

Out of a number of known polymers known to exhibit this behavior, water-soluble poly(N-isopropylacrylamide) (PNIPAAM) is a very common and extensively studied material. It has a lower critical solution temperature (LCST) of about 32°C [1–6], and below this temperature, the chains exhibit chain-extended conformations and random coil structure. The intermolecular hydrogen bonding with the water molecules due to the chain extended morphology generates the hydrophilic nature of the chains. The chains transform into a more collapsed globular form above the lower critical

229

solution temperature owing to the domination of intramolecular bonding between the CO and NH groups over the external hydrogen bonding. The highly reversible responses of PNIPAAM chains to temperature as well as other stimulants have led to its extensive use for the synthesis of stimulus-sensitive surfaces. Due to these characteristics, PNIPAAM chains grown on various surfaces have been used for protein adsorption, responsive gels, biological separations, etc. [7–17].

Figure 10.1 shows the schematic of thermal responsiveness of the PNIPPAM film grafted on a surface [18]. Such a shell of thermally responsive PNIPAAM grown on the surfaces shows extremely fast and reversible changes in wetting characteristics, and hence the size of the chains bound on the surface similar to the behavior seen in free polymer chains. Such modified surfaces have been observed to be useful for temperature-controlled adsorption and desorption of proteins and viruses, as functional filters and textiles as well as controlled drug release [19]. Sun et al. [19] also reported super hydrophobic and hydrophilic behavior reversible in nature when such chains were grown from very rough surfaces. The nature of the surface used to graft the environmentally responsive brushes has also been found to affect the wetting characteristics of the surfaces functionalized with PNIPAAM. Superhydrophobous and superhydrophilic surfaces were reported when the PNIPAAM chains were grown on extremely rough surfaces [19]. Contact angles with water of roughly 0° and 150° were reported at 25°C and 40°C respectively, while values of roughly 64° and 93° were found for the same temperatures when PNIPAAM was grown on flat surfaces. Based on this result, one would expect that polymeric porous materials could exhibit similarly enhanced reversible

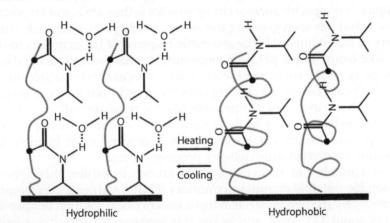

FIGURE 10.1

Schematic of thermal responsiveness of the PNIPAAM shell grafted onto a surface. (Reproduced from Mittal, V., and Matsko, N. B., *Open Surface Science Journal*, 1, 14–19, 2009.)

hydrophobic-to-hydrophilic transitions. You et al. [20] also showed the similar behaviors for the dendritic shell morphologies.

Many techniques to functionalize the polymer surfaces have been reported in the literature, like physical adsorption, grafting to surface, and grafting from surface. The grafting from surface approach has enjoyed more success because of its ability to generate densely packed polymer functionalities in the absence of any kinetic hindrance along with the advantage of covalently bound polymer chains on the surface. Out of all the polymerization techniques used to graft polymer brushes, controlled polymerization techniques such as ionic, atom transfer radical polymerization (ATRP) and reversible addition fragmentation transfer (RAFT) have received widespread attention. Moreover, ATRP has been observed to be more successful than other controlled polymerization techniques. For such purposes, the ATRP initiator needs to be chemically immobilized on the surface first.

Most of the studies were undertaken for the flat substrates, totally neglecting the spherical particles. The spherical particles, being of engineering applications in separation technologies, also need to be analyzed. As the polymer latex particle surfaces in water are very different from the silica or other flat surfaces commonly used, because of surface charges, charge concentration, wetting properties, etc., their behavior in the above-mentioned functionalization processes has to be investigated in detail. The following paragraphs describe few examples of grafting of PNIPAAM brushes from surfaces of polymer particles.

Emulsion polymerization is the commonly employed method to generate surface functionalized spherical polymer particles. Different methods can be used to achieve such a functionalized particle. In a representative example, styrene and N-isopropylacrylamide (NIPAAM) could be copolymerized and, owing to the different nature of monomers, the hydrophilic monomer forms the shell in the polymerized particle. As an example, to generate more ordered particles, the functionalizing moiety, e.g., ATRP or RAFT initiator, can be covalently immobilized onto seed polymer latex particles, and the particles can be subsequently grafted with functional polymers—PNIPAAM, for example—for thermal responsiveness. Other polymers with functional groups on the backbone that can be used for specific interactions with foreign media can also be similarly grafted. The process thus requires first the formation of polymer seed with the desired characteristics, followed by the immobilization of the initiator on the surface as a thin shell. This process forms a well-defined core-shell morphology; however, a mixed morphology can also be generated by a single-step process in which the monomer carrying the initiator (e.g., for NMP or ATRP process) is added to the polymerizing seed particles, once a desired degree of their polymerization is reached. In both cases, various surface morphologies can be achieved depending on various thermodynamic and kinetic factors.

10.2 Generation of PNIPAAM Brushes on Particles

Although the polymer particles are not exactly the same as other spherical substrates, like silica particles, in terms of their colloidal interactions, many reported surface modifications of silica or other inorganic particles with poly(N-isopropylacrylamide) can also be applied to polymer particles. Kanazawa et al. reported a new concept of chromatographic separations using temperature-responsive silica stationary surfaces [8]. The silica surfaces were modified with temperature-responsive polymers to exhibit temperature-controlled hydrophilic/hydrophobic changes. To achieve this functionalization, poly(N-isopropylacrylamide) (PNIPAAM) was grafted onto (aminopropyl) silica using an activated esteramine coupling method (Figure 10.2). The results showed that the temperature-dependent hydrophilic and hydrophobic character of PNIPAAM chains could be observed on the silica surfaces, as grafted silica surfaces showed hydrophilic properties at lower temperatures and were transformed to hydrophobic surfaces once the temperature was raised above the lower critical solution temperature of PNIPAAM. It could be successfully shown that the steroids separation could be controlled by the temperature of the aqueous phase in high-performance chromatography.

Go et al. reported the generation of functional chromatographic separation media based on PNIPAAM grafting [21]. In this approach, the polymerizable functional groups were immobilized on the surfaces of silica particles. These polymerizable methacryloylpropyl groups were generated by treatment of silica beads with silane coupling agents. NIPAAM was then copolymerized with the monomer units by the addition of initiator. The PNIPAAM chains start to form away from the surface of the beads; however, during polymerization, the growing radical attacks the double bond immobilized on the silica surfaces and transfers the radical to the surface. Subsequently, the chains start to grow from the surface. This way, the whole surface is grafted with PNIPAAM chains covalently bound to the surface. It also produces a large amount of free polymer in the solution, which is required to be properly cleaned if any useful application from the grafted particles is to be expected. The above functionalized particles were very efficient in separation of dextrans with different molecular weights. As a result of PNIPAAM functionalization, the elution times were also observed to shorten in general. Figure 10.3 shows the process of PNIPAAM grafting on silica particles.

In another interesting study, Hosoya et al. used porous polystyrene beads of roughly 1 µm diameter as base material, and PNIPAAM was polymerized and selectively grafted on the surface of the beads using porogens [22,23] as shown in Figure 10.4. By using cyclohexanol as porogen, it was possible to graft the PNIPAAM chains inside and outside of the porous polymer particles owing to the solubility of propagating PNIPAAM radical chains in cyclohexanol. On the other hand, it was only possible to graft PNIPAAM on the outer surface of porous particles when toluene was used as porogen, as

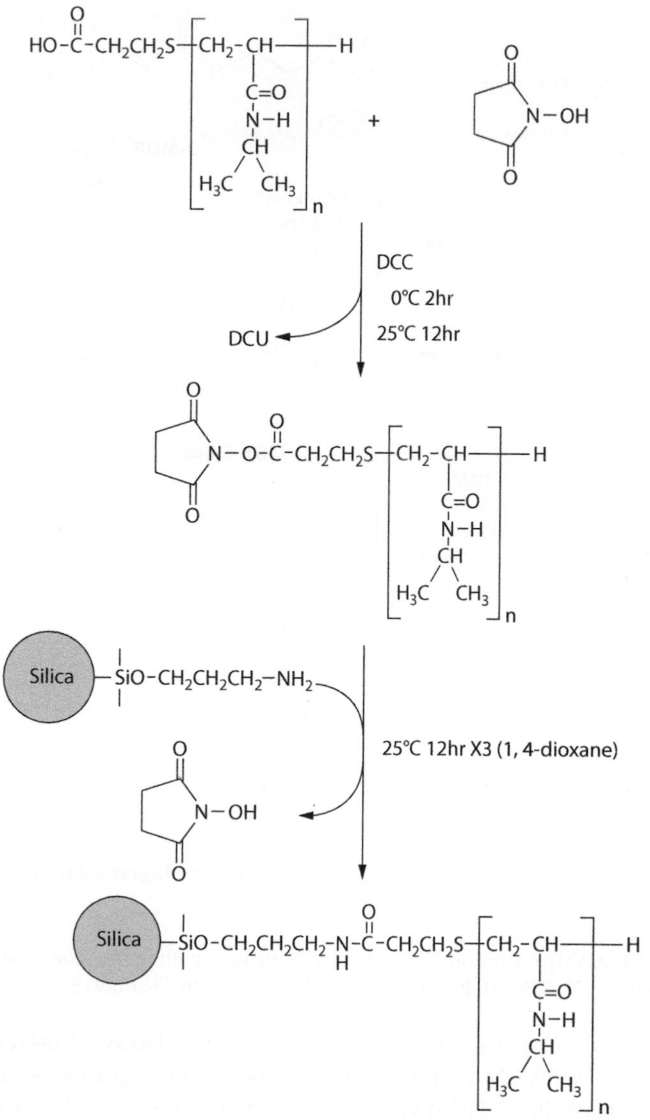

FIGURE 10.2
Modification of aminopropyl silica with PNIPAAM. (Reprinted with permission from Kanazawa, H., et al., *Analytical Chemistry*, 68, 100–58, 1996, copyright (1996) American Chemical Society.)

the propagating radical chains could not enter the inner voids of porous particles due to their insolubility in toluene. The functionalized particles could be successfully shown to be applicable for a number of chromatographic separations that could be controlled by the use of temperature, indicating that the PNIPAAM functionalization was useful in avoiding the harsh separation conditions of salt and pH used conventionally.

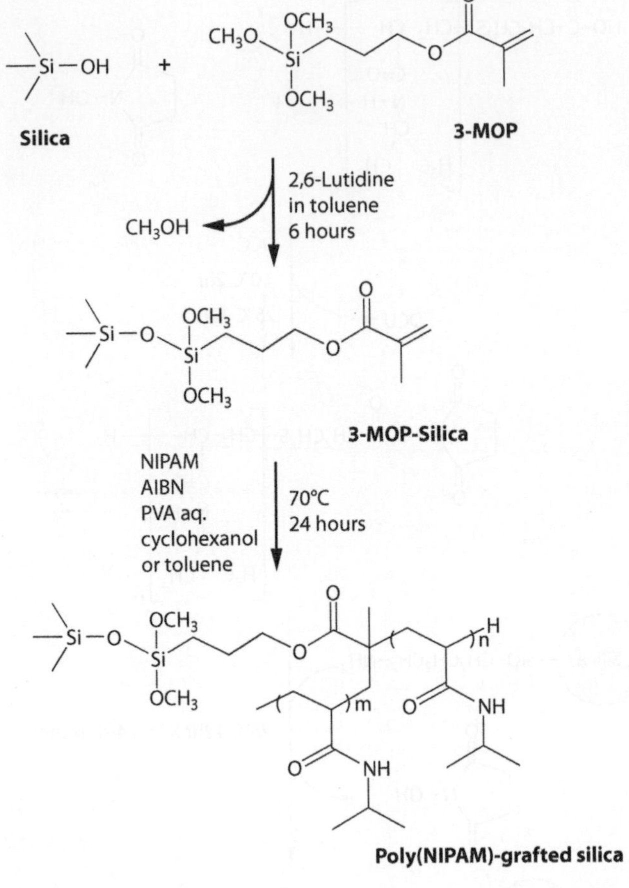

FIGURE 10.3

Schematic of PNIPAAM grafting on silica particles. (Reprinted with permission from Go, H., et al., *Analytical Chemistry*, 70, 4086–93, 1998, copyright (1998) American Chemical Society.)

   PNIPAAM brushes on a polystyrene particle were also synthesized by copolymerization of NIPAAM with styrene [24–26]. According to this process, initiation of hydrophilic (NIPAAM) monomer in the beginning is followed by polymerization of hydrophobic (styrene) monomer, which typically takes place inside the forming particles. Thick layers of PNIPAAM on the particles' surface as well as rough surface morphology can be achieved in this case. These particles were generated by initiating first the batch polymerization of styrene and N-isopropylacrylamide (NIPAAM) using 2,2′-azobis(2-amidinopropane) dihydrochloride as a cationic initiator. After 70% polymerization was achieved, a second batch of monomers containing NIPAAM, N,N′-methylenebisacrylamide (MBA) and aminoethyl methacrylate hydrochloride (AEMH), along with the initiator, were added to form a shell containing a majority of PNIPAAM. The adsorption-desorption behavior of bovine serum albumin (BSA) protein was

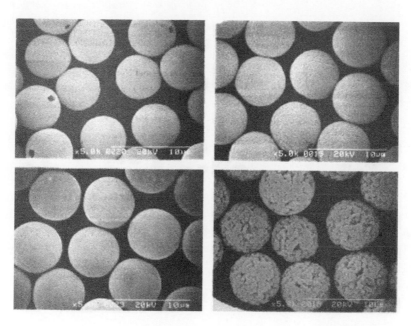

FIGURE 10.4
Top left and right: Scanning electron micrographs of the unmodified and modified particles using cyclohexanol as porugen, respectively. Bottom left and right: Scanning electron micrograph of the unmodified and modified particles using toluene as porogen, respectively. (Reprinted with permission from Hosoya, K., et al., *Macromolecules*, 27, 3973–76, 1994, copyright (1994) American Chemical Society.)

studied on the PNIPAAM functionalized particles. Adsorption of BSA protein onto such cationic and thermosensitive particles was found negligible below the LCST of PNIPAAM, whereas it was much higher above it, indicating that the thermally responsive behavior of PNIPAAM could be translated onto the particles. Figure 10.5 shows the transmission electron microscopy (TEM) micrographs of the generated core-shell particles.

Mittal et al. also reported similar polystyrene (PS)-PNIPAAM copolymer articles [26] as shown in Figure 10.6 and compared the swelling-deswelling behavior of these particles with PNIPAAM-grafted PS particles by using the atom transfer radical polymerization approach. PNIPAAM-grafted particles were observed to show a swelling degree of more than three times, whereas the copolymer particles were swollen to less than 100% at the same conditions. The accurately defined brush morphology was suggested to be responsible for the much better response of the grafted particles. In particular, it was believed that the copolymer structure did not allow PNIPAAM to fully exhibit its potential in switching between hydrophilic and hydrophobic behavior, nor did the presence of cross-linking.

This also indicated that much better control over the process can only be achieved if a grafting route involving controlled polymerization is selected to functionalize the particles. Thus, it was confirmed by these studies that, in

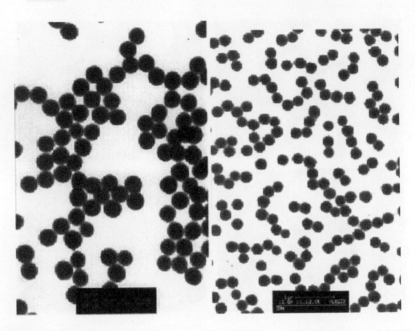

FIGURE 10.5
Transmission electron micrographs of core-shell microspheres. (Reproduced from Duracher, D., et al., *Polymer International*, 53, 618–26, 2004. With permission from Wiley.)

spite of the simplicity of the copolymerization process, more tunable properties of the functionalized material can be achieved using grafting techniques, as they allow the independent tuning of the support characteristics and of the grafting points. Moreover, the control over the density of the grafted brushes can also be achieved by tuning the initial density of the initiator on the grafting surface.

Physical adsorption of PNIPAAM on surfactant-free polystyrene nanoparticles was also reported by Gao and Wu [5], and methods like static and dynamic laser light scattering were employed to quantify the adsorption. The adsorption was achieved by mixing polystyrene latex particles with preformed PNIPAAM chains with controlled molecular weights. Figure 10.7 shows the distributions of the hydrodynamic size of the polymer particles before and after the polymer adsorption. By using these light scattering techniques, the temperature dependence of the hydrodynamic radius of the nanoparticles adsorbed with PNIPAAM was monitored to reveal the coil-to-globule transition of PNIPAAM on the particle surface. It was also reported that the adsorbed chains had a lower LCST of 29°C than the free chains in water (32°C), indicating slight changes in their physical characteristics after adsorption. It was also observed by the authors that, for a given temperature below the LCST, on increasing the amount of the added PNIPAAM, the thickness of the adsorbed PNIPAAM layer increased, but the average density of the adsorbed PNIPAAM layer decreased, suggesting an extension of the adsorbed chains.

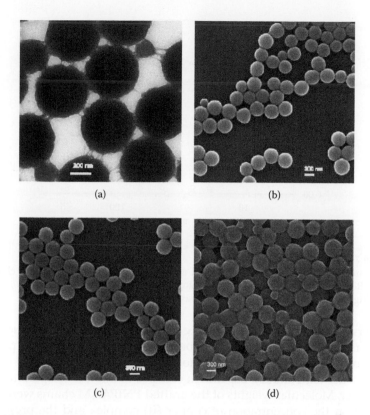

(a) (b)

(c) (d)

FIGURE 10.6
Microscope images in (a) - (d) of the NIPAAM functionalized particles formed by PS-PNIPAAM emulsion copolymerization using different reaction conditions and monomer amounts. (Reproduced from Mittal, V., *Advances in Polymer Latex Technology*, Nova Science Publishes, New York, 2009.)

Kizhakkedathu et al. reported a series of studies on the grafting of thermally responsive polymers from the latex particles using controlled ATRP methods [17,28]. A polystyrene shell latex (PSL) was synthesized, which was then modified by forming a shell of 2-(methyl-2'-chloropropionato) ethyl acrylate (HEA-Cl) on the polystyrene particles. HEA-Cl contained two functionalities in the molecule. By one terminal double bond, it could be polymerized along with styrene to form a thin shell on the seed particles, whereas by the other terminal ATRP initiator moiety, it could be used to subsequently graft the poly(N,N'-dimethylacrylamide) chains from the surface of the spherical particles. Different densities of the ATRP initiator could be achieved on the surface of the particles by changing the feed ratio of monomers in the shell-forming batch. It was observed by the authors that molecular weight of the grafted poly(N,N'-dimethylacrylamide) chains varied linearly with monomer concentration, and grafting density was roughly independent of monomer concentration except at the highest initiator concentration. Grafting density

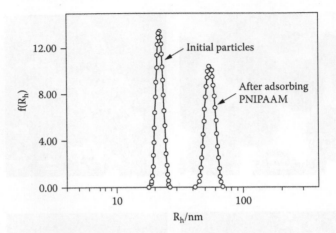

FIGURE 10.7

Hydrodynamic radius distribution before and after the adsorption of PNIPAAM chains on particles. (Reprinted with permission from Gao, J., and Wu, C., *Macromolecules*, 30, 6873–76, 1997, copyright (1997) American Chemical Society.)

was observed to vary (as initiator surface concentration). In another study from the authors, PNIPAAM brushes could be grafted from the surface of polymer particles by using similar ATRP methods, and the grafted brushes showed a second-order temperature-dependent collapse at a lower temperature than free polymer. Molecular weights of the grafted PNIPAAM chains were found to depend on the concentration of copper (II) complex and the presence of external initiator in the reaction medium along with the monomer concentration. A comparison of the ligands was also reported for their effect on the molecular weight of the grafted PNIPAAM chains, and hexamethyltriethylene-tetramine (HMTETA)/CuCl catalyst was observed to produce higher molecular weight chains than pentamethyldiethylenetriamine PMDETA/CuCl. The authors also studied the block copolymerization of N,N′-dimethylacrylamide from PNIPAAM-grafted latex. It confirmed the successful application of ATRP methods on the spherical particles. The hydrodynamic thickness of generated brushes could also be calculated by dynamic light scattering, and these values were found to be sensitive to temperature and salt concentration. The particles remained stable even when the PNIPAAM chains on the surface collapsed at a higher temperature, as shown in Figure 10.8.

Control of surface morphology of the particles has always been a challenge, as the homogenous surface distribution of the functionalizing entities (e.g., ATRP initiator) is of utmost importance in order to achieve subsequently homogenous polymer brushes. Reaction conditions such as the weight ratio of the functionalizing molecule (with terminal double bond and ATRP initiator moieties) to the seed polymer, preswelling of the seed particles with the functionalizing molecule, mode of addition of the monomer to the seed, etc., dramatically influence the resulting particles' size, shape, and morphology.

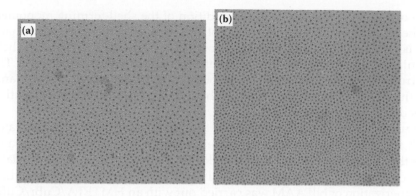

FIGURE 10.8
Optical micrographs of PNIPAAM brush in water above and below LCST: (a) 20°C and (b) 40°C. (Reprinted with permission from Kizhakkedathu, J. N., et al., *Macromolecules*, 37, 734–43, 2004, copyright (2004) American Chemical Society.)

Depending on the nature of the initiator and its ability to immobilize on preformed polymer latex particles, the resulting morphology in terms of initiator surface concentration and distribution can dramatically change [29,30]. Based on the kinetic and thermodynamic factors affecting the course of the polymerization, it is often challenging to find the optimum conditions leading to the uniform distribution of an ATRP initiator on the latex particles. In order to characterize the quality of the functionalization, the characterization of particle size—along with its distribution before and after brush growth, as well as changes in brush dimension because of thermal response—is essential. This is mostly carried out by laser light scattering. However, small amounts of aggregation can sometimes be observed, especially in the particles synthesized with surfactant-free emulsion polymerizations, thus making the use of light scattering difficult. In such cases, other qualitative or quantitative investigation tools are required. Mittal et al. [29,30] studied the conditions to optimize the surface functionalization of the polymer particles with ATRP initiator 2-(2-bromopropionyloxy) ethyl acrylate (BPOEA) by extensive use of microscopy. It was observed that a number of operating conditions led to the generation of secondary nucleation of the shell-forming monomers. Apart from that, the surface morphology of the ATRP initiator-modified particles was always different, depending upon the synthesis technique. It was reported that using the one-step process, i.e., by addition of the shell-forming particles to the seed particles when the seed particles have 70% monomer conversion, led to the elimination of the secondary nucleation. It was also observed that the addition of a small amount of cross-linker divinylbenzene (DVB) led to the total elimination of secondary nucleation, and particles could be successfully functionalized with ATRP initiator by following any method of synthesis. In the study it was concluded, therefore, that to achieve cross-linked material, the functionalization procedure should be a one-step process with starved addition of the shell-forming monomers at the end stages of polymerization of the seed. If non-cross-linked material is required, then good results are obtained by the shot addition of shell-forming monomers to the 70% polymerized polystyrene seed particles. The control of the density of functional groups could also be achieved by changing the styrene/BPOEA mole ratio in the shell-generating monomer mixture added to the polystyrene seed.

Figure 10.9 shows the examples of the polystyrene particles (generated with surfactant free emulsion polymerization) modified with a thin layer of either pure ATRP initiator molecules or a copolymer of ATRP initiator molecules with styrene (with and without DVB). The atom transfer radical polymerization of NIPAAM was carried out with an HMTETA/CuBr/CuBr$_2$ system, along with additionally added Cu powder to latex particles prefunctionalized with ATRP initiator. The presence of a PNIPAAM layer was also confirmed by electron energy loss spectroscopy (EELS), which indicates the presence of a layer containing oxygen and nitrogen around the particles. The particles were observed to be thermally responsive, and their thermally responsive character was quantified by measuring the amount of the swelling of the grafted chains below and above the LCST. A potential use of

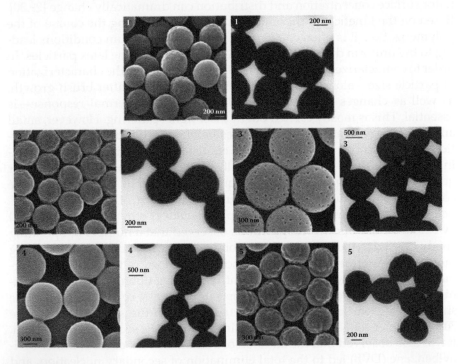

FIGURE 10.9
SEM (left) and TEM (right) images of the PNIPAAM-modified particles generated at different reaction conditions. (Reproduced from Mittal, V., et al., *European Polymer Journal*, 43, 4868–81, 2007. With permission from Elsevier.)

these materials as stationary phases for bioseparations was suggested, and it was reported that the adsorption behavior of tobacco mosaic virus could successfully be achieved in these particles. The process was also observed to be fully reversible with temperature and fast in response. Figures 10.10 and 10.11 show the particles after grafting with PNIPAAM. The particles in Figure 10.10 were generated without the surfactant, whereas the particles in Figure 10.11 were generated using surfactant [31]. Figure 10.12 also shows the effect of increasing the amount of initial amount of poly(N-isopropylacrylamide) monomer in the system. A corresponding amount of polymer could be grafted on the surface of particles [27]. Figure 10.13 also shows the EELS analysis coupled with transmission electron microscopy. The EELS image series was recorded with a Zeiss EM 912 Omega TEM operated at 120 kV and 10,000 magnification. A 90 µm objective diaphragm was used, corresponding to a collection of semiangles of 12.4 mrad. A series of energy-filtered images were recorded using a slow-scan charge-coupled device (CCD) camera

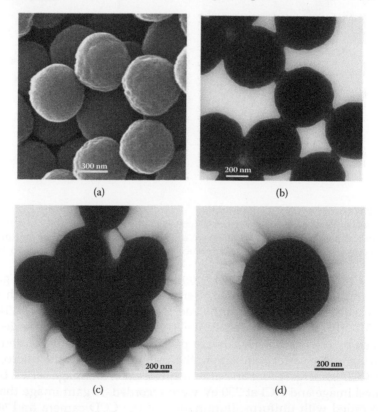

(a) (b)

(c) (d)

FIGURE 10.10
(a, b) SEM and TEM micrographs of the latex particles functionalized with a copolymer layer carrying ATRP initiator. (c, d) The grafted brushes of poly(N-isopropylacrylamide) from the surface of the functionalized particles using ATRP. (Reproduced from Mittal, V., *Polymers*, 2, 40–56, 2010.)

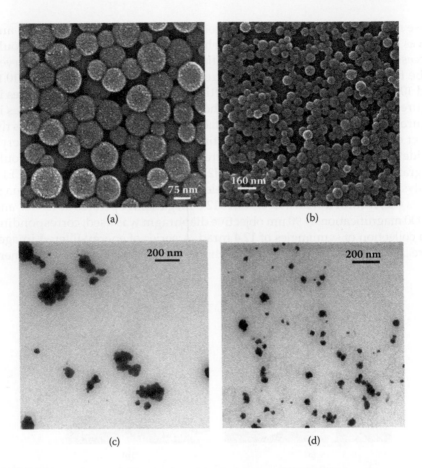

FIGURE 10.11
(a) Cross-linked polystyrene latex particles generated by the use of surfactant. (b) Particles of figure a surface modified with ATRP initiator. (c, d) The particles of figure b after grafting with poly(N-isopropylacrylamide) brushes. (Reproduced from Mittal, V., *Polymers*, 2, 40–56, 2010.)

(1 k × 1 k pixels and 19 μm pixel size) with a Yttrium aluminum garnet (YAG) scintillator screen (Proscan, Germany). An energy slit width of 15 eV was used. The series covered the energy range 258–303 eV(6C), 454–544 eV(8O), 317–410 eV(7N), and 1742–1970 eV(74W). Recording of the series was automated by the ESI vision software package (SIS, Germany) hosted on a PC. During image series acquisition, the program controls the microscope and the CCD camera. Before and after acquisition of an image series, both a black-level image and HCl at 250 eV were recorded. A gain image that has been recorded with uniform illumination of the CCD camera and stored before was used to correct for the pixel-to-pixel gain variations of the CCD camera [30]. The dots in the EELS image in Figure 10.13 confirm the presence of oxygen around the particles. The presence of oxygen is attributed to the grafting of PNIPAAM chains on the surface.

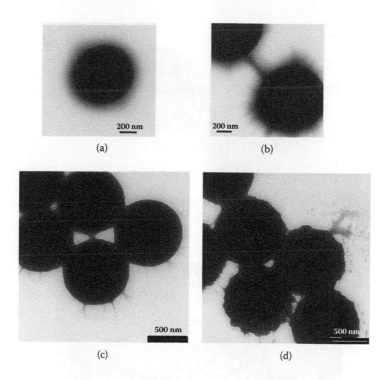

FIGURE 10.12
(a, b, d) Effect of increasing the amount of initial amount of poly(N-isopropylacrylamide) monomer in the system taken after staining. (c) The particles as shown in (d) but observed without the use of staining agent. (Reproduced from Mittal, V., *Advances in Polymer Latex Technology*, Nova Science Publishes, New York, 2009.)

10.3 Applications of PNIPAAM-Grafted PS Particles

Polymer particles are extensively used to generate chromatographic supports in which, based on the pore size of the supports, the separations of various molecules could be achieved. As seen in the earlier references, PNIPAAM chains could be grafted within the pores of the particles by free radical polymerization, and then these particles could be used to separate the mixture by making use of the thermally responsive character of PNIPAAM chains. However, these particles were packed to form the chromatographic column. Recently, a new technique named reactive gelation was reported [32], by which macroporous monolithic materials were generated by controlled aggregation of emulsion latexes that were initially swollen with monomers. The aggregated latexes were then repolymerized to fix the so-obtained macroporous particle network. As an additional functionalization, these particles could first be functionalized with an ATRP initiator and could then be formed onto macroporous monoliths. PNIPAAM chains

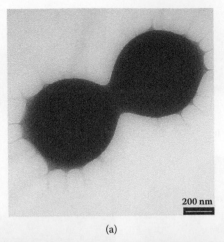

(a)

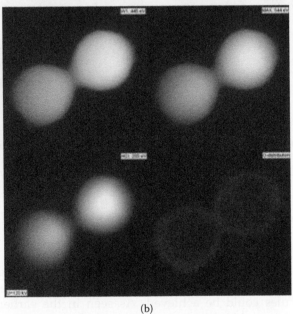

(b)

FIGURE 10.13

EELS analysis of the particles: (a) particles seen in the microscope and (b) the same particles analyzed for the presence of oxygen atoms corresponding to isopropylacrylamide (the image with dots on the bottom right side). (Reproduced from Mittal, V., *Advances in Polymer Latex Technology*, Nova Science Publishes, New York, 2009.)

could subsequently be grafted from these particles bound into monolithic structures [33]. The resulting structures showed similar behavior to free particles once they were functionalized with PNIPAAM brushes. The reduced swelling capabilities and the slower kinetics in swelling and deswelling were, however, observed, probably due to the constrained environment of the monolith, as opposed to free particles. An aggregate grafting process was also used to graft the PNIPAAM chains from particles in the reported study [26]. In this method, which is slightly different from the reactive gelation method mentioned above, the latex particles functionalized with an ATRP initiator were first shear mixed to form physical networks. These particles were then mixed with NIPAAM solution containing a small amount of cross-linker and the catalysts. The cross-linked material was highly stable and porous. The same behavior of thermoreversible hydrophobicity and hydrophilicity of the PNIPAAM chains was observed in the monoliths, though the extent and rate of swelling are hindered owing to the large degree of cross-linking among PNIPAAM brushes and space constraints on the chains. The solid fractions were also altered in these monoliths in order

(a) (b)

(c) (d)

FIGURE 10.14

SEM pictures of polymeric monoliths generated by reactive gelation process: (a) low- and (b) high-magnification images of the monolith produced from the particles functionalized with a thin layer of polymerized STY, functional monomer, and DVB; (c) low- and (d) high-magnification images of the monolith synthesized from the original cross-linked PSTY particles without subsequent functionalization. (Reproduced from Mittal, V., et al., *Macromolecular Reaction and Engineering*, 2, 215–21, 2008. With permission from Wiley.)

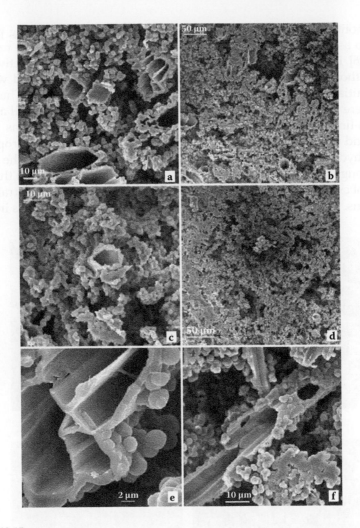

FIGURE 10.15

(a, b) High- and low-magnification SEM image of the monolith generated from aggregated particles using higher solid fraction and dried at room temperature. (c, d) High- and low-magnification SEM micrographs of the monolith prepared using lower solid fraction and dried at room temperature. (e, f) SEM images representing the tubes or channels as seen in images (a) and (b). (Reproduced from Mittal, V., and Matsko, N. B., *Journal of Porous Materials*, 16, 537–43, 2009. With permission from Springer.)

to estimate the effect of availability of free volume to the PNIPAAM chains in the monolith on their swelling-deswelling properties. It was observed that a higher extent of swelling-deswelling character occurred in the monolith with higher free volume, confirming the effect of the above-mentioned space constraints. Figure 10.14 shows an example of the monolithic structure generated by reactive gelation process [33]. Particles surface modified

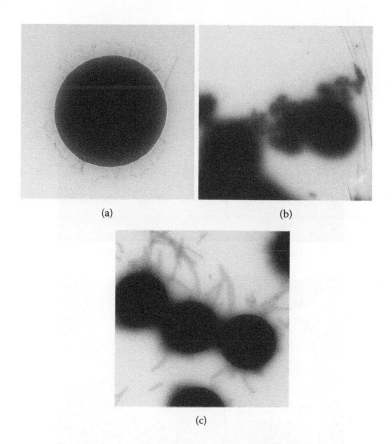

(a)   (b)

(c)

FIGURE 10.16
(a) Micrograph of the particles at 10°C. (b, c) Micrographs depicting the adsorption of protein and virus molecules, respectively, at 37°C. (Reproduced from Mittal, V., *Polymers*, 2, 40–56, 2010.)

by different methodologies were networked into the monolith structures, leading to different morphologies.

Another thing to note is that there are a number of factors that may influence the final morphology of the monolith. In Figure 10.15 are shown two examples of monoliths generated from PNIPAAM grafting on aggregated particles with different solid fractions [34]. The monolith generated from higher solid fraction aggregate particles had a large number of channels or tubes of varying dimensions, whereas such features were absent in the monolith of the same particles but with lower solid fraction, indicating the impact of solid fraction on the morphology.

PNIPPAM-modified particles can also be used directly for temperature-controlled adsorption and desorption of viruses, proteins, etc. [31]. To demonstrate the application of the PNIPAAM-grafted polymer particles for the temperature-controlled adsorption and desorption of proteins and viruses, IgG protein supported on gold particles and tobacco mosaic virus were

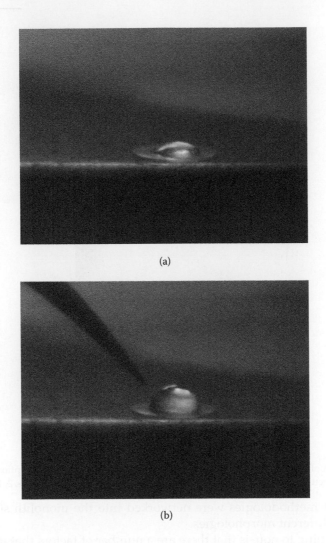

FIGURE 10.17
Behavior of the water droplets on the PNIPAAM-grafted sapphire discs at (a) 37°C and (b) 10°C. (Reproduced from Mittal, V., *Polymers*, 2, 40–56, 2010.)

chosen. The particles mixed with either protein or virus at 10°C (lower than the LCST) were observed to have extended morphology of the brushes on the particles and no adsorption of protein or virus molecules, as shown in Figure 10.16a. However, when the mixture was equilibrated at 37°C (higher than LCST), the brushes collapsed and the nature of the surface became suitable for the adsorption of the biological entities. As observed in Figure 10.16b and c, the particles tend to aggregate owing to hydrophobic interactions and the protein-supported gold particle (observed as tiny dots around the particles in the micrograph) or virus (seen as long tubes in the micrograph) molecules adsorb on the surface of the particles. The same mixture when recooled to 10°C showed desorption of the protein or virus molecules, and the brush morphology was again extended, as shown in Figure 10.16a. This behavior confirms the application potential of these polymer materials in advanced applications.

The ATRP initiator functionalized particles can also be adsorbed on macrosurfaces followed by PNIPAAM grafting to generate thermally reversible macrosurfaces. Figure 10.17 shows the behavior of the PNIPAAM-grafted sapphire discs [31]. At 10°C, the droplet of water placed on the disc spread out quickly, indicating the hydrophilic nature of the surface. On the other hand, when the disc was heated to 37°C, the droplet of water did not spread out at all, indicating the change in the nature of the surface with temperature. The hydrophilic-hydrophobic transition was also observed to be reversible, thus confirming the successful translation of reversible behavior to the flat surface also.

References

1. Ringsdorf, H., Sackmann, E., Simon, J., and Winnik, F. M. 1993. Interactions of liposomes and hydrophobically modified poly-(N-isopropylacrylamides): An attempt to model the cytoskeleton. *Biochimica Biophysica Acta* 1153:335–44.
2. Bae, Y. H., Okano, T., and Kim, S. W. 1990. Temperature dependence of swelling of crosslinked poly(N,N''-alkyl substituted acrylamides) in water. *Journal of Polymer Science, Part B: Polymer Physics* 28:923–36.
3. Heskins, M., and Guillet, J. E. 1968. Solution properties of poly(N-isopropylacrylamide). *Journal of Macromolecule Science: Chemistry* A2:1441–45.
4. Yoshida, R., Uchida, K., Kaneko, Y., Sakai, K., Kikuchi, A., Sakurai, Y., and Okano, T. 1995. Comb-type grafted hydrogels with rapid deswelling response to temperature changes. *Nature* 374:240–42.
5. Gao, J., and Wu, C. 1997. The "coil-to-globule" transition of poly(N-isopropylacrylamide) on the surface of a surfactant-free polystyrene nanoparticle. *Macromolecules* 30:6873–76.
6. Park, T. G., and Hoffman, A. S. 1993. Sodium chloride-induced phase transition in nonionic poly(N-isopropylacrylamide) gel. *Macromolecules* 26:5045–48.
7. Janzen, J., Le, Y., Kizhakkedathu, J. N., and Brooks, D. E. 2004. Plasma protein adsorption to surfaces grafted with dense homopolymer and copolymer brushes containing poly(N-isopropylacrylamide). *Journal of Biomaterials Science, Polymer Edition* 15:1121–35.
8. Kanazawa, H., Yamamoto, K., Matsushima, Y., Takai, N., Kikuchi, A., Sakurai, Y., and Okano, T. 1996. Temperature-responsive chromatography using poly(N-isopropylacrylamide)-modified silica. *Analytical Chemistry* 68:100–5.
9. Kikuchi, A., and Okano, T. 2002. Intelligent thermoresponsive polymeric stationary phases for aqueous chromatography of biological compounds. *Progress in Polymer Science* 27:1165–93.

10. Cunliffe, D., Heras Alarcon, C., Peters, V., Smith, J. R., and Alexander, C. 2003. Thermoresponsive surface-grafted poly(N-isopropylacrylamide) copolymers: Effect of phase transitions on protein and bacterial attachment. *Langmuir* 19:2888–99.

11. Okano, T., Yamada, N., Okuhara, M., Sakai, H., and Sakurai, Y. 1995. Mechanism of cell detachment from temperature-modulated, hydrophilic-hydrophobic polymer surfaces. *Biomaterials* 16:297–303.

12. Chen, G., Ito, Y., and Imanishi, Y. 1997. Regulation of growth and adhesion of cultured cells by insulin conjugated with thermoresponsive polymers. *Biotechnology and Bioengineering* 53:339–44.

13. Okajima, S., Yamaguchi, T., Sakai, Y., and Nakao, S. 2005. Regulation of cell adhesion using a signal-responsive membrane substrate. *Biotechnology and Bioengineering* 91:237–43.

14. Hosoya, K., Watabe, Y., Kubo, T., Hoshino, N., Tanaka, N., Sano, T., and Kaya, K. 2004. Novel surface modification techniques for polymer-based separation media: Stimulus-responsive phenomena based on double polymeric selectors. *Journal of Chromatography A* 1030:237–46.

15. Collier, T. O., Anderson, J. M., Kikuchi, A., and Okano, T. 2002. Adhesion behavior of monocytes, macrophages, and foreign body giant cells on poly (N-isopropylacrylamide) temperature responsive surfaces. *Journal of Biomedical Materials Research* 59:136–43.

16. Jones, D. M., Smith, R. R., Huck, W. T. S., and Alexander, C. 2002. Variable adhesion of micropatterned thermoresponsive polymer brushes: AFM investigations of poly(N-isopropylacrylamide) brushes prepared by surface-initiated polymerizations. *Advanced Materials* 14:1130–34.

17. Kizhakkedathu, J. N., Norris-Jones, R., and Brooks, D. E. 2004. Synthesis of well-defined environmentally responsive polymer brushes by aqueous ATRP. *Macromolecules* 37:734–43.

18. Mittal, V., and Matsko, N. B. 2009. New method to generate reversible hydrophobic and hydrophilic surfaces. *Open Surface Science Journal* 1:14–19.

19. Sun, T., Wang, G., Feng, L., Liu, B., Ma, Y., Jiang, L., and Zhu, D. 2004. Reversible switching between superhydrophilicity and superhydrophobicity. *Angewandte Chemie, International Edition* 43:357–60.

20. You, Y. Z., Hong, C. Y., Pan, C. Y., and Wang, P. H. 2004. Synthesis of a dendritic core-shell nanostructure with a temperature-sensitive shell. *Advanced Materials* 16:1953–57.

21. Go, H., Sudo, Y., Hosoya, K., Ikegami, T., and Tanaka, N. 1998. Effects of mobile-phase composition and temperature on the selectivity of poly(N-isopropylacrylamide)-bonded silica gel in reversed-phase liquid chromatography. *Analytical Chemistry* 70:4086–93.

22. Hosoya, K., Sawada, E., Kimata, K., Araki, T., and Tanaka, N. 1994. *In situ* surface-selective modification of uniform size macroporous polymer particles with temperature-responsive poly-N-isopropylacrylamide. *Macromolecules* 27:3973–76.

23. Hosoya, K., Kimata, K., Araki, T., Tanaka, N., and Frechet, J. M. J. 1995. Temperature-controlled high-performance liquid chromatography using a uniformly sized temperature-responsive polymer-based packing material. *Analytical Chemistry* 67:1907–11.

24. Duracher, D., Veyret, R., Elaissari, A., and Pichot, C. 2004. Adsorption of bovine serum albumin protein onto amino-containing thermosensitive core-shell latexes. *Polymer International* 53:618–26.
25. Duracher, D., Sauzedde, F., Elaissari, A., Perrin, A., and Pichot, C. 1998. Cationic amino-containing N-isopropyl-acrylamide-styrene copolymer latex particles. 1. Particle size and morphology vs. polymerization process. *Colloid and Polymer Science* 276:219–31.
26. Mittal, V., Matsko, N. B., Butte, A., and Morbidelli, M. 2008. Swelling-deswelling behavior of PS-PNIPAAM copolymer particles and PNIPAAM brushes grafted from polystyrene particles and monoliths. *Macromolecular Materials and Engineering* 293:491–502.
27. Mittal, V. 2009. *Advances in polymer latex technology.* New York: Nova Science Publishers.
28. Kizhakkedathu, J. N., Takacs-Cox, A., and Brooks, D. E. 2002. Synthesis and characterization of polymer brushes of poly(*N,N*-dimethylacrylamide) from polystyrene latex by aqueous atom transfer radical polymerization. *Macromolecules* 35:4247–57.
29. Mittal, V., Matsko, N. B., Butte, A., and Morbidelli, M. 2007. Functionalized polystyrene latex particles as substrates for ATRP: Surface and colloidal characterization. *Polymer* 48:2806–17.
30. Mittal, V., Matsko, N. B., Butte, A., and Morbidelli, M. 2007. Synthesis of temperature responsive polymer brushes from polystyrene latex particles functionalized with ATRP initiator. *European Polymer Journal* 43:4868–81.
31. Mittal, V. 2010. Synthesis of environmentally responsive polymers by atom transfer radical polymerization: Generation of reversible hydrophilic and hydrophobic surfaces. *Polymers* 2:40–56.
32. Marti, N., Quattrini, F., Butte, A., and Morbidelli, M. 2005. Production of polymeric materials with controlled pore structure: The "reactive gelation" process. *Macromolecular Materials and Engineering* 290:221–29.
33. Mittal, V., Matsko, N. B., Butte, A., and Morbidelli, M. 2008. PNIPAAM grafted polymeric monoliths synthesized by reactive gelation process and their swelling deswelling characteristics. *Macromolecular Reaction and Engineering* 2: 215–21.
34. Mittal, V., and Matsko, N. B. 2009. PNIPAAM grafted particle monoliths: Parameters affecting structure and morphology. *Journal of Porous Materials* 16:537–43.

11

Well-Defined Concentrated Polymer Brushes of Hydrophilic Polymers: Suppression of Protein and Cell Adhesions

Chiaki Yoshikawa

WPI Research Center, International Center for Materials Nanoarchitectonics, National Institute for Materials Science, Ibraki, Japan

Hisatoshi Kobayashi

Biomaterials Center, National Institute for Materials Science, Ibraki, Japan

CONTENTS

11.1 Introduction

11.1.1 Biomaterials

In 1986, during a consensus conference of the European Society for Biomaterials held in Chester, United Kingdom, more than 50 top scientists discussed the definition of the term *biomaterial*. The term had been previously defined by the U.S. Food and Drug Administration (FDA) as follows: "any substance, other than a drug, or combination or substances, synthetic or natural in origin, which can be used for any period of time, as a whole or as a part of a system which treats, augments, or replaces any tissue, organs, or functions of the body." The new definition became "a nonviable material used in a biomedical device intended to interact with biological systems." Even today, this definition remains little changed [1].

Synthetic materials utilized in medical devices are mainly composed of metals, ceramics, polymers, or a combination of these. In general, metals and ceramics are used to replace body parts that endure high loads, such as bone and teeth, and polymers replace all other body parts. Polymers offer many significant advantages: (1) they are easy to mold into various shapes, including particle, film, and fiber; (2) their chemical and physical properties are varied and controllable; and (3) most body parts, other than lipids and hard tissues such as bone and teeth, are composed of polymers. Therefore, from ex vivo to in vivo materials, polymers are widely used for medical applications. Table 11.1 shows examples of medical devices composed of synthetic polymers [2].

11.1.2 Biocompatibility

A biomaterial, even if composed of natural rather than synthetic materials, is still considered a foreign body unless it was created by cloning or tissue regeneration technology involving induced pluripotent stem cells (iPS). Thus, the performance of a biomaterial should optimally be described in terms of its biofunction and biocompatibility, where *biofunction* is the inherent function performed by an organ such as the heart and lung or by tissue such as cornea and skin, and *biocompatibility* is the capability to respond appropriately to a specific bioapplication. However, this concept of biocompatibility is ambiguous, since even the same material exhibiting the same response can be categorized as either biocompatible or nonbiocompatible, depending on how the material is used. For example, material that induces blood coagulation is considered biocompatible if it is used for curing an external wound (hemostasis) but nonbiocompatible if it is used as an artificial blood vessel. Thus, in biomaterial design, it is important to specify a material's biocompatibility for specific applications.

Figure 11.1 shows the accepted classification of biocompatibility for medical devices under Japanese regulation. Although biocompatibility is represented by the total balance of bulk and surface properties, in most cases, surface properties dominate during initial bioreactions at the material-body interface and bulk properties become evident relatively in the long term. The bulk properties of a device are typically determined by mechanical or design factors. For example, a hip joint should be continuously usable for at least a decade without abrasion (mechanical/design compatibility), and an artificial vessel

TABLE 11.1

Examples of Medical Devices of Polymers

Contact Part	Period	Use	Device Example	Polymer Materials
Skin	<1 day	Examination, surgery	Adhesive skin electrode, urethra catheter, surgical drape	Acrylic electroconductive gel silicone, polyvinyl chloride, etc.; polyethylene, polyurethane, cellulose, alginic acid gel, etc.
Skin	<1 month	Fixation, adjusting eyesight	Adhesive bandage, disposable contact lens	Acrylic adhesive, etc.; polyvinyl alcohol, poly (2-hydroxyethyl methacrylate), etc.
Skin	>1 month	Cure, adjusting eyesight	Wound skin cover, contact lens	Silicone, polyurethane, cellulose, collagen, etc.; poly(methyl methacrylate), polysufone, silicone, etc.
Tooth	>1 month	Filling	Resin	Photocuring acrylic resin
Tissue, blood, connection to the outer body	<1 day	Surgery	Surgical rubber gloves, infusion set, syringe, etc.	Natural rubber, latex, silicone, polypropylene, etc.
Tissue, blood, connection to the outer body	<1 month	Cure	Tracheotomy tube, dura mater catheter	Silicone, polyvinyl chloride, polyethylene, polypropylene, etc.
Tissue, blood, connection to the outer body	>1 month	Cure	Catheter, adjuvant artificial heart	Silicone, polyethylene, polypropylene, polyurethane, polycarbonate, etc.
In vivo	~1 month	Surgery	Suture	Nylon, polypropylene, silk, etc.
In vivo	>1 month	Alternative organ	Joint, bone cement, artificial blood vessel	High-density polyethylene, poly(methyl methacrylate), Teflon, polyurethane, etc.

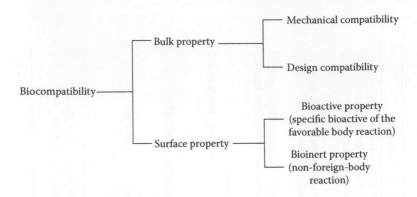

FIGURE 11.1
Biocompatibility of polymer materials.

should flexibly bend with body movement for a designated period of time without compressing other organs (mechanical compatibility).

The surfaces of most biomaterials implanted in the body come in direct contact with the body's blood or other liquids. Thus, surface biocompatibility plays a crucial role in determining device performance (surface biocompatibility) during initial bioreactions at the material-body interface. Surface biocompatibility is either bioactive (specifically bioactive of the favorable body reaction) or bioinert (non-foreign-body reactions).

When a material is implanted into our body, the material surface first contacts blood or body liquids and adsorption of proteins occurs. If the surface has sufficient biocompatibility (bioinert), reversible adsorption desorption would occur. Cell adhesion involves interaction between cell membrane proteoglycan and extracellular matrix, such as between a cell's integrin receptors and Arg-Gly-Asp (RGD) motif proteins [3], so irreversible protein adsorption often triggers cell adhesion. Cell adhesion, as well as other changes, such as spreading, proliferation, differentiation, and apoptosis, is thus affected by the external biological environment. Even if the bulk property of the material would show an excellent biocompatibility as long as our life span, the insufficient surface bioinertness can cause the material to undergo foreign body reactions, significantly reducing its effective expiration date.

The strategy to create an ideal bioinert surface is to suppress the initial bioreaction such as irreversible protein adsorption and cell adhesion, thus preventing consequent reactions. On the other hand, in order to create an ideal bioactive surface, it is the best way to introduce specific recognition or binding interactions on the ideal bioinert surface. By using the bioinert surface, i.e., zero-noise background, we can achieve 100% selectivity and specificity of the expected biological interactions. Unfortunately, however, such an ideal bioinert surface has not yet been achieved; the insufficient bioinertness of the substrate hampers the selective/specific recognition and binding reactions. Therefore, one of the top priorities of biomaterial researchers is to create an ideal bioinert surface, namely, to precisely understand the interactions of the initial stage at the interface between the given material and biomolecules in our body.

To better understand initial bioreactions at the material-body interface [4], self-assembly monolayers (SAMs) [5] and polymer brushes [6] have been extensively used. The SAM technique enables fabrication of a well-defined layer of organic molecules on an inorganic substrate by means of interactions between gold and thiol groups (–SH) or between hydroxylated surfaces and alkoxysiryl groups (–Si(OR)$_3$, e.g., R = CH$_3$, Cl, etc.). This technique offers several advantages; it is easy, simple, and produces well-ordered molecules. The polymer brushes are typically prepared by the grafting-to or grafting-from method. The grafting-to method physically or chemically immobilizes an end-functionalized polymer or a block polymer on the substrate using the terminal group or one side of the block. In the grafting-from method, polymers grow from the initiating sites, which are chemically fixed on the substrate surface. These grafting-to/from techniques provide a water-swollen layer that is relatively thick compared to a SAM. Moreover, the covalently tethered polymers are physically and chemically stable.

The use of SAMs and polymer brushes for biological applications is well reviewed in the literature. Hence, this chapter focuses mainly on a "concentrated" polymer brush prepared by surface-initiated living radical polymerization (SI-LRP) [7].

11.2 Concentrated Polymer Brush Prepared by SI-LRP

11.2.1 Living Radical Polymerization (LRP)

Over the past two decades, LRP has attracted great attention as a useful tool for preparing well-defined, low-polydispersity polymers [8–11]. The basic concept of LRP is a reversible activation-deactivation process (Scheme 11.1), where P-X is a dormant (end-capped) chain and P is a polymer radical. Due to this process, LRP allows fine control of polymer architectures retaining the advantages of conventional free radical polymerization such as simplicity, robustness, and versatility. For further details, see *Reference* [11h]. Here, we briefly describe the mechanism of LRP.

Mechanistically, LRP is distinguished from conventional radical polymerization by the reversible activation process. P-X is activated to the polymer radical P by thermal, photochemical, or chemical stimuli. In the presence of monomer M, P will propagate until it is deactivated back to P-X. For practically important systems, it usually holds that [P]/[P-X] < 10^{-5}, meaning that a living chain spends most of its polymerization time in the dormant state (here, *living chain* denotes the sum of the dormant and active chains). The activation-deactivation cycles will be repeated enough times to allow all

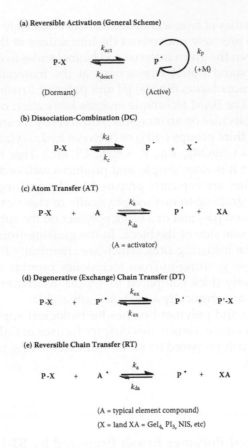

(a) Reversible Activation (General Scheme)

$$P\text{-}X \underset{k_{deact}}{\overset{k_{act}}{\rightleftharpoons}} P^{\cdot} \overset{k_p}{\curvearrowright} (+M)$$

(Dormant)   (Active)

(b) Dissociation-Combination (DC)

$$P\text{-}X \underset{k_c}{\overset{k_d}{\rightleftharpoons}} P^{\cdot} + X^{\cdot}$$

(c) Atom Transfer (AT)

$$P\text{-}X + A \underset{k_{da}}{\overset{k_a}{\rightleftharpoons}} P^{\cdot} + XA$$

(A = activator)

(d) Degenerative (Exchange) Chain Transfer (DT)

$$P\text{-}X + P'^{\cdot} \underset{k_{ex}}{\overset{k_{ex}}{\rightleftharpoons}} P^{\cdot} + P'\text{-}X$$

(e) Reversible Chain Transfer (RT)

$$P\text{-}X + A^{\cdot} \underset{k_{da}}{\overset{k_a}{\rightleftharpoons}} P^{\cdot} + XA$$

(A = typical element compound)

(X = I and XA = GeI$_4$, PI$_3$, NIS, etc)

SCHEME 11.1
(a) General scheme of reversible activation and (b–e) four main mechanisms of reversible activation.

polymer chains to grow equally, yielding a low-polydispersity product. The reversible activation reactions are mechanistically classified into four types: dissociation–combination (DC), atom transfer (AT), degenerative (exchange) transfer (DT), and reversible chain transfer (RT) processes.

- In the DC process, P-X is activated by a thermal or photochemical stimulus to produce P and the stable or persistent radical X, which is stable enough to undergo no reaction other than the combination with P (i.e., neither initiates polymerization or reacts with itself). Nitroxides such as 2,2,6,6-tetramethylpiperidinyl-1-oxy (TEMPO) are the typical X currently utilized (nitroxide-mediated polymerization (NMP)) [9,10].

- In the AT process, P-X is activated by the catalysis of activator A to produce P and the complex XA. The most successful example of this system uses a halogen like Cl and Br as a capping agent X, and a halide complex of transition metal like Cu and Ru as an activator A [12]. The LRP of this type is often termed atom transfer radical polymerization (ATRP).

- In the DT process, P-X is attacked by a polymer radical to form another polymer radical. The X in this process involves iodide [13], dithioesters [14], organotellurides [15], and organostibine [16]. The polymerization using dithioester as X is often called reversible addition fragmentation chain transfer (RAFT), and the polymerization using a tellurium compound as X is termed organotellurium-mediated radical polymerization (TERP).

- In the RT process, P-X is activated by an activator A to produce polymer radical and deactivator XA. XA plays a chain transfer agent, and P-X plays a catalyst of activation via the RT process. This polymerization process, in which iodine is used as X and Ge, Sn, O, or N compound is used as RT catalyst, XA, is termed the RT-catalyzed polymerization (RTCP) [17].

All LRPs such as NMP, ATRP, RAFT, TERP, or RTCP have already been applied to surface-initiated polymerization by fixing dormant species or conventional radical initiators. For details, readers are referred to other relevant chapters.

11.2.2 Concentrated Polymer Brush

Polymers end-grafted on a solid surface play an important role in many areas of science and technology, e.g., colloidal stabilization, adhesion, lubrication, tribology, and rheology [18–23]. The properties of polymer-grafted surfaces especially in a solvent are strongly related to the conformation of graft polymers, which can dramatically change with the graft density. When a polymer chain is isolated on a surface at a low graft density, it forms a coil-like structure, i.e., a "mushroom" structure (Scheme 11.2). When the graft density is high enough for a polymer chain to interact with its neighbors, the chain is obliged to stretch away from the surface in order to avoid the chain overlap, forming a so-called polymer brush (Scheme 11.2).

The polymer brushes can be classified into two groups: semidilute and concentrated brushes by the graft density. In a semidilute brush, polymer chains overlap with each other but their volume fraction is so low that the free energy of interaction is approximated by a binary interaction, and the elastic free energy distribution is approximately Gaussian. The structure and properties of semidilute brushes are relatively well studied and understood in both experimental and theoretical aspects [24–28]: scaling theory gives low-power dependence of the equilibrium thickness (L_e) of the semidilute brush in a good solvent on N, the number of monomers per chains, and σ, the surface density of the graft chains:

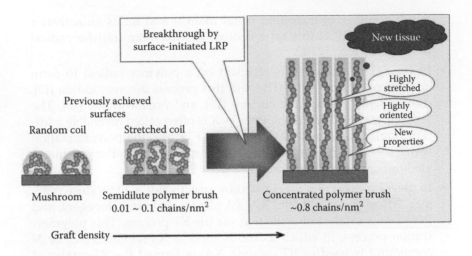

SCHEME 11.2
Development of concentrated polymer brush. (Redrawn from Tsujii et al. [7].)

$$L_e \sim N\sigma^{1/3}$$

It should be noted that L_e depends on N in a linear way, while the dimension of an isolated chain in a good solvent is scaled as $N^{3/5}$, meaning that the graft chains adopt stretched conformations. This theoretical prediction has been extensively verified with the semidilute brushes prepared by the grafting-to methods [27–33]. However, in a concentrated brush, these approximations are no longer valid and higher-order interactions must be considered [34]. Therefore, concentrated brushes are expected to exhibit different properties from semidilute ones. However, they had previously been little studied because of the difficulty to obtain the concentrated brushes by conventional methods.

Recently, LRP has been applied to surface-initiated graft polymerization, allowing controlled grafting of well-defined polymers from various solid surfaces with extremely high surface densities [7,35–38]. The surface density reached as large as 0.7 chains/nm² for common polymers like poly(methyl methacrylate) (PMMA) and polystyrene (PS). This density was more than one order of magnitude higher than those of typical semidilute brushes, going deep into the concentrated brush regime. Yamamoto et al. were the first to succeed in controlled synthesis of concentrated PMMA brushes by SI-ATRP and to reveal their structure and properties quite differently and even unpredictably from those of semidilute brushes. Most strikingly, concentrated PMMA brushes swollen in a good solvent (toluene) exhibited an equilibrium film thickness as large as 80–90% of the full (contour) length of the graft chains, indicating that the chains are extended to a similarly high degree [39]. Reflecting the high-extension structure, the concentrated PMMA brushes showed strong resistance against compression, super lubrication between swollen brushes, and so on. All these characteristic structures and properties are originated from the balance between a large elastic stress (by the conformational entropy) of the chains in the swollen brush layer and a huge osmotic pressure (by the mixing entropy) in a solvent.

Notable observations for dry concentrated brushes include that PMMA brushes in the melt had a glass transition temperature significantly higher [40], and a plate compressibility markedly smaller [41], than those of the equivalent cast films. More interestingly, they were immiscible even with free PMMA of an oligomeric chain length [42]. This last observation means that concentrated brushes in the melt have a size-exclusion effect of (conformational) entropic origin. A similar size-exclusion effect of concentrated brushes was also confirmed for molecules in solution. This will be detailed in Section 11.3.

More details of the concentrated brush are described in an excellent review by Tsujii et al. [7].

11.3 Protein Repellency of Concentrate" Polymer Brush: Size-Exclusion Effect

Considering the unique structure and properties of the concentrated polymer brush, one of the most attractive applications of this brush is directly toward a biointerface, especially bioinert surface, which prevents biologically important molecules such as protein and cells.

11.3.1 Protein Adsorption Model: Size-Exclusion Effect

As noted in Section 11.2, the initial biological event at the interface is a nonspecific protein adsorption. Accordingly, much effort has been made to modify surfaces with polymer brushes to prevent protein adsorption.

To understand the process of protein adsorption, the interactions between proteins and brush surfaces can be modeled by the three generic modes illustrated in Scheme 11.3 (after Currie et al. [43] with some modifications): (1) primary adsorption, in which a protein diffuses into the brush layer and adsorbs onto the substrate surface; (2) secondary adsorption at the brush-solvent interface, i.e., the outermost brush surface; and (3) tertiary adsorption, in which protein interacts with polymer segments in the brush layer. The model treats protein as a dense, rigid object with a nonadsorbing surface, that is, a structureless colloidal particle. For relatively small proteins, the primary and tertiary adsorptions would be particularly important, but they should become less important with increasing protein size and increasing graft density, since a larger protein would be more difficult to diffuse against the concentration gradient formed by the polymer brush, and this gradient, clearly, is a function of graft density.

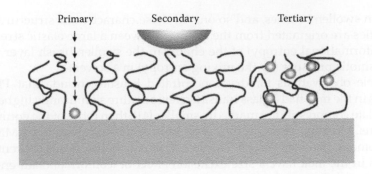

SCHEME 11.3
Three generic models of protein adsorption.

The size and density dependence of protein adsorption would manifest itself much more clearly for concentrated brushes due to a different mechanism. As already noted, the graft chains in a concentrated brush are highly extended and hence highly oriented so that the entire brush layer, from the substrate surface to the outermost surface throughout, could have a size-exclusion effect (Scheme 11.4a). By the term *size exclusion*, we emphasize the *physical* aspect of the phenomenon, meaning that the protein (or probe molecule) is excluded from the brush layer to avoid the large (mainly conformational) entropy loss caused on the highly extended chains by the entrance of the large molecule, as illustrated in Scheme 11.4a. Since the degree of chain extension is much less significant in semidilute brushes, this effect should be minor for them, and thus even a larger protein will partly diffuse into the brush layer, depending on its size (Scheme 11.4b). Thus, the concentrated brushes are expected to have a protein repellency effect by this new mechanism of size exclusion, and hence much better bioinertness.

11.3.2 Size-Exclusion Effect of the Concentrated Polymer Brush

In order to demonstrate size exclusion of the concentrated polymer brush by chromatography, a well-defined, concentrated poly(2-hydroxyethyl methacrylate) (PHEMA) brush with $M_{n,\text{theoretical}} = 10,700$ and $\sigma = 0.4$ chains/nm^2 was prepared on the inner surface of a silica monolithic column with ca. 80 nm mesopores by SI-ATRP [44,45]. A series of pullulans were eluted using phosphate-buffered saline (PBS) as eluent through the monolithic columns with or without PHEMA brush. Figure 11.2a shows size-exclusion chromatograms of standard pullulans with the columns. The chromatogram for the PHEMA grafted column (A) appeared in the lower region compared with the nonbrush column (B). This elution difference was ascribed to the volume of the PHEMA brush. More interestingly, the elution behavior of pullulans through the PHEMA-grafted column suggested the existence of two modes of size exclusion denoted by regions (a) and (b). Region (a) was

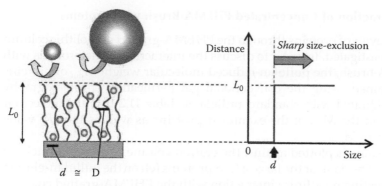

(a) Concentrated polymer brush

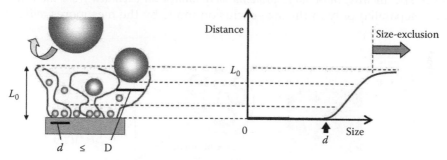

(b) Semidilute polymer brush

SCHEME 11.4

Schematic illustration of a size-exclusion effect for (a) concentrated brush and (b) semidilute brush. L_0 is the swollen brush thickness and d is the average distance between the nearest-neighbor graft points. The vertical and horizontal axes show the distance from the substrate and the protein size, respectively.

ascribed to the size exclusion by the mesopores from the similarity with the elution curve for the column without the PHEMA brush (A). On the other hand, the v value was sharply shifted in a rather small interval of M in region (b), which was ascribed to the size exclusion by the brush phase. Notably, the horizontal difference corresponding to region (b) is approximately equal to the volume of solvent in the brush layer, meaning that almost the whole brush layer is unavailable for the pullulans, with an M larger than the critical molecular weight of about 1,000. Here, we calculated the size of pullulan $2R_g$ at $M = 1,000$ to be about 1.6 nm, where R_g is the radius of gyration evaluated by using the known relation between the R_g and molecular weight of pullulan. Interestingly, this $2R_g$ value is almost the same as the average distance between the nearest-neighbor graft points, d ($s^{-1/2} = 1.6$ nm). This exactly supports our expectation, namely, solutes larger than the distance between the nearest-neighbor graft points d are *sharply* excluded from the entire concentrated brush layer as illustrated in Scheme 11.4a.

11.3.3 Interaction of Concentrated PHEMA Brush with Proteins

Elution behavior of proteins through the PHEMA-grafted monolithic column was also investigated. In order to discuss the interaction of the proteins with the PHEMA brush, the pullulan-reduced molecular weight M_{pul} of proteins was determined using the conventional gel permeation chromatography columns calibrated with standard pullulans. Table 11.2 lists the molecular weight M and the M_{pul} of the examined proteins as well as the $2R_g$ value evaluated from M_{pul}.

The M_{pul} value is plotted against the elution volume v for each protein in Figure 11.2b. The data for the largest four proteins fell on the pullulan-elution curve, suggesting no affinity interaction with the PHEMA-grafted column. More specifically, *those large proteins were almost all excluded from the brush layer*, separated only in the size-exclusion mode by the mesopores *without*

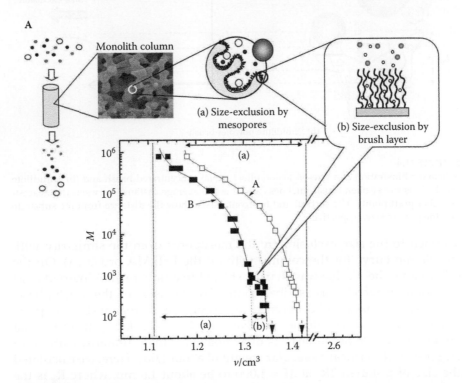

FIGURE 11.2
(a) Plot of molecular weight M vs. elution volume v for standard pullulans eluted through the (2-Bromo-2-methyl) propionyloxyhexyltriethoxysilane (BHE)-immobilized (open squares and curve A) and PHEMA-grafted (filled squares and curve B) monolithic columns: flow rate 0.2 ml/min, eluent PBS, room temperature. (b) Plot of pullulan-reduced molecular weight M_{pul} vs. elution volume v for proteins (circles) eluted through the PHEMA-grafted column. The solid curve is for the pullulans (curve B in (a)). The molecular weights M_{pul} of the proteins are the reduced values independently determined by pullulan-calibrated GPC.

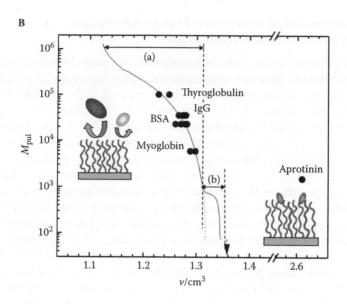

FIGURE 11.2B (CONTINUED)

TABLE 11.2

Absolute and Pullulan-Calibrated Molecular Weights and $2R_g$ for Studied Proteins

Protein	Molecular Weight M	Pullulan-Calibrated M_{Pul}	$2R_g{}^a$/nm
Aprotinin	6,500	1,500	2.0
Myoglobin	17,000	5,900	4.5
Bovin serum albumin (BSA)	67,000	22,800	9.9
Immunoglobulin G (IgG)	146,000	35,000	12.7
Thyroglobulin	669,000	100,000	23.4

^{a} Calculated from the known relation between R_g and molecular weight of pullulan.

affinity interaction with the brush surface. On the other hand, the smallest protein, aprotinin, was eluted much behind the pullulan of the equivalent size, suggesting a *strong affinity interaction* with the brush. The size of aprotinin ($M_{pul} = 1500$) is similar to the above-mentioned size-exclusion limit of the brush layer ($M \approx 1,000$), and proteins are typically anisotropic (ellipsoidal) in shape; thus, aprotinin should be able to penetrate the brush by an end-on approach, that is, by setting its long axis normal to the brush surface. For this reason, even aprotinin larger than the critical size would get in the brush layer and have affinity interaction within the brush layer to be eluted later.

11.3.4 Suppression of Irreversible Protein Adsorption on Concentrated PHEMA Brush

By varying protein sizes, protein adsorptions on poly(2-hydroxyehtl methacrylate) (PHEMA) brushes with different graft densities (σ = 0.007 ~ 0.7 chains/nm²) and chain length (dry thicknesses L = 2 ~ 10 nm) were systematically investigated [45,46]. Scheme 11.5 illustrated the preparation of concentrated PHEMA brushes (σ = 0.7 chains/nm²) on silicon wafers and quartz crystal microbalance (QCM) chips by SI-ATRP and semidilute brushes (σ = 0.007, 0.06 chains/nm²) on the substrates by grafting-to method. As shown in Figure 11.3, QCM measurements confirmed that the protein adsorption depends on the correlation between the protein sizes (the effective diameter of proteins ($2R_g$)) and the graft density (σ) of the concentrated brush, namely, when the diameter of a protein ($2R_g$) was sufficiently larger than the distance between the neighbor graft points (= $\sigma^{-1/2}$), protein adsorption was suppresed. Furthermore, the concentrated PHEMA brushes suppressed protein adsorptions independent of the graft thicknesses. This means that the concentrated brush holds a size-exclusion effect from its bottom to the outer surfaces throughout. This is the very feature expected for a concentrated brush in a good solvent (the highly extended brush structure).

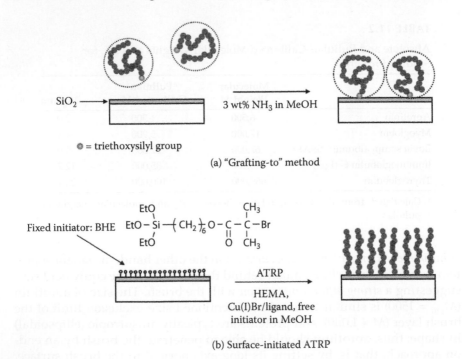

$\circledcirc$ = triethoxysilyl group

(a) "Grafting-to" method

Fixed initiator: BHE   EtO—Si—(CH₂)₆—O—C—C—Br

(b) Surface-initiated ATRP

SCHEME 11.5
Schematic illustration of grating polymers on an inorganic substrate by (a) grafting-to method and (b) SI-ATRP.

Protein	Molecular weight (M)	Pullulan reduced molecular weight	$2R_g^a$/nm
Aprotinin	6500	1500	2
Myoglobin	17000	5900	4
BSA	67000	22800	10
IgG	146000	35000	13

[a]Estimated from the known relation with molecular weight

(a) Molecular weight and sized of proteins

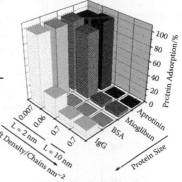

(b) Protein adsorption on PHEMA brushes
with different graft densities
and thicknesses. *L*:
graft thickness. [protein]$_0$ = 1.0 mg/ml

FIGURE 11.3
Protein adsorption on PHEMA brushes. (a) Molecular weights and sizes of examined proteins. (b) Irreversible protein adsorption on PHEMA brushes with different graft densities and graft thicknesses after 1 h soaking at 25°C. The vertical axis shows the adsorption amount normalized by that adsorbed on the low-density brush for each protein.

11.4 Suppression of Cell Adhesion on Concentrated Polymer Brushes

Since nonspecific adsorption of proteins often triggers cell adhesion, the concentrated polymer brush is expected to suppress cell adhesion. Thus, we investigated cell adhesions on polymer brushes of PHEMA, poly(2-hydroxyethyl acrylate) (PHEA), and poly(poly(ethyrene glycol)methyl ether monomethacrylate) (PPEGMA) with different graft densities and thicknesses [47]. The concentrated polymer brushes of all three, regardless of differences in chemical structure and hydrophilicity, almost completely suppressed the adhesion of L929 fibroblast cell to an almost undetectable level, while the corresponding semidilute brushes exhibited a cell adhesion similar to that of the reference substrates (Figure 11.4). These results are reasonable because the preadsorption of proteins that trigger cell adhesion was effectively inhibited on these concentrated brush samples. Until now, the protein repellency was demonstrated only on the concentrated PHEMA brush, but this would be the case with other concentrated brushes because this property comes from the entropically driven size-exclusion effect of a solvent-swollen concentrated brush (herein, water for biointerface).

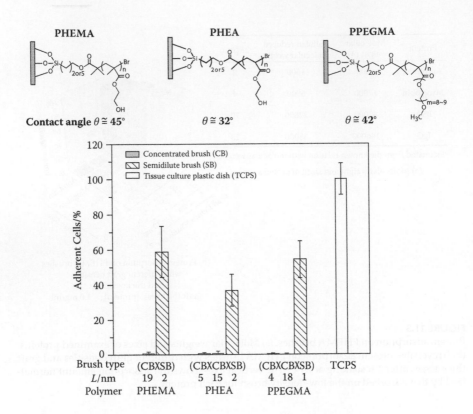

FIGURE 11.4
Amount of adhered L929 cells on semidilute (SD) and concentrated (CB) polymer brushes of PHEMA, PHEA, and PPEGMA as well as TCPS substrate. Incubation time = 24 h. $[L929]_0 = 5.0 \times 10^4$ cells/cm^2. The vertical axis was normalized with the number of adherent cells on TCPS.

11.5 Nonbiofouling Polymer Brushes Prepared by SI-LRP

Among all LRPs, ATRP is the most widely used for fabricating nonbiofouling surfaces because it can be performed on a wide range of hydrophilic monomers, and in addition, the necessary chemicals for ATRP, e.g., metal halides and ligands, are commercially available. To minimize secondary adsorption, it is best that neutral hydrophilic polymers do not have specific interactions with biomolecules. Figure 11.5 shows hydrophilic monomers (polymers) used previously for SI-ATRP.

Many researchers have reported biological applications, i.e., a nonbiofouling surface, with polymer brushes prepared by SI-ATRP [48]. To date, however, few have noted the aforementioned unique structure and properties of concentrated polymer brush obtainable by SI-LRP.

Feng et al. prepared poly(2-methacryloyloxyethyl phosphorylcholin) (PMPC) brushes on silicon wafers with different graft densities (0.10–0.39 chains/nm^2)

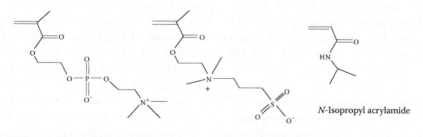

FIGURE 11.5

Examples of hydrophilic monomers (polymers) for SI-ATRP for bioapplications.

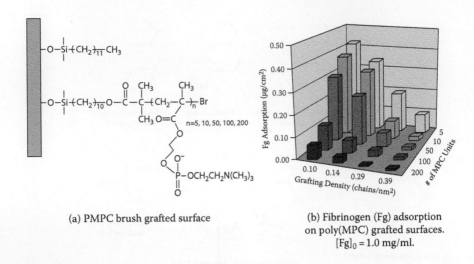

(a) PMPC brush grafted surface

(b) Fibrinogen (Fg) adsorption
on poly(MPC) grafted surfaces.
$[Fg]_0 = 1.0$ mg/ml.

FIGURE 11.6
Protein (fibinogen (Fg)) adsorptions on PMPC brushes prepared by SI-ATRP. (Redrawn and reproduced from Feng, W., et al., *Biomaterials*, 27, 847, 2006, with permission. Copyright 2006 Elsevier.)

and chain lengths (5–200 monomer unit) by SI-ATRP to evaluate protein (fibrinogen) adsorption on the brushes [49]. Figure 11.6 shows fibrinogen adsorption as a function of graft density and chain length. The amount of adsorbed proteins becomes more suppressed with increasing graft densities and chain lengths. This tendency corresponds with the observations on the concentrated PHEMA brush described in Section 11.3.4.

Iwata et al. reported micropatterned PMPC brushes prepared by SI-ATRP, and then investigated the resulting protein and cell arrays [50]. Figure 11.7 shows how a surface pattern of fixed initiators on a silicon wafer was prepared by ultraviolet (UV) irradiation through a photomask, after which SI-ATRP was preformed with the patterned substrate (graft density = 0.17 chains/nm²). On the patterned brush surface, L929 fibroblast cells significantly adhered onto the nonbrush region (the UV-irradiated region), but almost none on the brush region. The cell array pattern was clearer for brush thickness >5 nm, indicating that cell adhesion depends on the brush thickness. They also investigated protein adsorptions (bovine serum albumin and fibronectin (FN)) on the patterned brush surface and obtained the protein arrays with the same pattern as the cell array.

Mei et al. prepared the gradient PHEMA brush on a silicon wafer by SI-ATRP and investigated protein/cell adhesions on the brush surface [51]. First, they immobilized the fixed initiator on a silicon wafer with gradient density: a SAM of octyltrichlorosilane (OTS), which did not have an initiating site for ATRP, was deposited on a silicon wafer by evaporation. Then the OTS-gradient substrate was placed upright in the reaction vessel, and

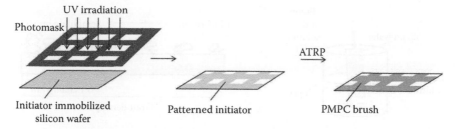

(a) Schematic representation for fabrication of a patterned PMPC brush.

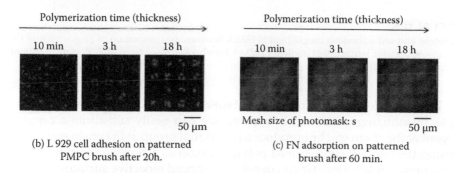

(b) L 929 cell adhesion on patterned
PMPC brush after 20h.

(c) FN adsorption on patterned
brush after 60 min.

FIGURE 11.7 (See color insert.)
Protein and cell adhesions on PMPC brushes prepared by SI-ATRP. (a) Schematic illustration for fabrication of a patterned PMPC brush. (b) L929 fibroblast cell adhesion on patterned PMPC brush after 20 h. (c) FN adsorption on PMPC brush after 60 min. (Images (b) and (c). (Reprinted with permission from Iwata, R., et al., *Biomacromolecules*, 5, 2308, 2004. Copyright (2004) American Chemical Society.)

subsequently, the solution of the fixed initiator for ATRP was pumped into the vessel at a specific pumping rate. On the obtained gradient initiator substrate, a PHEMA brush with gradient graft density was prepared (Figure 11.8). Even though the maximum graft density of the PHEMA brush ($\sim$0.08 chains/nm^2) was still in the semidilute brush regime, the adhesions of protein (FN) and fibroblast cell (NIH3T3) were more suppressed on the brush surface with increasing the graft density.

11.6 Conclusions

Concentrated polymer brushes obtainable by SI-LRP prevented proteins and cells compared with the corresponding semidilute brushes, owing to the unique size-exclusion effect (of entropic origin) of the concentrated polymer brushes. This means that the physical structure (possibly including chain dynamics) of graft chains on the surface plays an essential role in bioinertness.

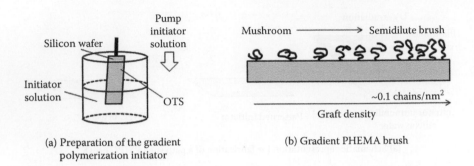

(a) Preparation of the gradient polymerization initiator

(b) Gradient PHEMA brush

FIGURE 11.8
Protein and cell adhesions on gradient PHEMA brushes. (a) Illustration of preparation of the gradient initiator surface. (b) Gradient PHEMA brush. (Redrawn from Mei, Y., et al., *J. Biomed. Mater. Res.*, 79A(4), 974, 2006.)

LRP provides a wide range of possibilities of chain architecture. For example, by utilizing chain-end livingness, we can easily substitute a capping agent at the chain end or introduce a functional group at the chain end. Thus, using the bioinert concentrated polymer brush as a zero-noise or detectably low-noise background, we can develop advanced bioactive surfaces that precisely detect/capture targeted biomolecules.

The research to use concentrated polymer brushes for biointerface application is on the way. We expect to see significant progress in research toward perfect biocompatiblity.

References

1. Williams, D. F. 1987. *Definitions in biomaterials*. Amsterdam: Elsevier.
2. The Ceramic Society of Japan. 2008. *Biomaterials*. Tokyo: Nikkan Kogyo Shimbun Ltd.
3. Niemeyer, C., and Mirkin, C. A. 2004. In *Nanotechnology—Concepts, applications and perspectives*. New York: Wiley-VCH.
4. (a) Ikada, Y. 1994. Surface modification of polymers for medical applications. *Biomaterials* 15: 725. (b) Kato, K., Uchida, E., Kang, E.-T., Uyama, Y., and Ikada, Y. 2003. Polymer surface with graft chains. *Prog. Polym. Sci.* 28: 209. (c) Klee, D., and Hocker, H. 2000. Polymers for biomedical applications: Improvement of the interface compatibility. *Adv. Polym. Sci.* 149: 1–57.
5. Ulman, A. 1991. *An introduction to ultrathin organic thin films: From Langmuir Blodget to self-assembly*. San Diego: Academic Press.
6. Advincula, R. C., Brittain, W. J., Caster, K. C., and Ruhe, J. 2004. *Polymer brushes*. Weinheim: Wiley-VCH.
7. Tsujii, Y., Ohno, K., Yamamoto, S., Goto, A., and Fukuda, T. 2006. Structure and properties of high-density polymer brushes prepared by surface-initiated living radical polymerization. *Adv. Polym. Sci.* 197: 1.

8. Otsu, T., and Yoshida, M. 1982. Role of initiator-transfer agent-terminator (iniferter) in radical polymerizations: Polymer design by organic disulfides as iniferters. *Makromol. Chem. Rapid Commun.* 3: 127.

9. Solomon, D. H., Rizzardo, E., and Cacioli, P. European Patent Application EP135280 (*Chem. Abstr.* 1985, 102, 221335q).

10. Georges, M. K., Veregin, R. P. N., Kazmaier, P. M., and Hamer, G. K. 1993. Narrow molecular weight resins by a free-radical polymerization process. *Macromolecules* 26: 2987.

11. For recent reviews, see: (a) Matyjaszewski, K., ed. 2000. *Controlled/living radical polymerization.* ACS Symposium Series 768. Washington, DC: American Chemical Society. (b) Matyjaszewski, K., and Davis, T. P., eds. 2002. *Handbook of radical polymerization.* Hoboken, NJ: John Wiley & Sons. (c) Matyjaszewski. K., ed. 2003. *Advances in controlled/living radical polymerization.* ACS Symposium Series 854. Washington, DC: American Chemical Society. (d) Matyjaszewski, K., and Xia, J. 2001. Atom transfer radical polymerization. *Chem. Rev.* 101: 2921. (e) Kamigaito, M., Ando, T., and Sawamoto, M. 2001. Metal-catalyzed living radical polymerization. *Chem. Rev.* 101: 3689. (f) Hawker, C. J., Bosman, A. W., and Harth, E. 2001. New polymer synthesis by nitroxide living radical polymerizations. *Chem. Rev.* 101: 3661. (g) Fukuda, T. 2004. Fundamental kinetic aspects of living radical polymerization and the use of gel permeation chromatography to shed light on them. *J. Polym. Sci. A Polym. Chem.* 42: 4743. (h) Goto, A., and Fukuda, T. 2004. Kinetics of living radical polymerization. *Prog. Polym. Sci.* 29: 329.

12. (a) Wang, J.-S., and Matyjaszewski, K. 1995. Controlled/"living" radical polymerization. Atom transfer radical polymerization in the presence of transition-metal complexes. *J. Am. Chem. Soc.* 117: 5614. (b) Kato, M., Kamigaito, M., Sawamoto, M., and Higashimura, T. 1995. Polymerization of methyl methacrylate with the carbon tetrachloride/dichlorotris- (triphenylphosphine)ruthenium(II)/methylaluminum bis(2,6-di-tert-butylphenoxide) initiating system: Possibility of living radical polymerization. *Macromolecules* 28: 1721.

13. (a) Yutani, Y., and Tatemoto, M. European Patent Application EP489370A1. (b) Kato, M., Kamigaito, M., Sawamoto, M., and Higashimura, T. 1994. Radical polimerization of styrene in the presence of alkyi lodides: Control of polymer molecular weight and end groups via long-lived growing species. *Polym. Prepr. Jpr. (Soc. Polym. Sci. Jpn.)* 43: 255. (c) Matyjaszewski, K., Gaynor, S., and Wang, J.-S. 1995. Controlled radical polymerizations: The use of alkyl iodides in degenerative transfer. *Macromolecules* 28: 2093.

14. (a) Kristina, J., Moad, G., Rizzardo, E., Winzor, C. L., Berge, C. T., and Fryd, M. 1995. Narrow polydispersity block copolymers by free-radical polymerization in the presence of macromonomers. *Macromolecules* 28: 5381. (b) Chiefari, J., Chong, Y. K., Ercole, F., et al. 1998. Living free-radical polymerization by reversible addition-fragmentation chain transfer: The RAFT process. *Macromolecules* 31: 5559.

15. (a) Takagi, K., Soyano, A., Kwon, TS., Kunisada, H., and Yuki, Y. 1999. Controlled radical polymerization of styrene utilizing excellent radical capturing ability of diphenyl ditelluride. *Polym. Bull.* 43: 143. (b) Yamago, S., Iida, K., and Yoshida, J. 2002. Organotellurium compounds as novel initiators for controlled/living radical polymerizations. Synthesis of functionalized polystyrenes and end-group modifications. *J. Am. Chem. Soc.* 124: 2874.

16. Yamago, S., Iida, K., Yoshida, J. et al. 2004. Highly versatile organostibine mediators for living radical polymerization. *J. Am. Chem. Soc.* 126: 13908.
17. Goto, A., Tsujii, Y., and Fukuda, T. 2008. Reversible chain transfer catalyzed polymerization (RTCP): A new class of living radical polymerization. *Polymer* 49: 5177.
18. Napper, D. H. 1983. *Polymeric stabilization of colloidal dispersions.* London: Academic Press.
19. Fleer, G. J., Cohen Stuart, M. A., Scheutjens, J. M. H. M., Cosgrove, T., and Vincent, B. 1993. *Polymers at interfaces.* London: Chapman & Hall.
20. Raphaël, E., and de Gennes, P. G. 1992. Rubber-rubber adhesion with connector molecules. *J. Phys. Chem.* 96: 4002.
21. Klein, J. 1996. Shear, friction, and lubrication forces between polymer-bearing surfaces. *Annu. Rev. Mater. Sci.* 26: 581.
22. Klein, J., and Kumacheva, E. 1995. Confinement-induced phase transitions in simple liquids. *Science* 269: 816.
23. Parnas, R. S., and Cohen, Y. 1994. A terminally anchored polymer chain in shear flow: Self-consistent velocity and segment density profiles. *Rheol. Acta.* 33: 485.
24. (a) Alexander, S. 1977. Adsorption of chain molecules with a polar head: A scaling description. *J. Phys. (Paris)* 38: 983. (b) de Gennes, P.-G. 1980. Conformations of polymers attached to an interface. *Macromolecules* 13: 1069.
25. Milner, S. T., Witten, T., and Cates, M. 1988. Theory of the grafted polymer brush. *Macromolecules* 21: 2610.
26. (a) Satija, S. K., Majkrzak, C. F., Russell, T. P., Sinha, S. K., Sirota, E. B., and Hughes, G. J. 1990. Neutron reflectivity study of block copolymers adsorbed from solution. *Macromolecules* 23: 3860. (b) Levicky, R., Koneripalli, N., Tirrell, M., and Satija, S. K. Concentration profiles in densely tethered polymer brushes. 1998. *Macromolecules* 31: 3731. (c) Cosgrove, T., Heath, T. G., Phipps, J. S., and Richardson, R. M. 1991. Neutron reflectivity studies of polymers adsorbed on mica from solution. *Macromolecules* 24: 94. (d) Field, J. B., Toprakcioglu, C., Ball, R. C., Stanley, H. B., Dai, L., Barfford, W., Penfold, J., Smith, G., and Hamilton, W. 1992. Determination of end-adsorbed polymer density profiles by neutron reflectometry. *Macromolecules* 25: 434. (e) Anastassopoulos, D. L., Vradis, A. A., Toprakcioglu, C., Smith, G. S., and Dai, L. 1998. Neutron reflectivity study of end-attached telechelic polymers in a good solvent. *Macromolecules* 31: 9369.
27. (a) Hadziioannou, G., Granick, S., Patel, S., and Tirrell, M. 1986. Forces between surfaces of block copolymers adsorbed on mica. *J. Am. Chem. Soc.* 108: 2869. (b) Watanabe, H., and Tirrell, M. 1993. Measurement of forces in symmetric and asymmetric interactions between diblock copolymer layers adsorbed on mica. *Macromolecules* 26: 6455. (c) Ansarifer, A., and Luckham, P. F. 1988. Measurement of the interaction force profiles between block copolymers of poly(2-vinylpyridine)/poly(t-butylstyrene) in a good solvent. *Polymer* 29: 329. (d) Taunton, H. J., Toprakcioglu, C., Fetters, L., and Klein, J. 1990. Interactions between surfaces bearing end-adsorbed chains in a good solvent. *Macromolecules* 23: 571.
28. (a) Courvoisier, A., Isel, F., François, J., and Maaloum, M. 1998. End-adsorbed telechelic polymer chains at surfaces: Bridging and elasticity. *Langmuir* 14: 3727. (b) Overney, R., Leta, D., Pictroski, C., Rafailovich, M., Liu, Y., Quinn, J., Sokolov, J., Eisenberg, A., and Overney, G. 1996. Compliance measurements of confined polystyrene solutions by atomic force microscopy. *Phys. Rev. Lett.*

76: 1272. (c) Kelley, T. W., Schorr, P. A., Johnson, K. D., Tirrell, M., and Frisbie, C. D. 1998. Direct force measurements at polymer brush surfaces by atomic force microscopy. *Macromolecules* 31: 4297. (d) O'Shea, S. J., Welland, M. E., Rayment, T. 1993. An atomic force microscope study of grafted polymers on mica. *Langmuir* 9: 1826.

29. Satija, S. K., Majkrzak, C. F., Russell, T. P., Sinha, S. K., Sirota, E. B., and Hughes, G. J. 1990. Neutron reflectivity study of block copolymers adsorbed from solution. *Macromolecules* 23: 3860.

30. Levicky R., Koneripalli N., Tirrell M., and Satija S. K. 1998. Concentration profiles in densely tethered polymer brushes. *Macromolecules* 31: 3731.

31. Cosgrove, T., Heath, T. G., Phipps, J. S., and Richardson, R. M. 1991. Neutron reflectivity studies of polymers adsorbed on mica from solution. *Macromolecules* 24: 94.

32. Field, J. B., Toprakcioglu, C., Ball, R. C., Stanley, H. B., Dai L., Barford, W., Penfold, J., Smith, G., and Hamilton, W. 1992. Determination of end-adsorbed polymer density profiles by neutron reflectometry. *Macromolecules* 25: 434.

33. Anastassopoulos, D. L., Vradis, A. A., Toprakcioglu, C., Smith, G. S., and Dai, L. 1998. Neutron reflectivity study of end-attached telechelic polymers in a good solvent. *Macromolecules* 31: 9369.

34. (a) Lai, P.-Y., and Halperin, A. 1991. Polymer brush at high coverage. *Macromolecules* 24: 4981. (b) Shim, D. F. K., and Cates, M. E. 1989. Smectic A-smectic B interface: Faceting and surface free energy measurement. *J. Phys. France* 50: 3535.

35. (a) Ejaz, M., Yamamoto, S., Ohno, K., Tsujii, Y., and Fukuda, T. 1998. Controlled graft polymerization of methyl methacrylate on silicon substrate by the combined use of the Langmuir-Blodgett and atom transfer radical polymerization techniques. *Macromolecules* 31: 5934. (b) Husseman, M., Malmström, E. E., McNamara, M., Mate, M., Mecerreyes, D., Benoit, D. G., Hedrick, J. L., Mansky, P., Huang, E., Russell, T. P., and Hawker, C. J. 1999. Controlled synthesis of polymer brushes by "living" free radical polymerization techniques. *Macromolecules* 32: 1424. (c) Huang, X., and Wirth, M. J. 1999. Surface initiation of living radical polymerization for growth of tethered chains of low polydispersity. *Macromolecules* 32: 1694. (d) Zhao, B., and Brittain, W. J. 1999. Synthesis of tethered polystyrene-*block*-poly(methyl methacrylate) monolayer on a silicate substrate by sequential carbocationic polymerization and atom transfer radical polymerization. *J. Am. Chem. Soc.* 121: 3557. (e) von Werne, T., and Patten, T. E. 1999. Preparation of structurally well-defined polymer–nanoparticle hybrids with controlled/living radical polymerizations. *J. Am. Chem. Soc.* 121: 7409. (f) Matyjaszewski, K., Miller, P. J., Shukla, N., et al. 1999. Polymers at interfaces: Using atom transfer radical polymerization in the controlled growth of homopolymers and block copolymers from silicon surfaces in the absence of untethered sacrificial initiator. *Macromolecules* 32: 8716.

36. Zhao, B., and Brittain, W. J. 2000. Polymer brushes: Surface-immobilized macromolecules. *Prog. Polym. Sci.* 25: 677.

37. Pyun, J., Kowalewski, Y., and Matyjaszewski, K. 2003. Synthesis of polymer brushes using atom transfer radical polymerization. *Macromol. Rapid. Commun.* 24: 1043.

38. Edmondson, S., Osborne, V. L., and Huck, W. T. S. 2004. Polymer brushes via surface-initiated polymerizations. *Chem. Soc. Rev.* 33: 14.

39. (a) Yamamoto, S., Ejaz, M., Tsujii, Y., Matsumoto, M., and Fukuda, T. 2000. Surface interaction forces of well-defined, high-density polymer brushes studied by atomic force microscopy. 1. Effect of chain length. *Macromolecules* 33: 5602. (b) Yamamoto, S., Ejaz, M., Tsujii, Y., and Fukuda, T. 2000. Surface interaction forces of well-defined, high-density polymer brushes studied by atomic force microscopy. 2. Effect of graft density. *Macromolecules* 33: 5608.

40. Yamamoto, S., Tsujii, Y., and Fukuda, T. 2002. Glass transition temperatures of high-density poly(methyl methacrylate) brushes. *Macromolecules* 35: 6077.

41. Urayama, K., Yamamoto, S., Tsujii, Y., Fukuda, T., and Neher, D. 2000. Elastic properties of well-defined, high-density poly(methyl methacrylate) brushes studied by electromechanical interferometry. *Macromolecules* 35: 9459.

42. Yamamoto, S., Tsujii, Y., Fukuda, T., Torikai, N., and Takeda, M. 2001–2002. The structure of high-density polymer brushes in contact with a chemically identical polymer matrix studied by neutron reflectometry. *KENS Rep.* 14: 204.

43. Currie, E. P. K., Norde, W., and Cohen Stuart, M. A. 2003. Tethered polymer chains: Surface chemistry and their impact on colloidal and surface properties. *Adv. Collide Interface Sci.* 100–102: 205.

44. Yoshikawa, C., Goto, A., Tsujii, Y., Ishizuka, N., Nakanishi, K., and Fukuda, T. 2007. Surface interaction of well-defined, concentrated poly(2-hydroxyethyl methacrylate) brushes with proteins. *J. Polym. Sci.* 45: 4795.

45. Yoshikawa, C., Goto, A., Ishizuka, N., Nakanishi, K., Kishida, A., Tsujii, Y., and Fukuda, T. 2007. Size-exclusion effect and protein repellency of concentrated polymer brushes prepared by surface-initiated living radical polymerization. *Macromol. Symp.* 248: 189.

46. Yoshikawa, C., Goto, A., Tsujii, Y., Fukuda, T., Kimura, T., Yamamoto, K., and Kishida, A. 2006. Protein repellency of well-defined, concentrated poly(2-hydroxyethyl methacrylate) brushes by the size-exclusion effect. *Macromolecules* 39: 2284.

47. Yoshikawa, C., Hashimoto, Y., Hattori, S., Honda, T., Zhang, K., Terada, D., Kishida, A., Tsujii, Y., and Kobayashi, H. 2010. Suppression of cell adhesion on well-defined concentrated polymer brushes of hydrophilic polymers. *Chem. Letters.* 39: 142–143.

48. (a) Senaratne, W., Andruzzi, L., and Ober, C. K. 2005. Self-assembled monolayers and polymer brushes in biotechnology: Current applications and future perspectives. *Biomacromolecules* 6: 2427. (b) Fristrup, C. J., Jankova, K., and Hvilsted, S. 2009. Surface-initiated atom transfer radical polymerization—A technique to develop biofunctional coatings. *Soft Matter* 5: 4623.

49. Feng, W., Brash, J. L., and Zhu, S. 2006. Non-biofoulingmaterials prepared by atom transfer radical polymerization grafting of 2-methacryloxyethyl phosphorylcholine: Separate effect of graft density and chain length on protein repulsion. *Biomaterials* 27: 847.

50. Iwata, R., Suk-In, P., Hoven, V. P., Takahara, A., Akiyoshi, K., and Iwasaki, Y. 2004. Control of nanobiointerfaces generated from well-defined biomimetic polymer brush for protein and cell manipulations. *Biomacromolecules* 5: 2308.

51. Mei, Y., Elliott, J. T., Smith, J. R., Langenbach, K. J., Wu, T., Xu, C., Beers, K. L., Amis, E. J., and Henderson, L. 2006. Gradient substrate assembly for quantifying cellular response to biomaterials. *J. Biomed. Mater. Res.* 79A(4): 974.

12

Polymer Brushes by Surface-Initiated Iniferter-Mediated Polymerization

Santosh B. Rahane

School of Polymers and High Performance Materials, University of Southern Mississippi, Hattiesburg, Mississippi

S. Michael Kilbey II

Departments of Chemistry and Chemical and Biomolecular Engineering, University of Tennessee, Knoxville, Tennessee; and Center for Nanophase Materials Sciences, Oak Ridge National Laboratory, Oak Ridge, Tennessee

CONTENTS

12.1 Introduction

The modification of organic and inorganic surfaces with polymeric materials is a conceptually attractive and widely practiced method of conferring desirable properties to the underlying substrate [1–6]. The ability to control the extent, topology, and surface density of chains as well as the number and type of chemical motifs displayed at the interface provides the capacity to tune interactions between the material and its environment without sacrificing bulk properties. As a result, surface properties such as wettability, adhesion, or lubricity can be tailored, or functional character such as biocompatibility through resistance to biofouling, sensing due to selective targeting and binding of analyte molecules, separation based on preferential sorbtion or gating of membrane channels, or catalytic activity through the directed deposition of absorbates can be imbued to the material.

Although perhaps not important for basic studies, in terms of designing interface-modifying polymer systems for various applications, it is often desirable to chemically bond or graft the polymeric material to the surface in order to impart permanence and stability to the surface coating. One particular class of surface-tethered polymer coatings that have garnered considerable attention over the last three decades is so-called polymer brushes [5–31]. Brushes are formed when polymer chains are tethered by one end at the solid-fluid interface at sufficiently high areal densities such that the chains stretch in the direction normal to the surface in order to alleviate lateral crowding. Because of the stretched configuration of the chains as well as the absence of translational motion in the direction normal to the surface, the structure and dynamics of polymer brushes differ from those of free polymer chains in solution. By controlling the extent and polydispersity of the brush chains, the range and "softness" of interactions across the interface can be manipulated [13,14,32–34]. Due to the development of various controlled (free) radical polymerization methods, a wide range of monomers can be utilized to produce functional and responsive brushes [9–11,15–18,24–29,35–45]. Coupled with the use of top-down lithographic patterning strategies as well as the application of various brush synthesis strategies to the modification of nanoparticles, soft colloids, or fibers, research in polymer brush systems has witnessed tremendous growth over the last decade.

12.1.1 Polymer Brushes by Surface-Initiated Polymerizations

In view of this growth and the potential to use brushes across a vast range of technologies, it is crucial to understand how the structure, chemical interaction potentials, and function can be optimized for a particular application or tuned for a range of working conditions. In these pursuits, polymer brushes attached to flat, low-area model substrates remain archetypical systems, in part because bench-top surface-sensitive instruments that can probe the chemical and physical properties of these interfacial layers are widely available, but also because the strategies for making brushes on these substrates have been well developed.

A variety of polymerization methods have been developed as a way to grow chains by surface-initiated polymerization (SIP) from interfaces decorated with various types of initiators: anionic polymerization, atom transfer radical polymerization (ATRP), nitroxide-mediated polymerization (NMP), reversible-addition fragmentation chain transfer (RAFT) polymerization, and iniferter-mediated photopolymerization (IMP) have all been adapted as surface-initiated approaches for making polymer brushes, and a recent review by Barbey et al. [11] provides a comprehensive view of the field of surface-initiated controlled radical polymerizations with thorough coverage given to the variety of monomeric species that have been used in those SIPs. The controlled (free) radical polymerization methods RAFT polymerization, NMP, and IMP rely on reversible activation and deactivation to create and maintain an equilibrium between reactive carbon radicals (growing chains) and stable persistent radical species and their combined, dormant form. In ATRP reactions, homolytic cleavage of a carbon-halogen bond creates an active, carbon-centered radical species, which leads to chain propagation by reaction with a monomer, and a higher oxidation state transition metal-ligand catalyst complex (rather than a stable, persistent free radical) [9,11,27,40]. The fast, reversible deactivation step involving transfer of the halogen back to the active carbon radical returns the growing chain to its dormant, deactivated state. As a result of this equilibrium between active and dormant species, these controlled (free) radical polymerization methods offer several advantages, including that they (1) allow the molecular weight of chains to be controlled, (2) yield polymer chains of low polydispersity, (3) are generally tolerant to impurities and are versatile in terms of the types of monomers that can be polymerized, and (4) are able to produce complex polymer architectures, such as multiblock copolymer brushes.

It is known that iniferter-mediated polymerizations can be catalyzed both photochemically and thermally, depending upon the type of the iniferter used [46,47]. However, in comparison to thermally activated iniferter-mediated polymerization, photoactivated iniferter-mediated polymerization (PA-IMP) is more attractive and widely used because of the additional advantages that arise due to the photoinitiation process (in addition to the typical aforementioned advantages of living radical polymerization (LRP) methods) [47]. For example, spatial and temporal control is enabled by controlling the location and duration of light exposure, which readily enables lithographic patterning. Also, the rate of photoinitiation can be easily manipulated by controlling the light intensity and because the process can be conducted at low temperatures (room temperature or physiological temperature, for example), IMP is well suited to thermally sensitive monomers, including conjugates formed with temperature-sensitive compounds or having unstable functional groups. Finally, because IMP processes use light as the catalyst (activating agent), postpolymerization purifications to remove catalyst compounds are obviated.

These advantages make the surface-initiated variant of PA-IMP, which is implemented mainly using the dithiocarbamate chemistry first discovered by Otsu et al. [48], a powerful and flexible method for creating polymer brushes. Throughout this chapter we will refer to surface-initiated (SI-IMP) as simply SI-IMP. Although this acronym does not distinguish between photoactivated versus thermally activated processes, to the best of our knowledge only PA-IMP has been adapted for the synthesis of polymer brushes. With this understanding, this chapter will focus on the synthesis and growth kinetics of polymer brushes created by surface-initiated PA-IMP, and those systems as technology platforms. Before reviewing synthesis and growth kinetics of SI-IMP, and presenting an overview of the variety of applications that have relied on the construction of well-defined brushes and hybrid layers by surface-initiated PA-IMP, a brief review of dithiocarbamate-based photoiniferter-mediated polymerization and its application in SI-IMP is provided.

12.1.2 Overview of Photoactivated Iniferter-Mediated Polymerization and Its Application to Surfaces

As mentioned previously, photoiniferter molecules are generally dithiocarbamate derivatives that can *ini*tiate upon exposure to light, act as trans*fer* agents, or *ter*minate during polymerization (hence the name *photoiniferter*). As shown in Figure 12.1, upon exposure to ultraviolet (UV) light, dithiocarbamate photoiniferter molecules undergo photolysis, yielding a carbon radical and a dithiocarbamyl radical. While the carbon-centered radical is reactive and can initiate polymerization by reacting with vinyl monomers, the dithiocarbamyl radical is generally viewed as stable, reacting weakly, if at all, with vinyl monomers [49,50]. However, dithiocarbamyl radicals can reversibly terminate the propagating chains, returning the chain to its dormant form. These reversible activation and deactivation processes provide the controlled (or pseudoliving) character to PA-IMP, and as implied in Figure 12.1, because of the rapid deactivation process, the equilibrium is shifted toward the dormant species.

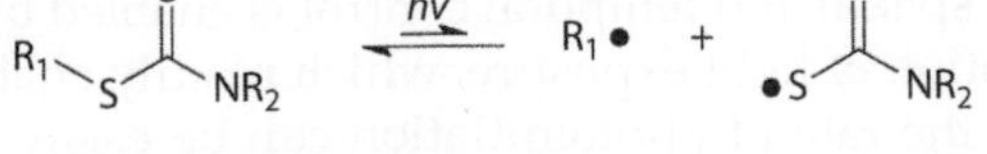

FIGURE 12.1
Typical reversible activation and deactivation of dithiocarbamate-based photoiniferters.

As shown by Otsu et al., photoiniferter-mediated polymerizations can exhibit the hallmarks of a "living" polymerization, including a linear increase in conversion and molecular weight with polymerization (irradiation) time, retention of appropriate end groups associated with reversible termination of the growing chain, and narrow molecular weight distributions [47]. Because disproportionation, coupling of radicals, or chain transfer reactions are difficult to eliminate completely, it is generally accepted that *controlled* and *pseudoliving* are better terms to describe these (free) radical polymerizations. Nevertheless, because of these characteristics, PA-IMP has been used to create block copolymers and topologically complex polymers.

SI-IMP was first performed by Otsu and coworkers, who devised a way to anchor benzyl-(*N,N*-diethyldithiocarbamyl)acetate to cross-linked polystyrene (PS) spheres by reactive modification [51]. These iniferter-modified particles were used to grow PS chains from the surfaces by photoactivated polymerization, and chain extension by reinitiation in the presence of methyl methacrylate was also demonstrated. Several years later, Nakayama and Matsuda used SI-IMP from flat, low-area substrates to synthesize surface-tethered polymer brushes [52]. In their work, *N,N*-diethyldithiocarbamate groups were immobilized onto polymeric surfaces by chemical reaction and by surface photografting. In addition, Nakayama and Matsuda demonstrated that photomasking could be applied to create patterned surfaces [52]. While these advances would be recognized as methods of making polymer brushes, because the dithiocarbamate-based photoiniferters were reactively coupled onto polymeric surfaces, two parameters typically used to assess whether a grafted layer exists as a brush—the grafting density (areal density) of chains and layer thickness—could not be measured. In 2000, de Boer et al. demonstrated the formation of self-assembled monolayers (SAMs) based on *N,N*-(diethylamino)dithiocarbamoylbenzyl(trimethoxy) silane (SBDC), which incorporates both the highly effective diethyldithiocarbamate motif and an alkoxysilane anchoring group that allows the photoiniferters to be chemically grafted onto hydroxyl-bearing substrates [53]. These photoiniferter SAMs were used to grow PS, and poly(methyl methacrylate) (PMMA) layers, and PS layers made by SI-IMP were reinitiated in the presence of methyl methacylate to create poly(styrene-*block*-methyl methacrylate) layers. These research endeavors paved the way for the development of SI-IMP as a robust and flexible method for creating polymer-modified and polymer brush-modified surfaces, and for full characterization of the polymer-modified interfaces via a variety of surface-sensitive, bench-top techniques.

Figure 12.2 shows a general scheme for surface-initiated PA-IMP based on SBDC. Here again, generation of surface-tethered carbon radicals is controlled by the intensity of incident light, allowing for the growth of polymer chains from the solid-fluid interface. The growing chains are capped by recombination with dithiocarbamyl radicals, and the reversibility of this

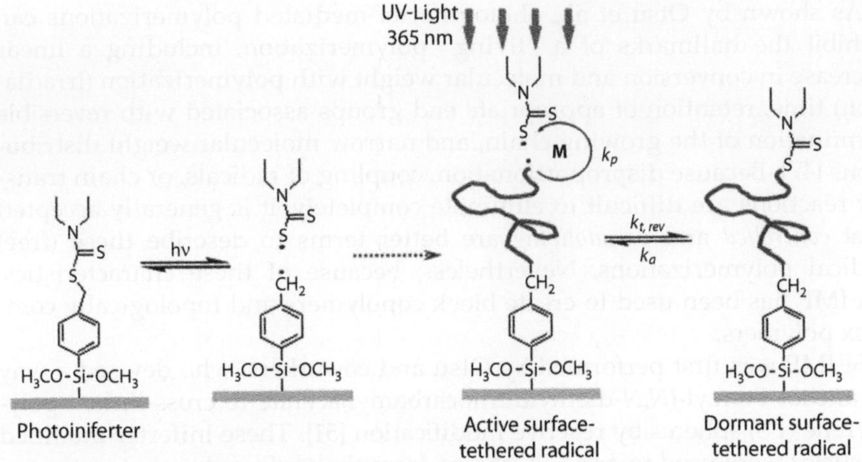

FIGURE 12.2
SI-IMP using dithiocarbamate-based photoiniferters.

activation and deactivation enables control of chain growth as well as the ability to reinitiate in the presence of a second monomer species. To enhance control in surface-initiated PA-IMP reactions, a source of deactivating dithiocarbamyl radicals is often added to the monomer solution in order to compensate for the exceedingly low surface concentration of tethered, photo-activated iniferters [54,55]. These matters of photoinitiation, manipulation of chain growth, and reversible and nonreversible termination reactions will be detailed more fully later in this chapter, as understanding these processes and how they are interconnected is central to applying SI-IMP as a means to tailor interfacial structure and properties using tethered polymer brushes.

It is worth noting that this overall strategy of installing well-defined inifer-ter molecules onto surfaces and activating them through photolytic cleavage of the C–S bond of the dithiocarbamate to create surface-tethered polymer brushes is general; it has been implemented using a variety of surface attachment strategies and direct attachment of iniferter fragments onto resin supports and iniferter designs [28,51–87]. Briefly, SI-IMP has been performed from flat surfaces, spherical or cylindrical capillaries, tubes or particles, porous surfaces such as membranes or polymer films/fibers, and even in microfluidic devices. While conducting SI-IMP from flat surfaces is generally straightforward, curved surfaces such as spherical or cylindrical capillaries, tubes, or particles require special equipment to ensure uniform exposure of the surface to light. Particular aspects and challenges in conducting SI-IMP using surfaces of various geometry are attenuated in this chapter in order to focus on the kinetics of polymer brush synthesis using SI-IMP. Nevertheless, SI-IMP has been shown to be a versatile and robust method of creating surface-tethered, interfacial polymer layers that have applications ranging from stimuli-responsive to bio-inspired surfaces. This subject is treated in the next section.

12.2 Surface-Initiated Iniferter-Mediated Polymerization as a Potential Technology Platform

Because brushes straddle the solid-fluid interface and alter surface properties, in his seminal publication in 1980, de Gennes envisioned five applications for polymer brushes: they were offered as coatings appropriate for adjusting wetting and adhesion, promoting biocompatibility and tailoring colloid stability, as well as for chromatographic applications [88]. In those pursuits, SI-IMP has proven to be a versatile and robust method of creating interfacial polymer layers suited for a variety of applications, including fabrication of stimuli-responsive coatings and polymer-modified surfaces relevant to biological and biomedical applications such as antibacterial coatings and control of cell-protein adhesion, and for creating molecularly imprinted coatings that have potential application in the fields of molecular recognition and separations. The suitability of SI-IMP in these and other areas springs from the advantages of the method described earlier, including the fact that SI-IMP can be conducted at room/physiological temperature in the absence of any added catalysts, and that the pseudoliving nature of SI-IMP offers formation of complex polymer architectures. Also, as will be seen, dithiocarbamate iniferters can be readily modified while preserving their inherent function, allowing them to be used in various ways. For the purpose of highlighting and summarizing a rather large volume of work, we have chosen to segment work into areas of stimuli-responsive layers, biofunctional coatings, and systems for chromatographic separations. We note that these distinctions are for convenience only, as many works cross these artificial boundaries.

12.2.1 Stimuli-Responsive Polymer Brushes

SI-IMP has been used for synthesis of different types of stimuli-responsive polymer brushes that are responsive to several external stimuli, such as pH, temperature, and ionic strength [28,58–65]. Because materials interact with their surroundings via their interfaces, the ability to fashion soft interfacial layers and tune the range, extent, and type of physicochemical interactions across interfaces is central to a variety of applications. Rahane et al. carried out sequential SI-IMP of two monomers to create bilevel poly(methacrylic acid)-*block*-poly(N-isopropylacrylamide) (PMAA-*b*-PNIPAM) block copolymer brushes that can respond to multiple stimuli [28]. They observed that each strata in the bilevel PMAA-*b*-PNIPAM brush retained its customary responsive characteristics: PMAA being a "weak" polyelectrolyte swells as pH is increased and the thermoresponsive PNIPAM block collapses as temperature is raised through the volume phase transition temperature due to its lower critical solution temperature (LCST) behavior. As a result of ions added to make buffer solutions of various pH and because of the effect of surface confinement, the swollen-collapse transition of the PNIPAM layer occurs at a lower temperature and over a wider range of temperature, as compared to the sharp transitions usually observed at 32°C for free PNIPAM chains in deionized water [89–92]. This study illustrates both the robustness of the SI-IMP method for creating multicomponent interfacial layers and the fact that the swelling behavior of these layers can be broadly manipulated by changing the pH, temperature, and ionic strength, either individually or in concert.

Vansco and coworkers have also used SI-IMP to create various stimuli-responsive polymer brushes, adapting the iniferter monolayer attachment strategy so that brushes could be grown from gold surfaces. In their work, monolayer-forming materials containing a dithiocarbamate motif were tethered to the gold surface via self-assembly of disulfide-derivatized dithiocarbamate-based photoiniferter molecules. Because gold-thiol bonds typically are not stable under UV irradiation conditions used frequently in SI-IMP processes (λ = 365 nm), SI-IMP was performed at lower wavelengths using a 280 nm cutoff filter. In this way, they were able to synthesize both PMAA and PNIPAM brushes [59–61,66]. Not only did they find that control of grafting density of the iniferter molecules (by mixing an inert component into the monolayer to dilute the surface-tethered iniferter) reduced recombination reactions that result in dead chains, but they also showed that photomasking provided spatial control, enabling micropatterned PNIPAM brushes to be made [59]. The response of these micropatterned PNIPAM brushes to changes in environmental temperature was followed using in situ atomic force microscopy. In addition to the anticipated swollen-collapse transition in thickness observed around the LCST, they observed an increase in the surface roughness of PNIPAM brushes at temperatures above LCST due to formation of hydrophobic polymer aggregates induced by shrinkage (dehydration) of the polymer chains.

Tagit et al. [60] extended their studies of temperature-induced changes on the conformation of surface-tethered PNIPAM chains synthesized by SI-IMP from gold surfaces by following changes in the luminescence properties of CdSe quantum dots attached covalently to the PNIPAM layer. With increases in the temperature of environment, PNIPAM chains collapsed, bringing the CdSe dots closer to the gold surface, causing a decrease in the luminescence intensity, supposedly due to nonradiative energy transfer. Decreasing the temperature reversed the conformation of PNIPAM chains to a swollen form, resulting in an increase in the luminescence intensity of quantum dots.

Sebra et al. [63] synthesized pH-responsive brushes of polyethylene glycol acrylate succinyl fluorescein, which can function as optical switches, using SI-IMP from custom-made dithiocarbamate-functionalized cross-linked polymer. These dithiocarbamate-functionalized cross-linked polymer films were synthesized by photopolymerization using a monomer formulation consisting of 48.75 wt.% aromatic urethane diacrylate (UDA) and 48.75 wt.% triethylene glycol diacrylate (TEGDA) mixed with 1.5 wt.% 2,2-dimethoxy-2-phenylacetophenone (DMPA) photoinitiator and 1 wt.% tetraethylthiuram disulfide (TED) as a source of deactivating dithiocarbamyl radicals. Their work demonstrated that the polymer brushes grafted from UDA/TEGDA platforms functioned as sensors that were able to monitor changes in pH upon mixing, reacting, or introducing new chemicals in microfluidic applications.

Ohya and Matsuda [64] used SI-IMP from biologically relevant gelatin-based substrate to synthesize PNIPAM-grafted gelatin, which was eventually used as a thermoresponsive three-dimensional (3D) artificial extracellular matrix for proliferation of bovine smooth muscle cells. In their work they investigated the effects of various formulation and structural parameters such as graft density and molecular weight of PNIPAM graft chains of PNIPAM-grafted gelatin with respect to gelation and cell proliferation. In general, they found that as the PNIPAM molecular weight and the grafting density of chains increased, the onset of gelation occurred at lower concentration. In addition, the ratio of PNIPAM to gelatin was found to be a crucial design parameter that along with the thermal responsiveness of PNIPAM, could be used to tailor mechanical properties, void and pore sizes of tissue scaffolds, and thus affect cell viability and proliferation [64].

In an effort to create adherent, stimulus-responsive bioconjugate materials, Matsuda et al. immobilized albumin onto the dithiocarbamate group of a photoiniferter that previously had been tethered to a glass substrate [65]. This surface-immobilized, albuminated dithiocarbamate-based photoiniferter was used to synthesize hydrophilic p(N,N-dimethylacrylamide) and PNIPAM brushes by SI-IMP. When the polymerizations were well controlled (displaying pseudoliving character), the tethered polymer chains contained albuminated chain ends that segregated to the periphery of the interfacial layer. They used the PNIPAM-based bioconjugates, which combined the thermoresponsive character of the chains with the control of colloidal osmotic pressure (of blood proteins) provided by albumin as a proof-of-concept example where SI-IMP could be harnessed to create bioactive surfaces that could regulate protein absorption through a temperature-induced self-cleaning function [65].

12.2.2 Biofunctional Surfaces

SI-IMP is particularly attractive for synthesis of biofunctional surfaces due to the ability to conduct polymerizations at physiological temperatures and without catalysts or other additives that may be detrimental to biological species or biosystems [67–74]. For example, Sebra et al. used their previously described UDA/TEGDA platforms decorated with dithiocarbamate groups to synthesize polymer layers by SI-IMP for use as highly sensitive and efficient assays for antigen detection [68,69]. The SI-IMP process allowed copolymer brushes comprising ethyleneglycol acrylate and acrylated antibodies to be made with facile control over the concentration of antibodies along the chains, and therefore the areal density of antibodies presented at the solid-fluid interface. The extension of the chains bearing grafted antibodies into the solution containing antigens, in combination with control over graft architecture imparted by the pseudoliving nature of the SI-IMP method and local environment provided by polyethylene glycol units of the grafted chains, facilitated accelerated detection, allowing quick detection of glucagon antigen, which has a short life.

Lawson et al. [70] also showed that the dithiocarbamate-decorated UDA/TEGDA platform could be used to grow polymer brush layers of photopolymerizable PEG acrylate and acrylamide derivatives containing the antibacterial glycoprotein, vancomycin. It was found that the structure of grafted polymer chains, which was controlled by SI-IMP, and the length of side chains that linked the repeat unit (either PEG or acrylamide) to vancomycin were crucial to the antibacterial activity of the vancomycin-functionalized polymer brushes. This same platform was used to grow polymer chains bearing protein antibodies that induce T-cell apoptosis, thereby creating a bioactive polymer coating designed to defeat the autoimmune system rejection of donor tissue [71]. Not only do the "levers" afforded through the controlled SI-IMP allow for tailoring of biofunctionality, but it was found that incorporation of T-cell-specific adhesion ligands enhances functionality (induction of T-cell apoptosis). This result demonstrates the broad potential of SI-IMP and polymer-based biomaterials in therapeutic applications.

Harris et al. [72] synthesized PMAA-modified substrates using SI-IMP of MAA and then modified the carboxylic acid groups with the cell-adhesive peptide sequence, arginine-glycine-aspartic acid (RGD), which is widely known to enhance cell adhesion. As expected, these RGD-functionalized materials were found to be cytocompatible and cell adhesive. Spatial control over SI-IMP was used to obtain surfaces having a gradient in areal density of RGD along the length of the substrate. The number of cells adhering to the substrate per unit area increased in accordance with the increasing RGD density. It is widely appreciated that cell adhesion depends on how RGD groups populate and orient at interfaces as well as the mechanical properties of the underlying film; thus, the facile control and robustness of SI-IMP provide an important pathway for tuning the function of the RGD-decorated interface. Navarro et al. [61] adapted the methodology described by Harris et al. [72] to incorporate RGD peptide sequences into PMAA brushes. However, Navarro et al. also harnessed the living character of SI-IMP to chain extend the RGD-functionalized PMAA layers with methacrylic acid, which buried the RGD groups within the layer [61]. Interestingly, while lactate dehydrogenase (LDH) assays revealed no differences when (human osteoblastic) cells were cultured on the two types of RGD-functionalized films, differences in cell shape and focal adhesion were observed between films having buried RGD sequences and those having RGD at the end of the brushes. These behaviors were speculated to arise from differences in how the chains respond (organize and structure) in the presence of the cells. They also point out the importance of understanding how brush chains, particularly functionalized brushes, rearrange structurally in the presence of absorbates. This line of research—understanding how the ability to tune the layer and its display of functionality affects biological response—demands considerable attention as applications for synthetic systems with biologically derived components increase.

Matsuda et al. [73,74] used a modified phosphorylcholine group functionalized dithiocarbamate-based photoiniferter to synthesize phosphorylcholine-end-capped oligomers and block oligomers as biomimetic surfaces. When synthesized appropriately and solvated in serum-containing media, these brush layers presented phosphorylcholine end groups to the environment that preferentially adsorbed nonadhesive proteins in order to prevent cell adhesion. In an analogous fashion, heparin was chemically transformed into a dithiocarbamyl-containing iniferter and used to grow heparin-capped PNIPAM chains [75]. This work is especially noteworthy because it demonstrates how SI-IMP can be compatible with biologically derived compounds. While the PNIPAM retained its customary thermoresponsive behavior in aqueous solution, becoming insoluble as temperature increased, molecular weight effects were observed in terms of the transition temperature, stability on polymeric surfaces (resistance to desorption), and ability to complex with antithrombin protein (AT III). These novel bioconjugates may find application as biomaterial coatings that provide anticoagulation activity.

12.2.3 Molecularly Imprinted Polymers for Molecular Recognition-Based Separations

Besides the typical syntheses of polymer brushes from flat, low-area substrates, SI-IMP can be conducted from cellulose membranes, carbon nanotubes, and silica or polymeric beads, which has allowed SI-IMP to be used for synthesis of molecularly imprinted polymers (MIPs) from a variety of supports [76–80]. The use of SI-IMP to fabricate MIPs is particularly advantageous over more common fabrication of MIPs by surface-initiated polymerization from surface-bound conventional initiators because the dithiocarbamyl radicals generated from surface-bound iniferters do not cause polymerization in solution, resulting in improved separation capacity [76–80]. In this section a few examples of fabrication of MIPs synthesized by SI-IMP from a variety of supports for molecular recognition or separation applications are briefly summarized with a focus on particular advantages enabled by the SI-IMP method.

Hattori and coworkers exploited both the living characteristics and photoactivation mechanism of SI-IMP to create MIPs consisting of methacrylic acid (MAA) and ethyleneglycol dimethacrylate (EDMA) synthesized from iniferter-modified cellulose membranes in the presence of theophylline, which served as the template molecule [76]. In addition to the ability to control polymer architecture, they observed that the SI-IMP photoactivation mechanism avoids template-monomer complexation, which is highly undesirable but usually observed in MIPs synthesized by thermally activated polymerization methods [76]. The MAA and EDMA comonomer system has also been used to synthesize MIPs from iniferter-modified carbon nanotubes

(CNTs) with theophylline as the template molecule [78] and from porous silica beads with thiabendazole as the print molecule [79], and these have been applied as molecular recognition agents. Qin et al. [80] made use of the stability of the dithiocarbamyl radicals generated upon photoactivation of iniferters to synthesize lysozyme-imprinted hydrophilic MIPs. These were created by SI-IMP of acrylamide and *N,N'*-methylenebisacrylamide from iniferter-derivatized mesoporous polystyrene beads (in the presence of the template protein lysozyme) and used in chromatographic separations that employed an aqueous mobile phase. They found that the control and stability imparted by SI-IMP are crucial elements that enhance the separation of lysozyme from other competing proteins.

These discussions illustrate how SI-IMP has enabled the design of new and interesting interface-modifying systems as well as potential technology platforms devised from the versatility and advantages of the SI-IMP process. The development of surface-modifying polymer layers by SI-IMP having particular structural characteristics (e.g., extent, grafting density), or properties tailored through control over the location, number, or arrangement of functional units, benefits from a thorough understanding of the synthesis of polymer brushes by SI-IMP and the impact of various reaction conditions on layer growth. We have previously described such a set of experimental and kinetic modeling studies aimed at providing a cohesive understanding of brush growth by SI-IMP. These efforts, augmented by findings from similar efforts, are reviewed in the following section.

12.3 Kinetics of Surface-Initiated Iniferter-Mediated Polymerization

To understand the kinetics of SI-IMP and the key parameters and events that govern the polymer brush growth, we first examine the basic processes and species that constitute SI-IMP. Figure 12.3 shows the idealized process for the formation of a tethered polymer chain by SI-IMP whereby growth proceeds by monomer addition to the carbon-centered radical at the end of the growing chain, which is initially produced by hemolytic cleavage of the carbon-sulfur bond of the surface-tethered photoiniferter molecule. This activation process concomitantly releases a dithiocarbamyl radical into solution. Throughout the text we identify these two radical species as surface-tethered radicals, STR•, and as DTC•, which refers to dithiocarbamyl radicals in solution, and their combined dormant form is represented as STR-DTC. The simultaneous growth of several such chains along the surface creates a surface-tethered polymer brush. The growth rate of the chains, and therefore of the polymer brush, depends on various parameters such as monomer concentration and light intensity.

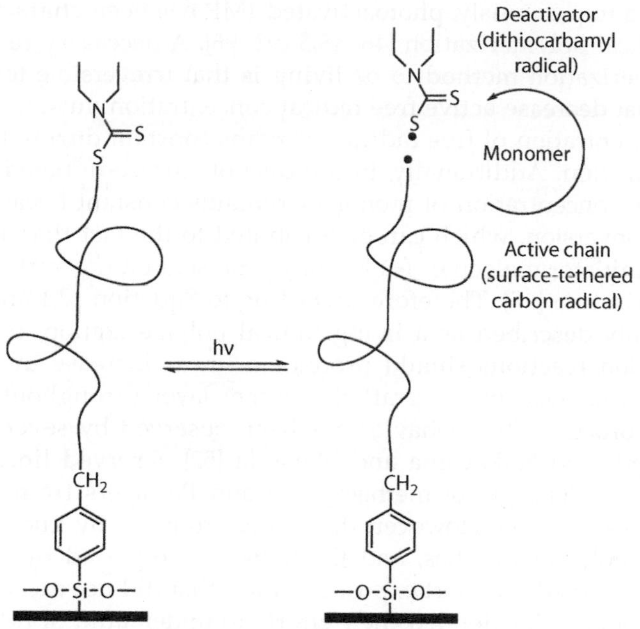

FIGURE 12.3

Idealized scheme for the growth of a polymer brush by surface-initiated photoiniferter-mediated photopolymerization. The stable dithiocarbamyl radical reversibly terminates the active carbon radical to produce the dormant species. When activated, the surface-bound carbon radical propagates by the addition of monomer to form a polymer chain. Several densely packed chains grow simultaneously to form a surface-tethered polymer layer or polymer brush.

As suggested by the figure, it is expected that the growth rate of the polymer brush depends on the concentration of surface-tethered radicals, STR•, and monomer, M. Equation 12.1 describes the rate of polymer brush growth:

$$\frac{dT}{dt} = k[STR\bullet][M] \tag{12.1}$$

Here $[STR\bullet]$ is the concentration of surface-tethered radicals, [M] is the monomer concentration, T is the ellipsometric thickness of the dry polymer layer, and t is the polymerization time. This expression uses k as a proportionality constant that lumps together the propagation rate constant for polymerization, a scaling constant between rate of SI-IMP and average molecular weight of polymer chains, and the proportionality constant assumed between the thickness of the polymer brush and the average molecular weight of the brush chains. As a result of this latter assumption of a linear relationship between brush thickness and molecular weight, Equation 12.1 is valid only in the brush regime.

As alluded to previously, photoactivated IMP has been characterized as a living radical polymerization [46–48,53,93–96]. A necessary requirement for a polymerization method to be living is that irreversible termination reactions that decrease active free radical concentration must be absent so that the concentration of free radicals remains constant during the course of polymerization. Additionally, in the case of surface-initiated polymerizations, the concentration of monomer remains constant because of low monomer conversion, which can be attributed to the fact that a very low number of initiator molecules ($<10^{15}$/cm^2) is present on the surface to initiate polymerization [97]. Therefore, according to Equation 12.1 and because IMP is ideally described as a living radical polymerization, the absence of termination reactions should produce a linear increase in molecular weight and thickness of the grafted polymer layer throughout the polymerization process. This behavior has been observed by several groups: Luo et al. [84] and Nakayama and Matsuda [52] observed linear growth of poly(polyethylene glycol methacrylate) and PS layers from polymeric substrates, respectively. However, the layers produced by Luo et al. were not strictly polymer brushes, and the behaviors reported by Nakayama et al. were observed for short exposure times (that did not exceed 20 min). Nakayama et al. [85] extended their efforts to understand SI-IMP kinetics from flat, low-area substrates using quartz crystal microbalance to track the change in mass with exposure time, demonstrating a linear increase in layer growth. Despite these studies supporting the notion of SI-IMP being a living polymerization process, other investigations of the SI-IMP process offer a contrasting view of the growth kinetics of SI-IMP: a variety of groups have shown that the growth of polymer brush layers produced by SI-IMP slows down and exhibits a plateau in the evolution of thickness vs. irradiation time [81–83,86,87].

To test the aforementioned hypothesis, Rahane et al. [81] synthesized PMMA brushes by SI-IMP of methyl methacrylate (MMA) using a wide variety of monomer concentrations and exposure times. The results of these studies, presented in Figure 12.4, show that there is a slow initial increase in grafted PMMA layer thickness followed by a rapid increase as the polymerization time increases for the three [M] examined. At early times, when the thickness is increasing slowly, we hypothesized that due to the small molecular weights the chains are in the "mushroom" regime. As the chains continue to grow, they begin to interact with one another and transition from the coiled mushroom regime to the extended "brush" regime [98]. This crossover behavior was also observed by Vansco et al. in their studies of PMAA brushes grown from mixed SAMs containing iniferter molecules and a nonphotoactive analog [66].

The dashed lines shown in Figure 12.4 represent the idealized case of linear growth behavior embodied by Equation 12.1. While the anticipated first-order dependence on the monomer concentration is observed during

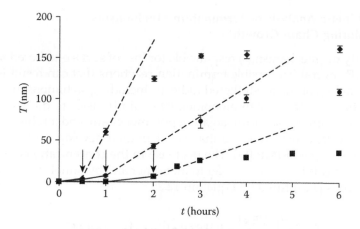

FIGURE 12.4

Dry poly(methyl methacrylate) layer thickness as a function of exposure time at a light intensity of 5mW/cm². Methyl methacrylate concentrations in toluene are (■) 1.17, (●) 2.34, and (♦) 4.68 M. Arrows indicate the apparent mushroom-to-brush transition for the concentrations studied, and the dotted lines show the behavior of an ideal, living photopolymerization. The slopes of these lines also represent the initial growth rates of the PMMA layer. (Reprinted with permission from Rahane, S. B., et al., Kinetics *Macromolecules*, 38, 8202–8210, 2005, copyright (2005) American Chemical Society.)

the initial phases of polymer layer growth in the brush regime, polymer layer growth deviates from the model at longer exposure times. This behavior indicates that SI-PMP of MMA under the described conditions did not proceed via a living free radical polymerization mechanism. The pseudo-living nature of SI-PMP of MMA was hypothesized to stem from the fact that very few initiator molecules are present on the surface to produce a sufficient concentration of deactivating dithiocarbamyl radicals in solution that reversibly terminate the surface-tethered radicals. In the absence of sufficient deactivator to establish and keep the equilibrium depicted in Figure 12.3 shifted toward the dormant species, irreversible termination reactions such as bimolecular termination or chain transfer of the radical from the surface will occur to a significant extent. Consequently and in consideration of Equation 12.1, a continuous decrease in [$STR•$] leads to a continuous decrease in the growth rate, as manifest in the decreasing rate of change in layer thickness with time during photopolymerization. That there is an insufficient DTC• to control the SI-IMP from flat, low-area surfaces is not surprising, as the release of dithiocarbamyl radicals from the two-dimensional surface upon photoactivation enables those fragments to diffuse away into the surrounding solution, leading to a decreased probability for reversible capping of the growing chain by a DTC•. As a result and as will be discussed in Section 12.3.2, it is often advantageous to add a source of deactivating dithiocarbamyl radicals.

12.3.1 Kinetic Analysis of Termination Mechanisms during Chain Growth

To identify the mechanism(s) responsible for loss of surface-tethered radicals in SI-PMP, several irreversible termination reactions that can result into the permanent loss of surface-tethered radicals, including (a) bimolecular termination, (b) chain transfer to monomer, (c) chain transfer to dithiocarbamyl radical, (d) chain transfer to an adjacent polymer chain, and (e) chain transfer to solvent, were considered by Rahane et al. in their development of a kinetic model to describe SI-IMP [81]. The decrease in the concentration of surface-tethered radicals by chain transfer to monomer and by bimolecular termination reactions is captured by Equation 12.2:

$$-\frac{d[STR\bullet]}{dt} = k_{bt}[STR\bullet]^2 + k_{ct}[STR\bullet][M] \tag{12.2}$$

Here, k_{bt} is the kinetic constant for bimolecular termination, and k_{ct} is the kinetic constant for chain transfer to monomer. Equation 12.2 suggests that loss of radicals by chain transfer to monomer should become more significant as monomer concentration increases, while bimolecular termination should become more significant as STR• increases.

To determine the prevailing termination mechanism in SI-PMP over a broad range of relevant reaction conditions, experimental data on film thickness evolution such as that shown in Figure 12.4 were fit in the brush regime (transitions marked by arrows in Figure 12.4) by the kinetic models that incorporate one or more termination mechanisms. For example, Rahane et al. combined Equation 12.1 with expressions for STR• based on either bimolecular termination or chain transfer to monomer to develop models for how layer thickness should evolve as a function of exposure time. These models, shown as Equations 12.3 and 12.4, respectively, can be compared to experimental data of polymer layer thickness as a function of time to deduce which irreversible termination mechanisms are prevalent.

$$T = \frac{k}{k_{bt}}[M]\ln\left\{1 + k_{bt}[STR\bullet]_0\left(t - t_{brush}\right)\right\} + T_{brush} \tag{12.3}$$

$$T = \frac{k}{k_{ct}}[STR\bullet]_0\left\{\exp\left(-k_{ct}[M]t_{brush}\right) - \exp\left(-k_{ct}[M]t\right)\right\} + T_{brush} \tag{12.4}$$

In these expressions $[STR\bullet]_0$ is the initial concentration of surface-attached radicals and T_{brush} is the thickness of the PMMA layer at the time it enters the brush regime, T_{brush}.

Figure 12.5 shows the comparison of model predictions with experimental data. The models are semipredictive, as unknown kinetic constants were determined by fitting the data obtained at high monomer concentration and then used to generate model curves at other concentrations [81]. While both termination mechanisms result in a decreasing growth rate with time and limit the film thickness, the patterns of PMMA layer growth were observed to be unmistakably different. As shown in Figure 12.5a, when bimolecular termination is the dominant termination mechanism, the initial growth rate and maximum layer thickness are functions of monomer concentration. Furthermore, the relative decrease in polymerization rate is independent of monomer concentration and only depends on exposure time. In contrast,

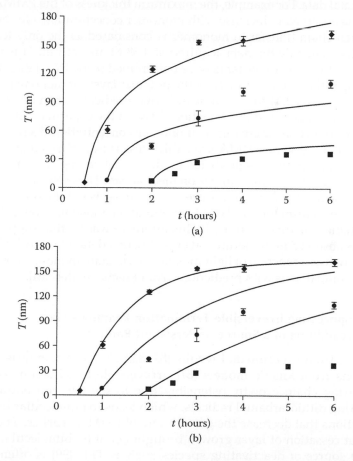

FIGURE 12.5
Comparison of the PMMA layer thicknesses measured using variable angle ellipsometry as a function of exposure time with model predictions (thin lines) for (a) bimolecular termination and (b) chain transfer to monomer. Irradiation intensity is 5 mW/cm² and methyl methacrylate concentrations in toluene are (■) 1.17, (●) 2.34, and (♦) 4.68 M. (Reprinted with permission from Rahane, S. B., et al., *Macromolecules*, 38, 8202–8210, 2005, copyright (2005) American Chemical Society.)

when chain transfer to monomer concentration is dominant, the maximum film thickness is independent of monomer concentration (Figure 12.5b). Equation 12.4 predicts that the limiting film thickness under these termination conditions is equal to the value of $k[STR\bullet]_0/k_{ct}$. As observed from Figure 12.5b, the time at which this limiting thickness is reached decreases as monomer concentration increases.

Figure 12.5a compares the measured film thicknesses to the predictions of the bimolecular termination model. As can be seen from this figure, the model predictions provide a reasonably good fit to all experimental data sets, including those obtained at 1.17 and 2.34 M monomer concentrations. Also, the essential features predicted by this model are exhibited by the experimental data. For example, the maximum thickness of the PMMA layer obtained is observed to increase with monomer concentration. On the other hand, when chain transfer to monomer is considered as the only termination mechanism, only the data acquired at 4.68 M are well fit. This is to be expected, as the kinetic constants were determined using this data set. More compelling is the fact that growth of the polymer layers at monomer concentrations of 1.17 and 2.34 M is significantly overpredicted when those kinetic constants are applied, and the deviation between the experimental data and the model predictions increases as monomer concentration decreases. It is also clear that the same PMMA layer thickness is not achieved at different monomer concentrations as predicted by the model having chain transfer to monomer as the dominant termination mechanism. These observations support the conclusion that chain transfer to monomer is not the dominant termination mechanism for the PMMA system at the monomer concentrations and irradiation conditions studied. This inference was further supported by the trends observed in the additional experimental data collected for PMMA brush growth as a function of light intensity at constant monomer concentration, and comparisons with predictions from kinetic models [81].

12.3.2 Suppressing Irreversible Termination Reactions via Preaddition of a Source of Persistent Radicals

As evident from experimental results [81–83,86,87] and comparison with predictions from kinetic models [81], irreversible termination reactions occur during chain growth, ostensibly due to the low concentration of deactivating dithiocarbamyl radicals, which leads to bimolecular termination reactions that decrease the growth rate of PMMA brushes. Therefore, to combat cessation of layer growth brought about by bimolecular termination, a source of deactivating species such as TED [99] is often added to the reaction mixture. As shown in Figure 12.6, TED undergoes homolytic cleavage when irradiated with UV light, yielding two dithiocarbamyl radicals. TED was used by Doi et al. [99] to prevent bimolecular termination during photoiniferter-mediated photopolymerization of methyl acrylate in bulk or in benzene. The presence of TED allowed Otsu et al. [51]

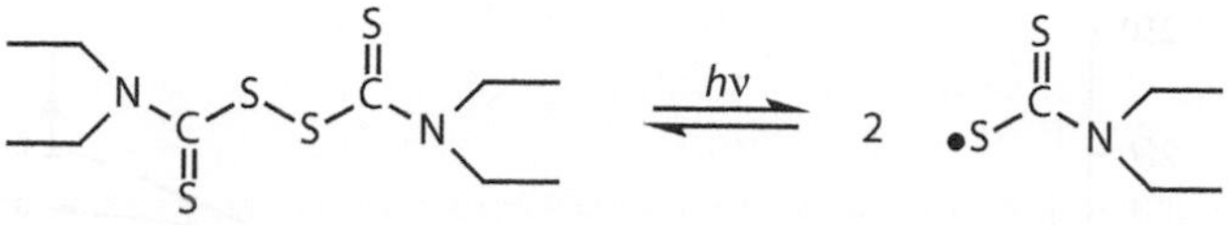

FIGURE 12.6

Formation of two identical dithiocarbamyl radicals by homolytic cleavage of tetraethylthiuram disulfide (TED) mediated by UV light.

to synthesize di- and triblock copolymers of PS and PMMA layers using surface-initiated photopolymerization from photoiniferter-modified PS beads.

However, in addition to suppressing the irreversible bimolecular termination reactions, preaddition of TED affects the SI-IMP process in a variety of ways. The main features can be gleaned from the sets of data presented in Figure 12.7, which demonstrate the impact of added TED on the SI-IMP of PMMA brushes (synthesized at $[M]$ = 4.68 M and a light intensity of 5 mW/cm^2). As seen by the curve labeled **a** in Figure 12.7, when no TED is added to the polymerization solution, the thickness of the PMMA layer increases rapidly after the initial lag period, but is followed by a sharp decline in the growth rate due to loss of active ends by bimolecular termination.

The data sets labeled **b, c, d,** and **e** in Figure 12.7 show the effect of increasing $[TED]$ on the growth of the PMMA brush layers. When $[TED]$ = 0.02 mM (data set **b**), the maximum growth rate (observed between 30 min and 1 h exposure time) is slower than when no TED is added; however, the thickness of this PMMA layer after 6 h of exposure exceeds that of the sample polymerized without TED by approximately 50 nm, suggesting that the extent of irreversible termination is lower in the presence of TED. At $[TED]$ = 0.2 mM (data set labeled **c** in Figure 12.7), a nonlinear increase in PMMA layer thickness is still observed. However, at $[TED]$ of 1 and 2 mM (data sets **d** and **e**, respectively), the measured thicknesses of PMMA layers increase linearly throughout the SI-IMP, suggesting that the extent of irreversible termination is less than at the lower $[TED]$ studied. The simultaneous decrease in the extent of irreversible termination and the PMMA layer growth rate is consistent with a shift in the equilibrium of the surface-tethered radicals toward the dormant state. Such a shift in the equilibrium toward the dormant surface-tethered radicals also suggests that bimolecular termination should be minimized as TED concentration is increased.

To support the inference that the extent of irreversible termination decreases upon preaddition of TED, a series of reinitiation studies in the presence of styrene monomer were conducted using PMMA layers synthesized in the presence of TED (at concentrations ranging from 0.02 to 2 mM) with 6 h of UV exposure. These studies showed that if no TED were added during SI-IMP of the PMMA layer, attempts to extend the chains by reinitiation in the presence of styrene failed. No changes in layer thickness or surface energy, as judged by ellipsometric and contact

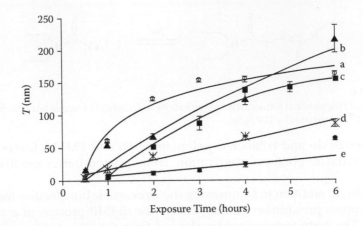

FIGURE 12.7
Dry PMMA layer thicknesses at TED concentrations of (a) 0 mM, (b) 0.02 mM, (c) 0.2 mM, (d) 1 mM, and (e) 2 mM. The thin lines are only to guide the eye. In these experiments, MMA concentration of 4.68 M and light intensity of 5 mW/cm² were used. Error bars represent the standard deviation calculated from repeat measurements using three identical samples (and five thickness measurements per sample). (Reprinted with permission from Rahane, S. B., et al., *Macromolecules*, 39, 8987–8991, 2006, copyright (2006) American Chemical Society.

angle measurements, respectively, were observed. On the other hand, when TED was preadded, increases in layer thickness and water contact angles consistent with the addition of PS blocks on the PMMA chains were observed. Additionally, as the [*TED*] used during growth of the PMMA layer was increased, the thickness of the PS block added upon chain extension increased [54].

While preaddition of TED offers the benefit of improving control as judged by linear growth of the layer in time and an increase in reinitiation ability, it comes at the cost of a decreased rate of growth of the brush layers. Also, it was noted that at high [*TED*], free chains were recovered from the solutions after SI-IMP, suggesting that at long irradiation times and high [*TED*], dithiocarbamyl radicals generated by hemolytic cleavage of TED are also responsible for radical initiation and polymerization [54]. These trade-offs highlight both the impact and importance of maintaining the concentration of active carbon-centered radicals low through addition of the deactivating species. Notable and inherent to SI-IMP processes is the fact that low concentrations of active radicals can also be achieved via decreasing the light intensity. However, the mechanistic impacts of low concentrations of the radicals brought about by preaddition of TED and by decreasing the light intensity are significantly different: while preaddition of TED reversibly caps the active surface-tethered radicals, decreasing the light intensity will simply decrease the rate at which active surface-tethered radicals are generated. This mechanistic difference, in turn, may affect the reinitiation ability of polymer brushes synthesized and the ultimate architecture of the polymer brush synthesized. In total, it is clear that important features of polymer brushes synthesized by surface-initiated SI-IMP such as brush thickness and reinitiation ability are interrelated and controlled by parameters including light intensity and exposure time, monomer concentration, concentration of surface-tethered radicals, and concentration of added deactivator (e.g., TED) [54,81–83]. As a result, it is efficient to use kinetic modeling to explore those interrelationships and impacts.

12.3.3 Application of a Comprehensive Kinetic Model for Simulating SI-IMP

To understand and effectively predict the effects of different reaction parameters on surface-initiated photochemically activated IMP, Rahane et al. [82] developed a rate-based kinetic model that supports and augments experimental findings of brush growth and reinitiation ability. This rate-based approach, which involves simultaneously solving rate equations that describe the individual processes (initiation, propagation, and termination) occurring during SI-IMP, is similar to the rate-based model developed for surface-initiated ATRP by Kim et al. [100]. As may be expected based on the scheme shown in Figure 12.3, the identification of bimolecular termination as the prevalent mode of irreversible radical loss, and the hemolytic cleavage of TED (shown in Figure 12.6), the following reactions comprise the surface-initiated photoactivated IMP process:

1. Reversible activation (termination) of surface-tethered photoiniferter molecules, forming (recombining) a surface-tethered carbon radical and a free dithiocarbamyl radical;
2. Reversible cleavage (termination) of TED molecules, each yielding two free DTC•;
3. Irreversible termination of surface-tethered radicals via bimolecular termination; and
4. Propagation via reaction of a STR• with monomer (MMA) to grow a surface-tethered PMMA chain.

The generalized kinetic model was validated by comparing predictions with experimental data (such as that shown in Figure 12.7) describing layer growth as a function of time and [*TED*] as well as by comparing predictions of the maximum PMMA layer growth rate of the full model with predictions resulting from a simpler, pseudo-steady-state model that is applicable during the initial stages of SI-PMP when PMMA layer growth rate is at its maximum [82]. Although the limited number of assumptions necessitate fitting of experimental data to determine a number of unknown rate constants and kinetic parameters specific to the monomer, the model is robust and flexible due to the fact that the trends predicted are not specific to a particular monomer type. The effects of [*TED*] and light intensity on SI-IMP were predicted and compared in terms of their impact on maximum growth rate of SI-IMP, thickness evolution as a function of exposure time and maximum achievable thickness, reinitiation ability, and the final architecture of the polymer brush.

Figure 12.8 shows the effects of TED concentration (12.8a) and light intensity (12.8b) on the maximum growth rates of the polymer brushes and maximum thickness attained by SI-IMP, respectively. These plots underscore the fact that while increasing TED concentration and decreasing light intensity qualitatively result in a decrease in the maximum polymer brush growth rate, the ways that these parameters affect the maximum growth rate and maximum thickness attainable (which for the purpose of the model was defined as the thickness at which 99% of the chains are irreversibly terminated) are markedly different. Predictions indicate that TED concentration

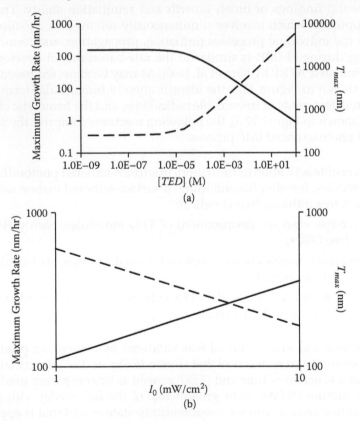

(a)

(b)

FIGURE 12.8
Predictions of maximum polymer growth rate (continuous lines) and maximum achievable thickness (broken lines) as a function of (a) TED concentration and (b) light intensity obtained from rate-based kinetic modeling.

does not affect the maximum growth rate significantly until the TED concentration exceeds the concentration of dithiocarbamyl radicals generated by photoactivated decomposition of surface-tethered photoiniferter molecules, $[STR-DTC]_0$. Beyond this concentration, the decrease in maximum growth rate is pronounced. Also observed is an increase in maximum thickness attainable as $[TED]$ increases. Both of these effects are a consequence of the equilibrium between active surface-tethered carbon-centered radicals (and free dithiocarbamyl radicals) and dormant dithiocarbamyl-capped radicals being shifted toward the dormant species, which slows the growth but preserves the ability of the surface-tethered iniferters to function. The trends predicted by simulations using the rate-based model [82] are consistent with experimental results discussed previously when TED is added prior to SI-IMP of MMA [54].

Because light intensity controls the rate of decomposition of surface-tethered photoiniferter molecules and the rate of formation of actively propagating surface-tethered radicals, it also may be used to tailor brush growth by SI-IMP. As shown in Figure 12.8b, decreasing the light intensity decreases the maximum growth rate and increases the maximum thickness attainable. Again, these changes can be understood in terms of the equilibrium between active growing chains and the dormant, dithiocarbamate-capped chains.

Increasing TED concentration and decreasing light intensity both result in the decrease in the maximum polymer brush growth rate by virtue of maintaining the low concentration of active surface-tethered radicals. In addition to a decrease in propagation rate (growth rate), the decrease in the concentration of propagating radicals also results in a decrease in irreversible termination reactions [101]. By virtue of this decrease in irreversible termination reactions with increasing $[TED]$ and decreasing light intensity, the average lifetime of growing PMMA chains increases with $[TED]$ and decreases with light intensity. As a result, the maximum thickness, T_{max}, defined as the thickness at which 99% of the tethered polymer chains are irreversibly terminated, achieved by SI-PMP should increase with increases in $[TED]$ and decreases in light intensity. Figure 12.8a and b show how T_{max} varies as a function of TED concentration and light intensity. As shown in Figure 12.8a, while T_{max} is approximately independent of $[TED]$ at concentrations less than $[STR-DTC]_0$, for $[TED] > [STR-DTC]_0$, T_{max} is proportional to $[TED]^{1/2}$. However, it should be noted that the exposure time required to reach T_{max} also increases with $[TED]$. These increases in T_{max} and exposure time required to reach T_{max}, both of which occur with increasing $[TED]$, are consequences of the previously noted trade-off between continued propagation and increased irreversible termination reactions. Similarly, as shown in Figure 12.8a, due to the low instantaneous concentration of active surface-tethered radicals, decreasing light intensity increases the T_{max} achievable by SI-IMP. Simulations also predicted that T_{max} is inversely proportional to $I_0^{1/2}$.

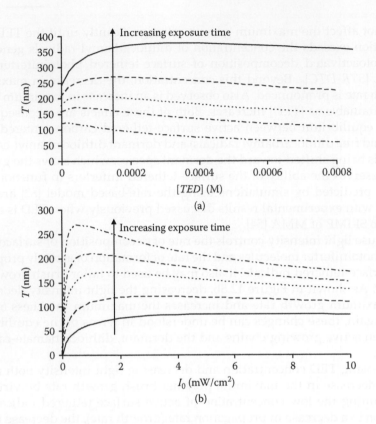

FIGURE 12.9
Evolution of simulated PMMA layer thickness as a function of (a) TED concentration and (b) light intensity at various exposure times. Lines for the TED concentration studies correspond to exposure times of 2 h (dashed-double-dotted line), 4 h (dashed-dotted line), 6 h (dotted line), 10 h (broken line), and 20 h (continuous line). The exposure times for light intensity studies are 1 h (continuous line), 2 h (broken line), 4 h (dotted line), 6 h (dashed-dotted line), and 10 h (dashed-double-dotted line). (Reprinted with permission from Rahane, S. B., et al., *Macromolecules*, 41, 9612–9618, 2008, copyright (2008) American Chemical Society.)

As shown in Figure 12.9a and b, at a given light intensity or [*TED*], polymer layer thickness increases with time. However, for a given exposure time, there is an optimum light intensity or [*TED*] that yields the maximum layer thickness. However, this optimum [*TED*] and light intensity that yield maximum thickness vary with exposure time: while optimum [*TED*] increases with exposure time, optimum light intensity decreases with exposure time. Thus, decreasing light intensity and increasing [*TED*] have similar effects on the polymer layer growth as described in terms of effects on evolution of layer thickness, maximum achievable thickness, and maximum growth rate observed during the initial period during the layer growth.

12.3.4 Application of a Comprehensive Model for Understanding Reinitiation Ability

While it appears that light intensity and added deactivator (TED) have similar impacts on SI-IMP, only light intensity affects the rate of initiation and decomposition of TED molecules. Therefore, as understood from Figures 12.8 and 12.9, TED and light intensity alter the formation, propagation, and termination of radicals in different ways. These differences will manifest in the ability to extend chains of the layer upon reinitiation. This subject was investigated by Rahane et al., who again used kinetic modeling to investigate how added TED and light intensity affect the reinitiation ability of PMMA layers. These studies are enabled by the ability to keep track of irreversibly terminated and "nonterminated" chains, which are those chains that can be reinitiated, within the framework of the rate-based kinetic model [83]. As a result, the fraction of nonterminated chains, f_{NT}, provides a measure of the reinitiation ability of the PMMA brush layer, as these chains can be reinitiated in the presence of a second monomer.

Figure 12.10 shows how the reinitiation ability of PMMA brushes, manifest in f_{NT}, is affected by [TED] and by light intensity as the layer grows [83]. In both situations, the predictions from the kinetic model show that f_{NT} increases at low thicknesses (which is equivalent to shorter exposure times), goes through a maximum, and then decreases as higher thicknesses (or longer exposure times) are reached. This pattern of behavior of f_{NT} as a function of PMMA layer thickness (or exposure time) was attributed to the balance between initiation and irreversible termination reactions involved in SI-IMP. Here again it is readily observed from the shapes of curves in Figure 12.10 that changes in [TED] and in light intensity have markedly different impacts on layer growth: the most significant difference is that while increasing [TED] improves the reinitiation ability of the tethered PMMA layers, decreasing the light intensity does not improve the reinitiation ability—the maximum f_{NT} is unchanged. Rather, decreasing light intensity results in an increase in layer thickness (or exposure time) at which maximum reinitiation ability is achieved [83].

Because of the differences observed in f_{NT} as a function of added TED and light intensity, it stands to reason that choices of light intensity and [TED] used during SI-IMP of the first monomer can lead to the formation of either bilevel block copolymer brushes or mixed brushes upon reinitiation in the presence of a second monomeric species. Again, kinetic modeling of the process provides a key ability because it allows the fraction of photoiniferter species that did not decompose to form active surface-tethered radicals and dithiocarbamyl radicals during polymer layer growth, f_{NI}, to be tracked as well as f_{NT}. As demonstrated by Rahane et al. [83], the ratio f_{NI}/f_{NT} establishes a guideline for the anticipated structure of polymer brushes formed by reinitiation of an existing polymer brush in the presence of another monomer: when $f_{NI}/f_{NT} \gg 1$, reinitiation with a second monomer results in formation of homopolymer chains

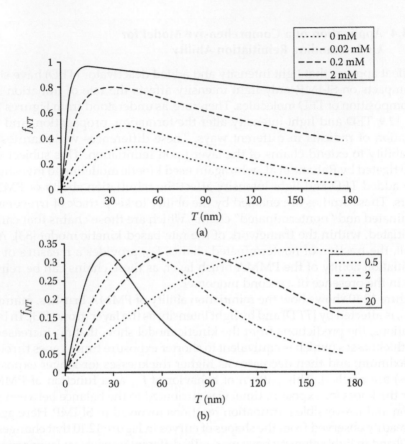

FIGURE 12.10

(a) Simulated fraction of nonterminated species as a function of thickness of PMMA layers at various TED concentrations. TED concentrations were 0 M (dotted line), 2×10^{-5} M (dashed-dotted line), 2×10^{-4} M (broken line), and 2×10^{-3} M (continuous line). The light intensity used for simulations was 5 mW/cm². (b) Simulated fraction of nonterminated species as a function of thickness of PMMA layers at various light intensities. The light intensities were 0.5 mW/cm² (dotted line), 2 mW/cm² (broken line), 5 mW/cm² (dashed-dotted line), and 20 mW/cm² (continuous line). TED concentration used for simulations was 0 M. (Reproduced from Rahane, S. B., et al., *J. Polym. Sci. A Polym. Chem.*, 48, 1586–1593, 2010. With permission from Wiley.)

of the second type of monomer tethered directly to the surface, resulting in a mixed brush structure. Conversely, if a brush layer with $f_{NI}/f_{NT} \ll 1$ is reinitiated in the presence of a second monomer, a block copolymer brush is formed. To identify the photopolymerization conditions that favor the formation of block copolymers rather than mixed polymer brushes (and vice versa), we examined the impact of various reaction conditions on f_{NI}/f_{NT}.

Figure 12.11 shows how f_{NI}/f_{NT} evolves with layer growth for layers synthesized at high intensity and with added TED (solid line) and at low intensity and with no added TED (broken line). For both sets of reaction conditions captured in Figure 12.11, f_{NI}/f_{NT} decreases as the layer thickness increases. However, f_{NI}/f_{NT} drops much more quickly as a function of thickness when

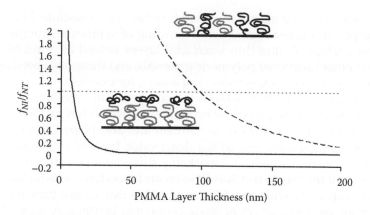

FIGURE 12.11

Comparison of simulated ratio of fraction of uninitiated photoiniferter species to fraction of nonterminated species as a function of thickness of the first brush layer synthesized in presence of [*TED*] and at light intensity of 5 mW/cm² (continuous line), and in absence of [*TED*] and lower light intensity of 0.5 mW/cm² (broken line). (Reproduced from Rahane, S. B., et al., *J. Polym. Sci. A Polym. Chem.*, 48, 1586–1593, 2010. With permission from Wiley.)

high intensities with TED are used than if low intensities and no TED are added, suggesting that conditions favor formation of block copolymers over a much broader range of thickness. On the other hand, lower intensities and no added TED favor formation of mixed polymer brushes. Interestingly, the dotted horizontal line represents a third situation in which after reinitiation in the presence of a second monomer, the layer consists of an equal number of tethered homopolymer chains (of the second monomer) and diblock copolymer chains [83]. In summary, and even though the kinetic model was parameterized based on PMMA being the first type of chain grown, kinetic modeling guides the selection of reaction conditions that can be used to tailor the design of the brush layer. This represents a useful and insightful tool that enhances the design of soft interfaces by SI-IMP.

12.4 Summary and Outlook

The adaptation of various controlled (free) radical polymerization methods to create polymer brushes that straddle interfaces and, as a result, modify surface properties has dramatically advanced the development of tailored soft interfaces. In this chapter, we have attempted to highlight commonalities among the various methods, as well as advantages of surface-initiated photoactivated IMP, which include facile spatial and temporal control and tolerance to thermally sensitive monomers or those with labile groups. Iniferters based on dithiocarbamate chemistry have been used to create single-component and multicomponent brushes and cross-linked layers that have deepened a fundamental understanding of synthesis-structure-property relationships of ultra-thin interfacial layers as well as novel bioconjugates and other functional polymeric materials, and these have been applied as biocompatible coatings, sensors, and assays, for example.

The design of such interfacial layers is facilitated by insights gained through kinetic modeling of the growth of brush chains, which has provided insight into the dominant termination mechanism and also helped to sort out the impact of light intensity and added deactivator, both of which are known to improve control, though in remarkably different ways. In concert, these findings point the way to use SI-IMP to create layers having particular thicknesses, which along with preaddition of added deactivator provides access to either mixed brush layers or block copolymer brushes. A general shortcoming of the kinetic model is that unknown kinetic rate constants must be extracted from a set of experimental data. While this makes the model pseudopredictive, guidance may be obtained through general trends to the extent that the same assumptions apply at other experimental conditions.

Finally, the examples reviewed in this chapter as well as the broader, growing interest in polymer brushes as useful constructs and devices are testaments to particular themes, many of which reside at the interface of biology and polymer chemistry and physics and demand continued attention. Because of their well-defined structure, polymer brushes remain important model systems. As interactions across interfaces are determined by the details of how chemical functionality is presented at that interface, it remains important to understand how interfacial structure is impacted by changes in the local environment as well as the presence of absorbates (e.g., proteins or nanocrystals). Along these lines, the ability to articulate design rules for modifying polymer brushes (or polymer-modified surfaces) by proper integration of biochemical motifs or biological compounds to achieve optimal function, whether it be targeting efficiency, sensing, cellular adhesion, or antifouling character, remains crucial for the development and application of polymer brushes as platforms in biotechnology. All of these efforts are advanced by and benefit from the ability to tailor interfacial polymer layers through control of composition, topology, and layer characteristics such as extent and grafting density, which are directly enabled through controlled surface-initiated polymerizations, including photoactivated SI-IMP.

Acknowledgments

The authors gratefully acknowledge helpful discussions and guidance from Dr. Derek Patton. Dr. Andrew Metters and J. Alaina Floyd are thanked as well for contributions made during completion of the research, which was completed at the Center for Advanced Engineering Fibers and Films, an engineering research center established by the National Science Foundation at Clemson University. SMKII also acknowledges financial support from the Interfacial Processes and Thermodynamics Program of the National Science Foundation (Award Number CBET-0840249).

References

1. Lahann, J., Balcells, M., Rodon, T., Lee, J., Jenson, K. F., Langer, R., Reactive Polymer Coatings: A Platform for Patterning Proteins and Mammalian Cells onto a Broad Range of Materials, *Langmuir* 2002, 18, 3632–3638.
2. Hammond, P. T., Form and Function in Multilayer Assembly: New Applications at the Nanoscale, *Adv. Mater.* 2004, 16, 1271–1293.
3. Dorrer, C., Rühe, J., Some Thoughts on Superhydrophobic Wetting, *Soft Matter* 2009, 5, 51–61.
4. Jones, R. A. L., Polymer Interfaces and the Molecular Basis of Adhesion, *Curr. Opin. Solid State Mater. Sci.* 1997, 2, 673–677.
5. Bhat, R., Tomlinson, M. R., Wu, T., Genzer, J., Surface-Grafted Polymer Gradients: Formation, Characterization, and Applications, *Adv. Polym. Sci.* 2006, 198, 51–124.
6. Ayres, N., Polymer Brushes: Applications in Biomaterials and Nanotechnology, *Polym. Chem.* 2010, 1, 769–777.
7. Uhlmann, P., Ionov, L., Houbenov, N., Nitschke, M., Grundke, K., Motornov, M., Minko, S., Stamm, M., Surface Functionalization by Smart Coatings: Stimuli-Responsive Binary Polymer Brushes, *Prog. Org. Coatings* 2006, 55, 168–174.
8. Xu, F. J., Neoh, K. G., Kang, E. T., Bioactive Surfaces and Biomaterials via Atom Transfer Radical Polymerization, *Prog. Polym. Sci.* 2009, 34, 719–761.
9. Boyes, S. G., Granville, A. M., Baum, M., Akgun, B., Mirous, B. K., Brittain, W. J., Polymer Brushes—Surface Immobilized Initiators, *Surf. Sci.* 2004, 570, 1–12.
10. Edmondson, S., Osborne, V. L., Huck, W. T. S., Polymer Brushes via Surface-Initiated Polymerizations, *Chem. Soc. Rev.* 2004, 33, 14–22.
11. Barbey, R., Lavanat, L., Paripovic, D., Schuewer, N., Sugnaux, C., Tugulu, S., Klok, H.-A., Polymer Brushes via Surface-Initiated Controlled Radical Polymerization: Synthesis, Characterization, Properties, and Applications, *Chem. Rev.* 2009, 109, 5437–5527.
12. Draper, J., Luzinov, I., Minko, S., Tokarev, I., Stamm, M., Mixed Polymer Brushes by Sequential Polymer Addition: Anchoring Layer Effect, *Langmuir* 2004, 20, 4064–4075.
13. Klein, J., Shear, Friction, and Lubrication Forces between Polymer-Bearing Surfaces, *Annu. Rev. Mater. Sci.* 1996, 26, 581–612.
14. Forster, A. M., Mays, J. W., Kilbey II, S. M., Effect of Temperature on the Frictional Forces between Polystyrene Brushes, *J. Polym. Sci. B Polym. Phys.* 2006, 44, 649–654.
15. Tugulu, S., Silacci, P., Stergiopulos, N., Klok, H.-A., RGD—Functionalized Polymer Brushes as Substrates for the Integrin Specific Adhesion of Human Umbilical Vein Endothelial Cells, *Biomaterials* 2007, 28, 2536–2546.

16. Tugulu, S., Arnold, A., Sielaff, I., Johnsson, K., Klok, H.-A., Protein-Functionalized Polymer Brushes, *Biomacromolecules* 2005, 6, 1602–1607.
17. Jain, P., Baker, G. L., Bruening, M. L., Applications of Polymer Brushes in Protein Analysis and Purification, *Annu. Rev. Anal. Chem.* 2009, 2, 387–408.
18. Jain, P., Sun, L., Dai, J., Baker, G. L., Bruening, M. L., High-Capacity Purification of His-Tagged Proteins by Affinity Membranes Containing Functionalized Polymer Brushes, *Biomacromolecules* 2007, 8, 3102–3107.
19. Raynor, J. E., Petrie, T. A., Garcia, A. J., Collard, D. M., Controlling Cell Adhesion to Titanium: Functionalization of Poly[Oligo(Ethylene Glycol)Methacrylate] Brushes with Cell-Adhesive Peptides, *Adv. Mater.* 2007, 19, 1724–1728.
20. Zou, Y. Q., Yeh, P. Y. J., Rossi, N. A. A., Brooks, D. E., Kizhakkedathu, J. N., Nonbiofouling Polymer Brush with Latent Aldehyde Functionality as a Template for Protein Micropatterning, *Biomacromolecules* 2010, 11, 284–293.
21. Xu, F. J., Liu, L. Y., Yang, W. T., Kang, E. T., Neoh, K. G., Active Protein-Functionalized Poly(Poly(Ethylene Glycol) Monomethacrylate)-Si(100) Hybrids from Surface-Initiated Atom Transfer Radical Polymerization for Potential Biological Applications, *Biomacromolecules* 2009, 10, 1665–1674.
22. Hucknall, A., Kim, D. H., Rangarajan, S., Hill, R. T., Reichert, W. M., Chilkoti, A., Simple Fabrication of Antibody Microarrays on Nonfouling Polymer Brushes with Femtomolar Sensitivity for Protein Analytes in Serum and Blood, *Adv. Mater.* 2009, 21, 1968–1971.
23. Iwata, R., Satoh, R., Iwasaki, Y., Akiyoshi, K., Covalent Immobilization of Antibody Fragments on Well-Defined Polymer Brushes via Site-Directed Method, *Coll. Surf. B Biointerfaces* 2008, 62, 288–298.
24. Tugulu, S., Barbey, R., Harms, M., Fricke, M., Volkmer, D., Rossi, A., Klok, H.-A., Synthesis of Poly(Methacrylic Acid) Brushes via Surface-Initiated Atom Transfer Radical Polymerization of Sodium Methacrylate and Their Use as Substrates for the Mineralization of Calcium Carbonate, *Macromolecules* 2007, 40, 168–177.
25. Wu, T., Gong, P., Szleifer, I., Vlek, P., Šubr, V., Genzer, J., Behavior of Surface-Anchored Poly(Acrylic Acid) Brushes with Grafting Density Gradients on Solid Substrates. 1. Experiment, *Macromolecules* 2007, 40, 8756–8764.
26. Husseman, M., Malmstrom, E. E., McNamara, M., Mate, M., Mecerreyes, D., Benoit, D. G., Hedrick, J. L., Mansky, P., Russell, T. P., Hawker, C. J., Controlled Synthesis of Polymer Brushes by "Living" Free Radical Polymerization Techniques, *Macromolecules* 1999, 32, 1424–1431.
27. Matyjaszewski, K., Miller, P. J., Shukla, N., Immaraporn, B., Gelman, A., Luokala, B. B., Siclovan, T. M., Kickelbick, G., Vallant, T., Hoffmann, H., Pakula, T., Polymers at Interfaces: Using Atom Transfer Radical Polymerization in the Controlled Growth of Homopolymers and Block Copolymers from Silicon Surfaces in the Absence of Untethered Sacrificial Initiator, *Macromolecules* 1999, 32, 8716–8724.
28. Rahane, S. B., Floyd, J. A., Metters, A. T., Kilbey, S. M., II, Swelling Behaviour of Multiresponsive Poly(Methacrylic Acid)-Block-Poly(N-Isopropylacrylamide) Brushes Synthesized Using Surface-Initiated Photoiniferter-Mediated Photopolymerization, *Adv. Funct. Mater.* 2008, 18, 1232–1240.
29. Ding, S., Floyd, J. A., Walters, K., Comparison of Surface Confined ATRP and SET-LRP Syntheses for a Series of Amino (Meth)acrylate Polymer Brushes on Silicon Substrates, *J. Polym. Sci. A Polym. Chem.* 2009, 47, 6552–6560.

30. Prucker, O., Rühe, J., Polymer Layers through Self-Assembled Monolayers of Initiators, *Langmuir* 1998, 14, 6893–6898.
31. Prucker, O., Rühe, J., Synthesis of Poly(Styrene) Monolayers Attached to High Surface Area Silica Gels through Self-Assembled Monolayers of Azo Initiators, *Macromolecules* 1998, 31, 592–601.
32. Alonzo, J., Mays, J. W., Kilbey, S. M., II, Forces of Interaction between Surfaces Bearing Looped Polymer Brushes in Good Solvent, *Soft Matter* 2009, 5, 1897–1904.
33. Huang, Z., Alonzo, J., Liu, M., Lay, M., Ji, H., Zhang, Y., Yin, F., Smith, G. D., Mays, J. W., Kilbey, S. M., II, Dadmun, M. D., Density Profile of "Looped" Triblock Copolymer Brushes at the Liquid-Solid Interfaces by Neutron Reflectivity Measurements, *Macromolecules* 2008, 41, 1745–1752.
34. Dan, N., Tirrell, M., Effect of Bimodal Weight Distribution on the Polymer Brush, *Macromolecules* 1993, 26, 6467–6473.
35. Rowe-Konopacki, M. D., Boyes, S. G., Synthesis of Surface Initiated Diblock Copolymer Brushes from Flat Silicon Substrates Utilizing the RAFT Polymerization Technique, *Macromolecules* 2007, 40, 879–888.
36. Li, Y., Benicewicz, B. C., Functionalization of Silica Nanoparticles via the Combination of Surface-Initiated RAFT Polymerization and Click Reactions, *Macromolecules* 2008, 41, 7986–7992.
37. Tsujii, Y., Ejaz, M., Sato, K., Goto, A., Fukuda, T., Mechanism and Kinetics of RAFT-Mediated Graft Polymerization of Styrene on a Solid Surface. 1. Experimental Evidence of Surface Radical Migration, *Macromolecules* 2001, 34, 8872–8877.
38. Zhai, G. Q., Yu, W. H., Kang, E. T., Neoh, K. G., Huang, C. C., Liaw, D. J., Functionalization of Hydrogen-Terminated Silicon with Polybetaine Brushes via Surface-Initiated Reversible Addition-Fragmentation Chain Transfer (RAFT) Polymerization, *Ind. Eng. Chem. Res.* 2004, 43, 1673–1680.
39. Ghannam, L., Parvole, J., Laruelle, G., Fancois, J., Billon, L., Surface-Initiated Nitroxide-Mediated Polymerization: A Tool for Hybrid Inorganic/Organic Nanocomposites "*In Situ*" Synthesis, *Polym. Int.* 2006, 55, 1199–1207.
40. Pyun, J., Kowalewski, T., Matyjaszewski, K., Synthesis of Polymer Brushes Using Atom Transfer Radical Polymerization, *Macromol. Rapid Commun.* 2003, 24, 1043–1059.
41. Tsujii, Y., Ohno, K., Yamamoto, S., Goto, A., Fukuda, T., Structure and Properties of High-Density Polymer Brushes Prepared by Surface-Initiated Living Radical Polymerization, *Adv. Polym. Sci.* 2006, 197, 1–45.
42. Ma, H. W., Hyun, J. H., Stiller, P., Chilkoti, A., "Non-Fouling" Oligo(Ethylene Glycol)-Functionalized Polymer Brushes Synthesized by Surface-Initiated Atom Transfer Radical Polymerization, *Adv. Mater.* 2004, 16, 338–341.
43. Ejaz, M., Tsujii, Y., Fukuda, T., Controlled Grafting of a Well-Defined Polymer on a Porous Glass Filter by Surface-Initiated Atom Transfer Radical Polymerization, *Polymer* 2001, 42, 6811–6815.
44. von Werne, T., Patten, T. E., Preparation of Structurally Well-Defined Polymer-Nanoparticle Hybrids with Controlled/Living Radical Polymerizations, *J. Am. Chem. Soc.* 1999, 121, 7409–7410.
45. Gauthier, M. A., Gibson, M. I., Klok, H.-A., Synthesis of Functional Polymers by Post-Polymerization Modification, *Angew. Chem. Int. Ed.* 2009, 48, 48–58.
46. Otsu, T., Iniferter Concept and Living Radical Polymerization, *J. Polym. Sci. A Polym. Chem.* 2000, 38, 2121–2136.

47. Otsu, T., Matsumoto, A., Controlled Synthesis of Polymers Using the Iniferter Technique: Developments in Living Radical Polymerization, *Adv. Polym. Sci.* 1998, 136, 75–137.
48. Otsu, T., Yoshida, M., Tazaki, T., A Model for Living Radical Polymerization, *Macromol. Chem. Rapid Commun.* 1982, 3, 133–140.
49. Lambrinos, P., Tardi, M., Polton, A., Sigwalt, P., The Mechanism of the Polymerization of n-Butyl Acrylate Initiated with N,N-Diethyl Dithiocarbamate Derivatives, *Eur. Polym. J.* 1990, 26, 1125–1135.
50. Kazmaier, P. M., Moffat, K. A., Georges, M. K., Veregin, R. P. N., Hamer, G. K., Free-Radical Polymerization for Narrow-Polydispersity Resins. Semiempirical Molecular Orbital Calculations as a Criterion for Selecting Stable Free-Radical Reversible Terminators, *Macromolecules* 1995, 28, 1841–1846.
51. Otsu, T., Ogawa, T., Yamamoto, T., Solid Phase Block Copolymer Synthesis by the Iniferter Technique, *Macromolecules* 1986, 19, 2087–2089.
52. Nakayama, Y., Matsuda, T., Surface Macromolecular Architectural Designs Using Photo-Graft Copolymerization Based on Photochemistry of Benzyl N,N-Diethyldithiocarbamate, *Macromolecules* 1996, 29, 8622–8630.
53. de Boer, B., Simon, H. K., Werts, M. P. L., van der Vegte, E. W., Hadziioannou, G., "Living" Free Radical Photopolymerization Initiated from Surface-Grafted Iniferter Monolayers, *Macromolecules* 2000, 33, 349–356.
54. Rahane, S. B., Metters, A. T., Kilbey, S. M., II, Impact of Added Tetraethylthiuram Disulfide Deactivator on the Kinetics of Growth and Reinitiation of Poly(Methyl Methacrylate) Brushes Made by Surface-Initiated Photoiniferter-Mediated Photopolymerization, *Macromolecules* 2006, 39, 8987–8991.
55. Bossi, A., Whitcombe, M. J., Uludag, Y., Fowler, S., Chianella, I., Subrahmanyam, S., Sanchez, I., Piletsky, S. A., Synthesis of Controlled Polymeric Cross-Linked Coatings via Iniferter Polymerisation in the Presence of Tetraethylthiuram Disulfide Chain Terminator, *Biosen. Bioelectron.* 2010, 25, 2149–2155.
56. Sebra, R. P., Anseth, K. S., Bowman, C. N., Integrated Surface Modification of Fully Polymeric Microfluidic Devices Using Living Radical Photopolymerization Chemistry, *J. Polym. Sci. A Polym. Chem.* 2006, 44, 1404–1413.
57. Kobayashi, T., Takahashi, S., Fujii, N., Silane Coupling Agent Having Dithiocarbamate Group for Photografting of Sodium Styrene Sulfonate on Glass Surface, *J. App. Polym. Sci.* 1993, 49, 417–423.
58. Duñer, G., Anderson, H., Myrskog, A., Hedlund, M., Aastrup, T., Ramström, O., Surface-Confined Photopolymerization of pH-Responsive Acrylamide/ Acrylate Brushes on Polymer Thin Films, *Langmuir* 2008, 24, 7559–7564.
59. Benetti, E. M., Zapotoczny, S., Vansco, G. J., Tunable Thermoresponsive Polymeric Platforms on Gold by "Photoiniferter"-Based Surface Grafting, *Adv. Mater.* 2009, 19, 268–271.
60. Tagit, O., Tomczak, N., Benetti, E. M., Cesa, Y., Blum, C., Subramaniam, V., Herek, J. L., Vansco, G. J., Temperature-Modulated Quenching of Quantum Dots Covalently Coupled to Chain Ends of Poly(N-Isopropyl Acrylamide) Brushes on Gold, *Nanotechnology* 2009, 20, 185501(1)–185501(6).
61. Navarro, M., Benetti, E. M., Zapotoczny, S., Planell, J. A., Vansco, G. J., Buried, Covalently Attached RGD Peptide Motifs in Poly(Methacrylic Acid) Brush Layers: The Effect of Brush Structure on Cell Adhesion, *Langmuir* 2008, 24, 10996–11002.

62. Geismann, C., Tomicki, F., Ulbricht, M., Block Copolymer Photo-Grafted Poly(Ethylene Terephthalate) Capillary Pore Membranes Distinctly Switchable by Two Different Stimuli, *Sep. Sci. Tech.* 2009, 44, 3312–3329.
63. Sebra, R. P., Kasko, A. M., Anseth, K. S., Bowman, C. N., Synthesis and Photografting of Highly pH-Responsive Polymer Chains, *Sensors Actuators B* 2006, 119, 127–134.
64. Ohya, S., Matsuda, T., Poly(*N*-Isopropylacrylamide) (PNIPAM)-Grafted Gelatin as Thermoresponsive Three-Dimensional Artificial Extracellular Matrix: Molecular and Formulation Parameters vs. Cell Proliferation Potential, *J. Biomater. Sci. Polym. Edn.* 2005, 16, 809–827.
65. Matsuda, T., Ohya, S., Photoiniferter-Based Thermoresponsive Graft Architecture with Albumin Covalently Fixed at Growing Graft Chain End, *Langmuir* 2005, 21, 9660–9665.
66. Benetti, E. M., Reimhult, E., de Bruin, J., Zapotoczny, S., Textor, M., Vansco, G. J., Poly(Methacrylic Acid) Grafts Grown from Designer Surfaces: The Effect of Initiator Coverage on Polymerization Kinetics, Morphology, and Properties, *Macromolecules* 2009, 42, 1640–1647.
67. Higashi, J., Nakayama, Y., Marchant, R. E., Matsuda, T., High-Spatioresolved Microarchitectural Surface Prepared by Photograft Copolymerization Using Dithiocarbamate: Surface Preparation and Cellular Responses, *Langmuir* 1999, 15, 2080–2088.
68. Sebra, R. P., Masters, K. S., Bowman, C. N., Anseth, K. S., Surface Grafted Antibodies: Controlled Architecture Permits Enhanced Antigen Detection, *Langmuir* 2005, 21, 10907–10911.
69. Sebra, R. P., Masters, K. S., Cheung, C. Y., Bowman, C. N., Anseth, K. S., Detection of Antigens in Biologically Complex Fluids with Photografted Whole Antibodies, *Anal. Chem.* 2006, 78, 3144–3151.
70. Lawson, M. C., Shoemaker, R., Hoth, K. B., Bowman, C. N., Anseth, K. S., Polymerizable Vancomycin Derivatives for Bactericidal Biomaterial Surface Modification: Structure-Function Evaluation, *Biomacromolecules* 2009, 10, 2221–2234.
71. Hume, P. S., Anseth, K. S., Inducing Local T Cell Apostosis with Anti-Fas-Functionalized Polymer Coatings Fabricated via Surface-Initiated Photopolymerizations, *Biomaterials* 2010, 31, 3166–3174.
72. Harris, B. P., Kutty, J. K., Fritz, E. W., Webb, C. K., Burg, K. J. L., Metters, A. T., Photopatterned Polymer Brushes Promoting Cell Adhesion Gradients, *Langmuir* 2006, 22, 4467–4471.
73. Matsuda, T., Kaneko, M., Ge, S., Quasi-Living Surface Graft Polymerization with Phosphorylcholine Group(s) at the Terminal End, *Biomaterials* 2003, 24, 4507–4515.
74. Matsuda, T., Nagase, J., Ghoda, A., Hirano, Y., Kidoaki, S., Nakayama, Y., Phosphorylcholine-Endcapped Oligomer and Block Co-Oligomer and Surface Biological Reactivity, *Biomaterials* 2003, 24, 4517–4527.
75. Magoshi, T., Ziani-Cherif, H., Ohya, S., Nakayama, Y., Matsuda, T., Thermoresponsive Heparin Coating: Heparin Conjugated with Poly(*N*-Isopropylacrylamide) at One Terminus, *Langmuir* 2002, 18, 4862–4872.
76. Hattori, K., Hiwatari, M., Iiyama, C., Yoshimi, Y., Kohori, F., Sakai, K., Piletsky, S. A., Gate Effect of Theophylline-Imprinted Polymers Grafted to the Cellulose by Living Radical Polymerization, *J. Membrane Sci.* 2004, 233, 169–173.

77. Oxelbark, J., Legido-Quigley, C., Aureliano, C. S. A., Titirici, M.-M., Schillinger, E., Sellergren, B., Courtois, J., Irgum, K., Dambies, L., Cormack, P. A. G., Sherrington, D. C., Lorenzi, E. D., Chromatographic Comparison of Bupivacaine Imprinted Polymers Prepared in Crushed Monolith, Microsphere, Silica-Based Composite and Capillary Monolith Formats, *J. Chromatogr. A* 2007, 1160, 215–226.
78. Lee, H.-Y., Kim, B. S., Grafting of Molecularly Imprinted Polymers on Iniferter-Modified Carbon Nanotube, *Biosens. and Bioelectron.* 2009, 25, 587–591.
79. Barahona, F., Turiel, E., Cormack, P. A. G., Martín-Esteban, A., Chromatographic Performance of Molecularly Imprinted Polymers: Core-Shell Microspheres by Precipitation Polymerization and Grafted MIP Films via Iniferter-Modified Silica Beads, *J. Polym. Sci. A Polym. Chem.* 2010, 48, 1058–1066.
80. Qin, L., He, X., Zhang, W., Li, W., Zhang, Y., Surface-Modified Polystyrene Beads as Photografting Imprinted Polymer Matrix for Chromatographic Separation of Proteins, *J. Chromatogr. A* 2009, 1216, 807–814.
81. Rahane, S. B., Kilbey, S. M., II, Metters, A. T., Kinetics of Surface-Initiated Photoiniferter-Mediated Photopolymerization, *Macromolecules* 2005, 38, 8202–8210.
82. Rahane, S. B., Kilbey, S. M., II, Metters, A. T., Kinetic Modeling of Surface-Initiated Photoiniferter-Mediated Photopolymerization in Presence of Tetraethylthiuram Disulfide, *Macromolecules* 2008, 41, 9612–9618.
83. Rahane, S. B., Metters, A. T., Kilbey, S. M., II, Modeling of Reinitiation Ability of Polymer Brushes Grown by Surface-Initiated Photoiniferter-Mediated Photopolymerization, *J. Polym. Sci. A Polym. Chem.* 2010, 48, 1586–1593.
84. Luo, N., Metters, A. T., Hutchison, J. B., Bowman, C. N., Anseth, K. S., A Methacrylated Photoiniferter as a Chemical Basis for Microlithography: Micropatterning Based on Photografting Polymerization, *Macromolecules* 2003, 36, 6739–6745.
85. Nakayama, Y., Matsuda, T., *In-Situ* Observation of Dithiocarbamate-Based Surface Photograft Copolymerization Using Quartz Crystal Microbalance, *Macromolecules* 1999, 32, 5405–5410.
86. He, D., Ulbricht, M., Tailored "Grafting-From" Functionalization of Microfiltration Membrane Surface Photo-Initiated by Immobilized Iniferter, *Macromol. Chem. Phys.* 2009, 210, 1149–1158.
87. Heeb, R., Bielecki, R. M., Lee, S., Spencer, N. D., Room-Temperature, Aqueous-Phase Fabrication of Poly(Methacrylic Acid) Brushes by UV-LED-Induced, Controlled Radical Polymerization with High Selectivity for Surface-Bound Species, *Macromolecules* 2009, 42, 9124–9132.
88. de Gennes, P. G., Conformations of Polymers Attached to an Interface, *Macromolecules* 1980, 13, 1069–1075.
89. Yim, H., Kent, M. S., Mendez, S., Balamurugan, S. S., Balamurugan, S., Lopez, G. P., Satija, S., Temperature-Dependent Conformational Change of PNIPAM Grafted Chains at High Surface Density in Water, *Macromolecules* 2004, 37, 1994–1997.
90. Balamurugan, S., Mendez, S., Balamurugan, S. S., O'Brien, II, M. J., Lopez, G. P., Thermal Response of Poly(N-Isopropylacrylamide) Brushes Probed by Surface Plasmon Resonance, *Langmuir* 2003, 19, 2545–2549.
91. Plunkett, K. N., Zhu, X., Moore, J. S., Leckband, D. E., PNIPAM Chain Collapse Depends on the Molecular Weight and Grafting Density, *Langmuir* 2006, 22, 4259–4266.

92. Douglas, J. F., Karim, A., Kent, M. S., Satija, S. K., in *Encyclopedia of Materials: Science and Technology, Buschow*, K. H. J., Cahn, R. W., Flemings, M. C., Ilschner, B., Kramer, E. J., Mahajan, S., eds., New York, Pergamon Press, 2001, vol. 8, pp. 7218–7223.

93. Otsu, T., Matsunaga, T., Doi, T., Matsumoto, A., Features of Living Radical Polymerization of Vinyl Monomers in Homogeneous System Using *N,N*-diethyldithiocarbamate Derivatives as Photoiniferters, *Eur. Polym. J.* 1995, 31, 67–78.

94. Kannurpatti, A. R., Lu, S., Bunker, G. M., Bowman, C. N., Kinetic and Mechanistic Studies of Iniferter Photopolymerizations, *Macromolecules* 1996, 29, 7310–7315.

95. Qin, S. H., Qiu, K. Y., A New Polymerizable Photoiniferter for Preparing Poly(Methyl Methacrylate) Macromonomer, *Eur. Polym. J.* 2001, 37, 711–717.

96. Ward, J. H., Shahar, A., Peppas, N. A., Kinetics of "Living" Radical Polymerizations of Multifunctional Monomer, *Polymer* 2002, 43, 1745–1752.

97. Gopireddy, D., Husson, S. M., Room Temperature Growth of Surface-Confined Poly(Acrylamide) from Self-Assembled Monolayers Using Atom Transfer Radical Polymerization, *Macromolecules* 2002, 35, 4218–4221.

98. Marques, C. M., Joanny, J.-F., Block Copolymer Adsorption in a Nonselective Solvent, *Macromolecules* 1989, 22, 1454–1458.

99. Doi, T., Matsumoto, A., Otsu, T., Radical Polymerization of Methyl Acrylate by Use of Benzyl *N,N*-Diethyldithiocarbamate in Combination with Tetraethylthiuram Disulfide as a Two-Component Iniferter, *J. Polym. Sci. A Polym. Chem.* 1994, 32, 2911–2918.

100. Kim, J. B., Huang, W., Miller, M. D., Baker, G. L., Bruening, M. L., Kinetics of Surface-Initiated Atom Transfer Radical Polymerization, *J. Polym. Sci. A Polym. Chem.* 2003, 41, 386–394.

101. Fischer, H., The Persistent Radical Effect in "Living" Radical Polymerization, *Macromolecules* 1997, 30, 5666–5672.

Index